Rob.

Purposes only

INDEX TO THE CURRAGH

Printed at the Ordnance Survey Office, Phœnix Park, Dublin. 1909

A MOST DELIGHTFUL STATION:

The British Army on the Curragh of Kildare, Ireland, 1855–1922

CON COSTELLO

THE COLLINS PRESS

First published in 1996 by
The Collins Press, Carey's Lane, The Huguenot Quarter, Cork

The generous sponsorship of Kildare County Council and the County Kildare Archaeological Society towards the publication of this book is acknowledged, and both indicate that their support was an expression of their appreciation of the author's many years of research into, and promotion of the history of county Kildare.

British Library cataloguing in publication data.

A CIP catalogue record for this book is available from the British library.

Printed in Ireland by Colour Books Ltd., Dublin, Ireland.

Designed by Elaine Shiels.
Typeset by Seton Music Graphics, Bantry, Ireland.
Jacket Design by Upper Case Ltd., Cork.

ISBN-189825608X

Contents

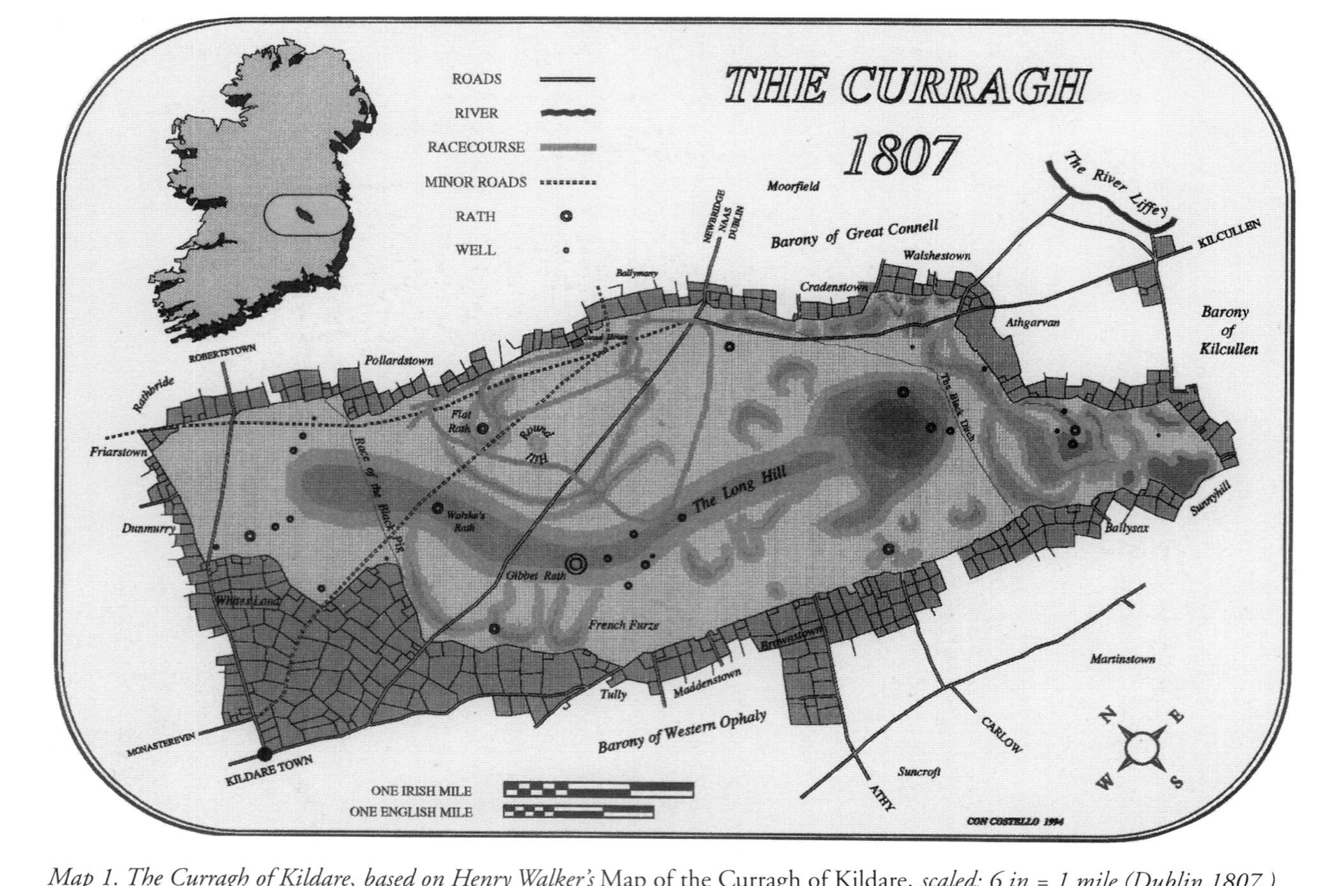

Map 1. The Curragh of Kildare, based on Henry Walker's Map of the Curragh of Kildare, *scaled: 6 in = 1 mile (Dublin 1807.). Drawn by C.W. Costello, 1994..*

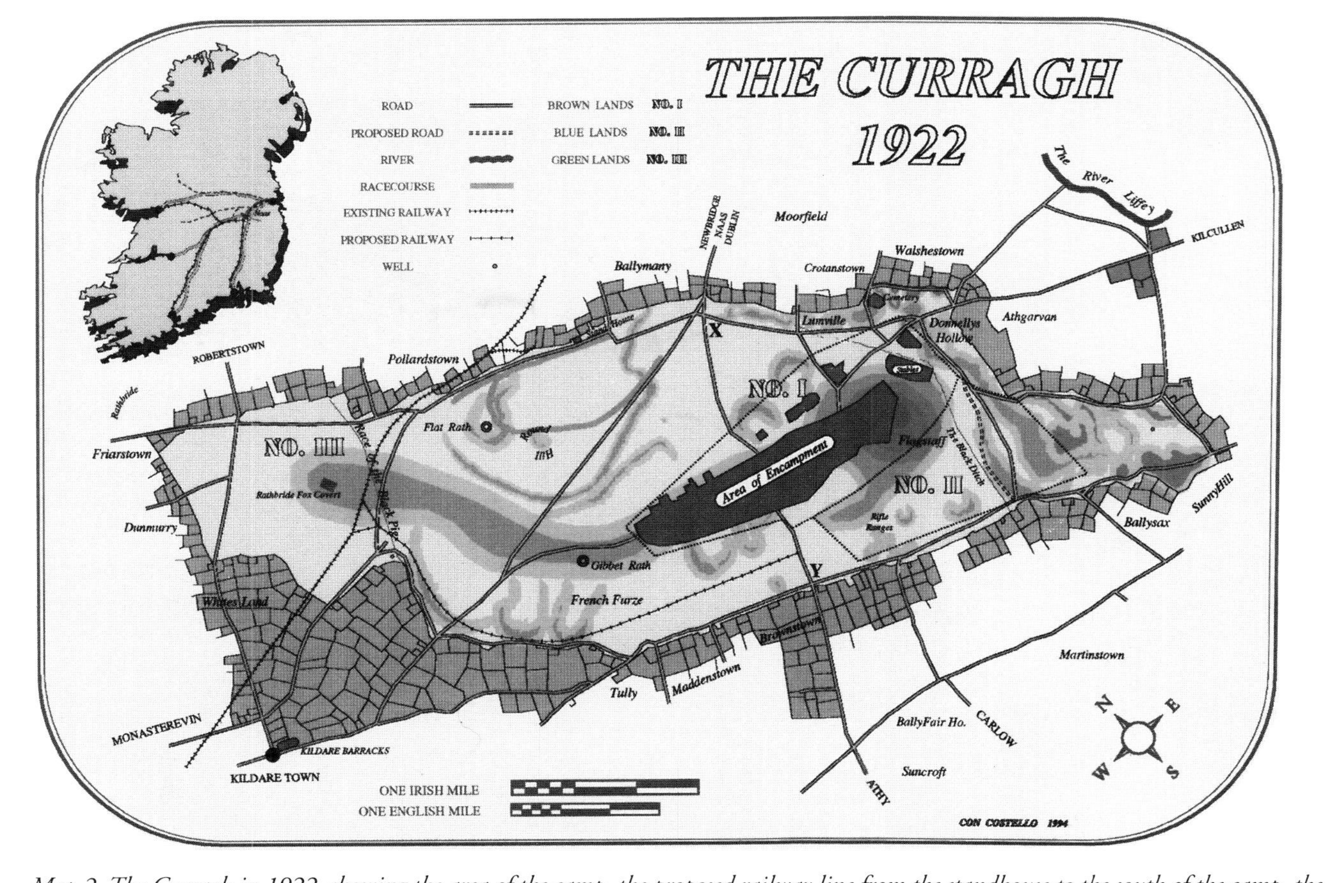

Map 2. The Curragh in 1922, showing the area of the camp, the proposed railway line from the standhouse to the south of the camp, the new roads, and the division of the plain into three 'lands'. Drawn by C.W. Costello, 1994..

Acknowledgements

Grateful appreciation is expressed to the following people and institutions for their help in the writing of this book, especially to Professor R. V. Comerford without whose interest and encouragement it would not have come to fruition. To Gregory Allen; Peter Bardon; Niall Brannigan; Pat Bell; Michelle Brown; Suzanne Byrne; Mary Carroll; Lieut. Col. C. M. L. Clements; John Cooke; Dr Mark Costello; Con W. Costello; Hugh Crawford; Reg Darling; Maeve Davison; Major John de Burgh; Áine Delahunt; Ken Dunne; Lieut. Col. Eugene Field; Anne Fitzsimons; Carmel Flahavan; Pat Foley; the late Tom Garrett; Comdt. J. C. Gallagher; Thomas Geoghegan; Lieut. Col. William Gibson; Daniel Gillman; Elizabeth Gleeson; the Knight of Glin; Mr and Mrs J. Gorry, Gorry Gallery; Lieut. Hubert Hamilton; Anne Marie Heskin; Stan Hickey; Brig. Gen. P. D. Hogan; Arnold Horgan; Capt. Rory Hynes; Vivien Igoe; Dr Keith Jeffery; Michael Kavanagh; Michael Kenny; Karel Kiely; Ronnie Kinane; Tony Kinsella; Elizabeth Kirwan; Dr Noel Kissane; Comdt. Victor Lang; Gerry Lyne; Ger McCarthy; Dr Patrick McCarthy; Nicholas MacDermott; Richard McGlynn; Philip Goldrick; Lieut. Col. Dermot McLoughlin; Brig. Gen. P. F. Nowlan; Colette O'Daly; Comdt. P. D. O'Donnell; Capt. Stephen Ryan; Jacinta Ruddy; Col. Desmond Swan; Glenn Thompson; Alex Ward; Bob Webster; Sgt. Joe White; Comdt. Peter Young.

The staff of the following institutions: Carlow County Library; the Central Statistics Office; College of Physicians Library; Curragh Command Quartermaster's Office, Press Office, and Command Engineers Draughtsman's Office; Maynooth College Library; Military College Library; Department of Defence, Property Management Section; Department of Irish Folklore, University College, Dublin; Diocese of Kildare and Leighlin Archive; FÁS projects at Celbridge, Naas and Newbridge; Freemasons' Hall; Irish Architectural Archive; Kildare County Library; Land Evaluation Office; Leinster Leader Ltd.; Mercer Library; Military Archives; Ministry of Defence, London; *Nationalist and Leinster Times* Ltd.; La Sainte Union des Sacres Coeurs, Killashee; National Army Museum, London;

National Library; *Nationalist and Leinster Times*; Office of the Local Registrar of Titles, County Kildare; Public Record Office; Public Record Office, Kew; Representative Church Body Library; Royal Archives, Windsor Castle; Royal Engineers Corps Library, Chatham; Royal Irish Academy.

Alan Sutton Publishing Ltd. for permission to quote from *The Army and the Curragh Incident 1914* by I. F. W. Beckett; Sinclair-Stevenson for permission to quote from *Vive Moi!* by Sean O'Faoláin. For sponsorship in the research for this book I am indebted to Bord na Mona, Curragh racecourse, the Department of Defence, Kildare County Library and Wyeth Medica Ireland.

The interest of the publisher and the enthusiasm and skill of Dr Colin Rynne brought the task to completion.

Con Costello
Tullig, Naas
April 1996

ABBREVIATIONS USED

A.D.C.	*aide-de-camp*
A.S.C.	Army Service Corps
Bn.	Battalion
Brig. Gen.	Brigadier General
By.	Battery
Capt.	Captain
Col.	Colonel
Coy.	Company
D.A.G.	Deputy Adjutant General
G.O.C.	General Officer Commanding
Gen.	General
Lieut. Col.	Lieutenant Colonel
M.C.	Military Cross
N.C.O.	Non-Commissioned Officer
O.C.	Officer Commanding
Pte.	Private
R.A.	Royal Artillery
R.A.M.C.	Royal Army Medical Corps
R.A.O.C.	Royal Army Ordnance Corps
R.D.F.	Royal Dublin Fusiliers
R.E.	Royal Engineers
R.H.A.	Royal Horse Artillery
R.M.F.	Royal Munster Fusiliers
Sgt. Maj	Sergeant Major
V.C.	Victoria Cross

List of Illustrations

MAPS

For Maeve, Mark, Denis, Sheila, Con

Chapter 1

The Historic Curragh of Kildare

The geographical, archaeological, historical and mythical importance of the 5,000 acre plain known as the Curragh of Kildare, and situated in the centre of the county Kildare, has been recognised by scholars since early in the nineteenth century. The townland of the Curragh is in the barony of East Offaly and in the parishes of Ballysax and Kildare.[1] Since the earliest times it has been a great common or unenclosed plain, and its ancient name of *Cuirreach Lifé*, or Curragh of the Liffey, suggests that long ago it stretched to the river's banks. In the early maps of the county the Curragh is shown as an open space. Both the Down Survey of 1654–5 and Sir William Petty's map[2] of 1685 show it is as an area of approximately four and a half by one and a half Irish miles. The eighteenth-century cartographers Noble and Keenan in their map of the county, published in 1752, give the same measurements of the plain, but include the acreage: Irish: 2,910 acres, 2 roods, 12 perches; English: 4,714 acres, 2 roods, 23 perches.[3] The most detailed map of the Curragh, prior to those of the Ordnance Survey, is that of Henry Walker and dated 1807 (Map 1, p. iv). It gives the circumference of the plain as 11 miles, 6 furlongs, 18 perches, Irish, with the greatest length 4 miles 6 furlongs 20 perches, Irish, and width 1 mile 4 furlongs Irish, covering in Irish measure, 3,022 acres 18 perches.[4] It also confirms that the 1791 Act of George III which made it lawful for the Grand Jury of county Kildare to make a road 'round the common by the name of the Curragh of Kildare [which] would not only prevent encroachments from being made thereon, but would also communicate with the several market towns adjacent, to the great advantage of the neighbourhood'[5] was not availed of by the Grand Jury.

John O'Donovan, engaged in the Ordnance Survey in 1837, was of the opinion that 'the Curragh of Kildare has been a plain from the most remote ages',[6] but he believed that the claims that it was once the site of druidicial groves should not be accepted until 'we discover actually existing monuments to bear us out'. He was also conscious of the local attitude to the earthworks, noting that 'the natives do not consider these mounds (at least the small ones) are ancient but that they were thrown up by the military in the year 1798 and before and since that period . . . whether these tumuli be the work of British soldiers who encamped on the Curragh at various periods or that of the pagan Irish can perhaps be ascertained by opening and examining a few of them'.[7]

Brewer, writing a decade before O'Donovan, referred to the then visible earthen works and he also quoted from earlier writings on the plain.[8] Though the assertion by the twelfth-century historian Giraldus Cambrensis that Stonehenge once stood on the plain has never been taken seriously, the Iron Age Hill Fort of *Dún Ailinne* [Knockaulin], *Suidhe Finn* on the Hill of Allen, the many earthworks on the plain itself and the early Christian site at Kildare give adequate testimony to the ancient importance of the area.

The geology of the plain has been described by Frank Mitchell[9] as a grey-brown podzolic soil, which the grazing of large flocks of sheep on the common, and the neglect of the soil 'removed all the calcium carbonate from the surface layers, which became acid and low in nutrients' and encouraged the proliferation of ling heather. Formed on rocks of the carboniferous limestone series, with a variable covering of boulder clay, sand and gravel the plain permits the percolation of the alkaline water to the adjacent Pollardstown fen which is a place of major national ecological importance.[10] John Feehan and Roland McHugh, noting that an important group of fungi was found there, believed that 'one aspect [of the plain] which certainly has not been appreciated is its natural interest and conservation value. The ecological interest of the Curragh is not immediately evident or obviously reflected in its botanical composition, and indeed its natural history has been virtually neglected'.[11]

The sward which covered the plain has been described as 'forming a more beautiful lawn than the hand of art ever made. Nothing can exceed the extreme softness and elasticity of the turf, which is of a verdure that charms the eye, and is still further set off by the gentle inequality of the surface . . .'[12] Gorse bushes are the only natural cover there, and when they are in bloom the full glory of the plain is evident. As the agriculturalist Thomas Rawson observed in 1807, 'what a turf it must be to bear twenty sheep an acre?'.[13] While cattle may have been grazed on the plain in ancient times, and the 1299 act aimed to prevent swine feeding on commonages to the detriment of the sward,[14] by the seventeenth century only sheep were recorded there. The Curragh Act of 1870 restricted depasturing of the plain to sheep, and the destruction of the furze was forbidden as it provided shelter for the sheep. The commission which adjudicated on the rights of claimants in 1868 found that 'the right of use of pasturage for any animals other than sheep' was unsustained, and that 'the pasturage on the Curragh is adapted for sheep only . . .' The prohibition on the grazing of cattle and horses would have prevented the poaching of the sward, while the rooting of pigs also damaged the turf.[15] As the surface of the plain does not retain water, in summer-time the dryness of the turf was ideal for the sheep, but it would not have suited cattle. The three draw-wells on the plain were not found adequate by the military prospectors in 1855, and served the needs of the local population.

With the publication of the *Sites and Monuments Record for County Kildare* by the Office of Public Works in 1988 the extent of the archaeological features on the Curragh of Kildare was definitively established. Forty-eight earthworks and barrows, or the sites of such monuments, were listed on the plain.[16] Numerous other ancient sites were identified in the vicinity of the plain, the most important being the early monastic and medieval remains at Kildare, the hill-fort at Knockaulin, another possible such site at Dunmurry, a possible tumulus and cist on the Hill of Allen[17] and a mound, well and castle site on the Chair of Kildare.[18]

In a survey of the Curragh in 1944 by Séan P. Ó Ríordáin thirty-two existing earthworks were identified and described. He was of the opinion that the belief that the plain was an ancient assembly place was the reason why it remained unenclosed land, 'since the earliest times; it was in fact, then, as now, state property and the horse fair and horse racing of the Curragh are probably as much due to continuing tradition as to the suitability of the venue'.[19] The sites he selected for investigation were in the northern part of the plain as 'it was hoped that these, being further from the camp, would have suffered to a lesser degree from disturbance due to military activities',[20] and he considered O'Donovan's opinion of 1837 that 'around these raths in every direction are traces of modern encampments which do not come under the head of antiquity'.[21] O Ríordáin and his team, which included soldiers from the camp, found the remains of a man who was believed to have been buried alive. It led the archaeologist to conclude that 'the presence there of sites of ritual purpose gives countenance to the contention that the Curragh is to be equated with one of the ancient assembly places named in early Irish literature . . . the archaeological evidence merely strengthens the general acceptance that the Curragh served as an ancient assembly place'.[22]

Modern excavations on the hill of Knockaulin confirmed that it was a major site of assembly during the Iron Age, and that a wood-henge and processional way stood there.[23] It has been proposed by an historical geographer that in Celtic Leinster 'the vast majority of the population lived in the Liffey plain of county Kildare and the valleys of the rivers Barrow and Slaney . . . the Curragh of Kildare and the plains of Meath formed the heartland of early Irish civilisation'.[24]

John O'Donovan, in 1837, had listed the only eight monuments of antiquity which he found on the plain. They included the Gibbet Rath and three other earthworks, including the Race of the Black Pig; a large stone with the impression of two feet, another large stone and, on the Hill of Grange, the large rock called the Chair of Kildare. Cambrensis' story that a stupendous stone circle once stood on the Curragh, O'Donovan dismissed (Fig. 1).[25] The Race of the Black Pig he

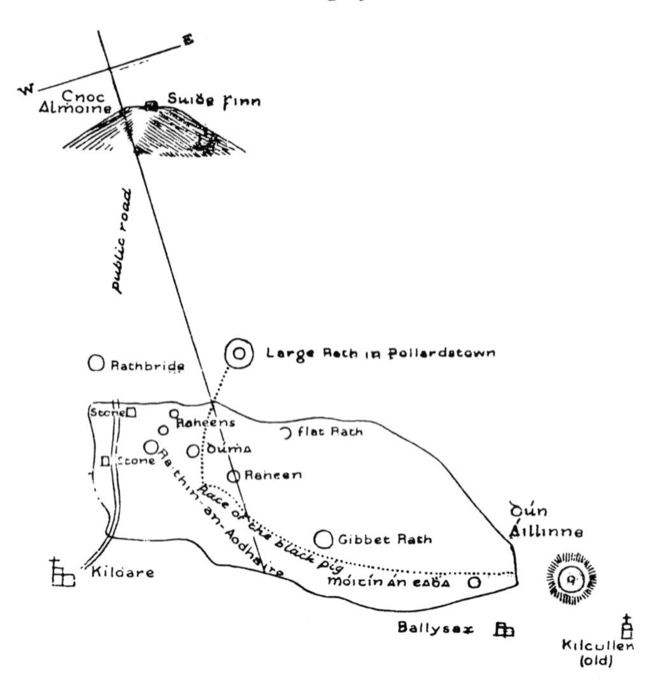

Fig. 1. *Map of the Curragh of Kildare, from* Ordnance Survey Letters 1837, *showing main road crossing the plain and the ancient sites and monuments.*

identified as the ancient royal road leading across the plain to the great rath of *Ailleann* [Knockaulin].[26] In his first letter from Kildare, O'Donovan said that he had found no reference to the plain previous to the period of St Brigid, and if tradition could be relied on 'it would go to prove that it was first formed in a common by the saintess'.[27] He retold the story of her miraculous cloak, that as a reward for banishing the ass' ears from the head of the Leinster king she was offered any reward she wanted. Replying that she wanted as much ground for her cell as her cloak would cover, she spread the garment and it covered the entire plain.[28] Thus, in popular belief, the saint's ownership of the sward was established. A seventeenth-century writer suggested that the legend of the cloak was based on the story of Dido purchasing so much ground as she could surround with ox-hide, and on which she built Carthage.[29] Giraldus Cambrensis in the twelfth century referred to 'the most delightful plains, which are called the pasturage of St Brigid', and when the rights of commonage were being considered in 1867 it was reported that since the saint's time, 'the dwellers round the Curragh claim and enjoy the rights of pasture which St Brigid and the angels won for them'.

Stories about cattle and dairying were told about Brigid, and as the wealth of Gaelic Ireland was counted in its herds of cattle, it was believed that the saint tended her cows on the short grass.[30] In his evidence to the commission in 1866, Lord Strathnairn, the general officer commanding in Ireland, said that 'the military and the sheep-owners benefited and accommodated each other, and the poor people of the district [who had rights on the plain] then believed that the Curragh is our own still, and St Brigid's doing is not undone'.[31]

In pre-christian times the Kildare area was in the territory of the *Uí Failge*,[32] and the Curragh has been associated with one of the fairs or *aonechs*, as the annual or triennial gatherings of king and people were called. Then the dead were mourned, markets were held, the ancients recited the genealogies and history of the tribe, and there were athletics and horse races.[33] The latter sport, according to P.W. Joyce, gave the Curragh its name as the Gaelic word *Cuirreach* meant a racecourse. It could also mean a morass.[34] Ossianic literature associated the plain and the Hill of Allen with Fionn MacCool and the *Fianna*,[35] and the name of the rock of *Suidhe Finn* on the hill commemorates that belief. In modern times when the Aylmer tower was being built on the summit the skeleton of a giant man was found and it immediately entered local folklore as being that of Fionn himself.[36]

The foundation by St Brigid in the fifth century of a convent which accommodated both nuns and monks in an oak grove on the edge of the Curragh is held to be the origin of the name Kildare, *cill-dara*, the cell of the oak.[37] But the advent of the new faith meant the suppression of the old, and so a pagan place of worship became a christian sanctuary. Even if it is not certain exactly when Brigid lived, S. Connolly and J.M. Picard, writing on the lives of the saint, have noted that in the seventh century Kildare had claimed to be head of almost all the Irish churches and with superiority over all other monasteries. Then the dynasty of *Uí Dunlainge* controlled the abbacy, and there was a fine church which was described as 'the treasury of kings'.[38] The transition of worship from paganism to christianity has been investigated by Mary Condren,[39] while the traditional life of the saint has been retold by D.D.C. Pochin Mould.[40] The cult of Brigid, and the eternal flame which her community tended until the reformation, brought pilgrims to Kildare for centuries, and it became the centre of a large diocese at the synod of Raith Bressail in 1111.[41]

St Brigid's cathedral, as constructed in Anglo-Norman times, and restored in the nineteenth century, retains much evidence of the past. A round tower, high cross, early-Christian grave stones, the reputed site of the fire-house and a collection of medieval monuments testify to the antiquity of the site. That the cathedral retained grazing rights on the Curragh until the Reformation was revealed at the Commission of Inquiry in 1866. Then the claim of the Dean and Chapter of St Brigid to rights of common of pasture for the tenants on their lands was allowed.[42] Giraldus Cambrensis, at the end of the twelfth century, learnt that the plain was in the possession of the Crown, and that it had never been ploughed. Its sward was so luxurious, the Welshman was told, 'that if all the cattle in the province should graze the herbage from morning 'till night, the next morning the grass would be as luxuriant as ever'.[43] The royal ownership of the Curragh was much later said to have come about through the marriage of Aoife, daughter of Dermot, king of Leinster, to Richard de Clare, called Strongbow, in 1170.[44] Strongbow's heiress, Isabel de Clare, was given by Henry II in marriage to William Marshal in 1189, and he thus became lord of all Strongbow's lands. However, by May 1207 the cantred of Offaly, which included the Curragh, had been taken into the king's hands.[45]

By the early thirteenth century the rights of pasture were no longer confined to people living in the vicinity of the plain. The Augustinian Canons of St Thomas of Dublin were granted free pasture of the Curragh in 1205,[46] and thus their flocks might be regarded as the first 'foreigners', as those of the '*rawgorrah* [from the Townland of Rathgorragh in west Wicklow] men', or shepherds from the Wicklow mountains, were called in evidence given to the commission in 1866.[47] Confirmation of the crown's ownership of the sward was demonstrated in 1299 when the statute of Edward I forbidding the feeding of swine on the Curragh was enacted.[48]

In the 1830s the Kildare Grand Jury sought to impose its authority on the Curragh by striking a rate in respect of the plain on the landowners fronting on the plain. Thus it was intended to raise a salary for a Conservator of the Curragh, to which office they appointed Graydon Medlicott of Dunmurry. But the objections of some local gentlemen and of the Office of Woods and Forests succeeded in having the innovation suppressed in 1836.[49]

The earliest reference to military activity on the Curragh is found in the *Annals of the Four Masters* for the year 777 AD when a battle between neighbouring tribal groups was fought there.[50] Another fatal engagement took place on the plain in 1234. Sir John Gilbert, writing in 1865, told of the bequeathing by William Marshal, earl of Pembroke and nominal lord of Ireland, of his Irish possessions to his brother Richard, and of the obstruction by Henry III of the bequest. Gilbert wrote that 'at this time the king was greatly influenced by foreigners from Potiou and Brittany. . . .' to suppress the nobility of England.[51] A few years later, on 1 April 1234, Earl Richard summoned a conference on the Curragh with the Irish barons with the intention of solving his proprietorial problems. Despite the efforts of the negotiating Knights Templar the meeting ended in violence and Richard was fatally wounded. He was a grandson of Strongbow, through whom the plain had come to the crown. A century and a half was to pass before the next recorded military engagement on the Curragh.[52] It was in 1406 when the Augustinian Prior of Great Connell, at the head of twenty soldiers, put to flight two-hundred, well-armed Irish. This achievement Lord Walter Fitzgerald believed should be 'taken *cum grano salis*'.[53]

Apart from the violent military encounters on the plain there are also records of more peaceful army presence there. In the month of May 1599 the Earl of Essex and his army camped on the turf. Then 'the lord lieutenant rode forth from Dublin to the champion fields between Kilrush and Kilcullen, where he had appointed to meet him twenty-seven ensigns of foot and three-hundred horse, which he proposed to divide into regiments, appointing colonels for the same . . . These champion fields are called by the natives "Curraghs" . . . A better place for deploying of an army I never beheld'.[54] During the 1641 Rebellion James Butler, the first duke of Ormond, billeted his army around the Curragh. Two years later the confederates were there and Lord Castlehaven, having taken the castle of Tully, 'encamped on a heath called the Curragh of Kildare, from whence I summoned all the castles thereabouts, and had them yielded'.[55]

That the plain in the seventeenth century was an accepted mustering place for the military was confirmed by Thomas Monk who in 1682 was assembling information for the map-maker Sir William Petty. This is an extract from Monk's account:

> Near the centre of this county is the Curragh of Kildare, a large spacious plain and common to all the adjacent neighbourhood, who find it a rich and commodious as well as a healthful pasturage, especially for sheep that bear a fine staple and the sweetest flesh of any in the kingdom, it being thronged with flocks all the year round. It is about nine miles in compass, and together with the adjoining grounds, is reckoned one of the most pleasant soils these kingdoms anywhere can show. The easy ascents, yielding noble and various prospects, and the gentle declinings give content to the wearied traveller as well as recreate and please the gentle horseman and keeper, it being a place naturally adapted to pleasure, and its vicinity to Dublin, being but 17 miles distance, occasions that hither repairs the lord lieutenant, or chief governor, when his majesty's important affairs will admit leisure to unbend and slacken from tiring cares. Hither are also seen to come all the nobility and gentry of the kingdom that either pretend to love, or delight in, hawking, hunting, or racing, for in this clearer and finer air the falcon goes to a higher pitch or mount so as often to be scarce visible; the hounds enjoy the scent more freely, and the courser in his swift career is less sensible of pressure or opposition than otherwhere. And upon any general meeting or rendezvous of the army or militia this is the place, and indeed it is no unacceptable sight upon such occasions to see what numbers of gentlemen with fair equipage, good mien and port, appear there to accompany and attend his excellency.[56]

In the years before the battle of the Boyne the Curragh was regularly used as a place of assembly; in 1687 Richard Talbot, earl of Tyrconnell, and from Carton near Maynooth, encamped his army at the Curragh; Col. Dan Bryan, writing in the *The Irish Sword*, believed that 'the Curragh was firmly established as a training camp under Richard Talbot, duke of Tyrconnell'. In 1688 three regiments of horse, one regiment of dragoons and seven regiments of foot formed a camp there for the month of July. It was announced that persons who so desired might sell 'bread, wine, drink, flesh, fish, butter, cheese or the like at the camp'. On 20 July Tyrconnell issued *A Declaration for the Good Government of the Army* at 'the camp'. It provided regulations for pay, clothing, subsistence, discipline, leave, etc. Receipts for utensils, necessaries and straw for use at the camp were issued in October of that year.[57]

Part of the army of James II trained 'in the camp specially prepared near Dublin not far from the Curragh in Kildare' in 1689, and by August there were nine regiments, with a strength of 4,400, on the plain.[58] The short grass was also to provide a resting place for the retreating Jacobites after the battle of the Boyne when Tyrconnell and the duke of Berwick 'went away as did all the rest to the Curragh of Kildare'.[59] The English traveller John Dunton came to Kildare in 1698 and he recorded this impression of his visit: 'We soon came to the Curragh so much noised here. It is a very large plain covered in most places with heath; it is said to be five and twenty miles round. This is the Newmarket of Ireland, where the horse races are run, and also hunting matches made there being great store of hares, and more game for hawking, all of which are carefully preserved.' He decided that it was 'indeed a noble plain'[60] (Fig. 2).

Fig. 2. *Capt. Robert Gore's 'Oakstick' wins the One Thousand Guineas at the Curragh in 1815. The Standhouse, built before 1777, was to be used as a social centre by military for many years.* Photo: *National Library of Ireland.*

Training camps continued to be formed on the Curragh in the early eighteenth century, and in September 1704 when Sir Richard Cox, lord chancellor, was writing to the lord lieutenant, he petulantly commented 'all the world and our secretary being at the Curragh'. There they might have been visiting a camp, or attending the races, or combining both. A couple of years later Lord Cutts, the commander-in-chief, found that in early April he was 'in such want of air and exercise that on Wednesday last [I] came down to this place [the Curragh] to see the plates and matches and to hunt a little'. He was familiar with the plain, and in the summer of 1705 he had assembled a board of general officers there to investigate complaints against a Captain Morgan, who failed to appear before the board.

That camps were held elsewhere than on the Curragh is evident from Lieutenant General Francis Langston's query, to the lord lieutenant in February 1705, as to where the camp for his regiment would be based. It was the general's opinion that 'the kingdom affords no better place than the Curragh'. He reminded Ormonde that it was in March that contracts were usually made for grass during the time of the camp for horse and dragoons. Such pasturage was a financial bonus for landowners in the vicinity of a camp, and in April 1706 grass was taken at the Curragh for the June encampment. The spring contract for grazing was to ensure a good crop for the horses, and even though the tents themselves would be pegged on the plain the pasturage for the chargers had to be on neighbouring farms. When Lord Cutts was informing the lord lieutenant of the arrangements for the camp of 1706 he wrote 'Langston has already by my order agreed forage, I mean grass, for the horse and dragoons during the encampment on the Curragh. The contract begins from 10 June, and is absolute for eighteen days, but so ordered that we can have grass for a longer time if your grace would not have us decamp so soon . . .' When the camp had been formed Cutts signalled his intention of inspecting the regiments, and 'to see them make their movements, and enquire of the commanders into the conduct and behaviour of their officers'. The Inniskilling Dragoons participated in the review that summer, and afterwards returned to quarters in Boyle and other barracks. Another advantage of using the Curragh that year was that the army had 'preserved our pumps there, which cost us money last year'.[61] This observation suggests that one of the draw-wells there was being used.

Later in the century the plain was again to be the scene of historic happenings. In the years 1778–9, when France and Spain supported the American colonists against the British, there was apprehension that Ireland could be invaded as most of the garrison had been withdrawn for service in America. For the defence of the country the Irish Volunteers were raised[62] and in July 1783 a review of the Volunteers was held on the Curragh. The men were transported there from Dublin on the Grand Canal, on which passenger boats had plied for the first time just three years before when the line was newly opened. It was found necessary to employ all of the boats of the Grand Canal Company (and for which the company provided 100 boards to make seats for the men) to bring the Volunteers to Osberstown.[63] Taking part in the review was a squadron commanded by Captain John Wolfe of Forenaghts, Naas. The review 'afforded one of the most glorious sights to upwards of 50,000 spectators, that was perhaps ever before beheld in this or any country'. The Curragh plain itself was seen as 'the most extensive and beautiful that can be found perhaps in any other part of Europe'.[64] Official concern for the integrity of the plain was demonstrated in April 1796 when a detachment of six field guns marched from the barracks in the lower castle-yard in Dublin to the Curragh in order to assist the High Sheriff of the county 'in prostrating the numerous cabins that had been illegally built on the common'.[65]

Twelve years later when Lord Edward Fitzgerald and his wife Pamela were living in Kildare town he had an unpleasant experience on the Curragh. One afternoon in 1795 when he was returning from the Curragh races, he encountered some dragoon officers who took exception to his wearing a green cravat. They accosted him and demanded that he should remove the cravat, but when Edward's companion proposed that a duel might be necessary, the officers retired. Soon the incident was widely known, and as Thomas Moore noted in his biography of Fitzgerald 'a significant verdict was passed [on the dragoons' conduct] at a Curragh ball shortly after, when it was agreed, as I have heard, by all the ladies in the room not to accept any of them as partners'.[66] The presence of the band of the Londonderry militia at the laying of the foundation stone of the seminary at Maynooth on 20 April 1796 might suggest that the regiment was encamped on the Curragh, but it is just as likely to have come from quarters in Dublin.[67]

More momentous were the events of 1798. General Dundas, the officer in command of the forces near Kildare, was quartered at Castlemartin, near the Curragh. On 29 May, in response to a proclamation offering free pardon to rebels who would give up their arms at the Gibbet Rath on the Curragh, several hundreds of United Irishmen assembled there. When the military under Maj. Gen. Duff approached the rath it was claimed that one of the rebels fired a shot, and the soldiers immediately opened fire. At the end of the day 350 United Irishmen lay dead.

Speaking at a monster repeal rally at Mullaghmast, near Athy, on Sunday 1 October 1843 Daniel O'Connell gave his verdict on that tragedy: 'In 1798 there were some brave men, some valiant men to lead the people at large; but there were many traitors who left the people in the power of their enemies. The Curragh of Kildare afforded an instance of the fate which Irishmen were to expect who confided in their Saxon enemies.'[68] Over half a century later Lord Walter Fitzgerald's understanding of the massacre was that 'some historians of the rebellion state that the troops were attacked on this occasion, which led to their returning the fire; but this is a deliberate lie, concocted to hush up a vast military murder'.[69] It is interesting that both commentators condemned the generals; Lord Walter himself had served in the King's Royal Rifles, and in the social life of the county he must have had frequent intercourse with officers of the garrison.

Early in the nineteenth century when England and France were again at war, and a French invasion of Ireland was feared, militia units camped on the Curragh. The Grand Canal Company operated an occasional service to Robertstown and Milltown to facilitate the encampments.[70] There was a camp in 1803,[71] and in 1804 about 13,000 (or 16,000 by one account) men were there 'in light marching order and ready to move at short notice'. Again the lack of water on the camp site had to be overcome, and it had to be brought by hand a distance of nearly a mile. This was the principal inconvenience. The health of the men in the camp was considered to be better than ever it was in city quarters, and morale was high. On Sunday 12 August, a gentleman who rode through the camp 'saw the men assembled for divine service in hollow squares round the chaplains and most laudably attentive'. Afterwards they formed up and fired a volley, in honour of the Prince of Wales' birthday.[72]

Other visitors to the camp were a team of graphic artists who were recording the assembly for a Dublin exhibition. It opened just before Christmas 1804 'in the large building near Carlisle Bridge', and it was advertised as 'Curragh of Kildare Panorama. An extensive panorama representing the late encampment on the above celebrated plain with the military spectacle on a canvas of 2,530 square feet, from drawings and sketches carefully taken upon the spot, and brings under the eye, as a view, a body of 16,000 troops, forming various evolutions over an extensive track of country, with numerous groups of spectators in the foreground, including portraits of distinguished characters'. An explanatory leaflet was available in the exhibition hall, and admission to the panorama was one British shilling.[73] It was open every day from 9 o'clock until dusk, and it was still being advertised in March of the following year.[74] [A panorama was a pictorial representation on a circular surface].[75] An official report sent to London on the camp stated that 'it has been of great use . . . it has served to convince the lower orders of people that . . . the troops can be moved to different points in a very short time'.[76]

In the summer of 1805 the camp on the Curragh was again a major military and social occasion. It was considered to be 'the military event of the year. The scheme provided for twenty-two infantry battalions being present, nine of which were militia. On the first of August, [luckily a fine day] the troops destined to occupy this camp appeared on the ground in six divisions. The order to take up positions being given by signal, the different points allotted to the several regiments were immediately occupied; and the command to pitch tents being announced by another signal, the whole camp was formed in little more than two minutes from the instant in which the signal was made'. The militia formed a large part of the light infantry, the movement of which to their posts on the sound of a bugle was very impressive: 'the celerity of the manoeuvres, the brilliancy of their uniforms and the gay diversity of their various standards glancing across the hills'[77] greatly delighted the spectators. Even Lieut. Col. William Blacker of the Armagh Militia was impressed by the action, as he noted in his journal: 'Great was the galloping hither and thither of the staff officers of every grade and denomination for nearly an hour, until 1 o'clock, bang went a signal gun . . . when the columns poured into the magnificent plain, or rather downs, of the Curragh by all the roads leading to it. Bands played, colours flying, drums beating, bugles screaming and all the pomp, pride and circumstance of military display.'[78] General Lord Cathcart, the commander of the forces, who had 'a great liking for' Blacker's regiment had his quarters near the standhouse until the camp dispersed in September.

Three years were to elapse before the next camp was made on the Curragh. In 1808 seven militia and six regular regiments, totalling some 11,000 men, were encamped there and a post office was operated for the duration of the camp. A letter marked in ink with a metal hand-stamp and the words Curragh Camp 1808 is preserved. Written in June to a relation in Derry by Major General George Vaughan Hart, an officer on the staff in Ireland, and a veteran of service in America, the West Indies and India, it gives a clear indication of the rigorous training programme of the camp in the Curragh. It tells that:

> . . . the weather here is very hot, and the duties of this camp [of instruction], extremely fatiguing. In order to be in time for guard mounting I am every morning obliged to be on horseback at a quarter before five. Drill from half after six until a little before ten, again after eleven until half after twelve. Dine exactly at two, and at exercise in the field again from six to half after eight, when I endeavour after a ride of three and a half miles, to get to bed as soon as possible . . .[79]

The high point of that season occurred on 4 July 1808 when Gen. Sir Arthur Wellesley, travelling to Cork to take ship for the Peninsular Wars, 'passed the line'. He got a great reception from the troops, causing one onlooker to remark that 'had it been permitted . . . the whole army on the Curragh would have willingly accompanied the gallant general'.[80] The summer camps provided a social season for the local gentry as, in 1808, 'balls were held every Thursday at the standhouse . . . and a public ordinary every Sunday at three o'clock for ladies and gentlemen. This was calculated to relieve the tedium which want of society of their fair countrywomen must give, even in camp'.[81] As recalled at the commission of 1866 by the London solicitor of the Commissioners of Woods and Forests, there had been a camp in 1812. He said camps were made on the Long Hill and 'on the Dunmurry side [of the plain]'.[82]

An engraving of *The Foot Camp on the Curragh* of Kildare from that period shows the lines of tents guarded by sentries, and with a mounted and an unmounted soldier in conversation in the foreground. A water-waggon and a coach are shown proceeding towards the encampment, the former indicating that the lack of water on the plain remained a source of nuisance.[83] Colonel Lugard later remarked that 'the difficulty of procuring a sufficient supply of that necessary of life caused very great inconvenience in former encampments on the Curragh, and finally led to its abandonment as a site for the assemblage of troops'.[84]

The Curragh, as crown lands, was then in the care of the Commissioner of Her Majesty's Woods and Forests and Land Revenues.[85] In 1807 when the Commander of the Forces in Ireland proposed that new barracks should be built around the country the open aspect of the Curragh made it suitable to be considered for a cavalry barracks, and it was surveyed. When news of the possible establishment of barracks on the plain reached the board of the Grand Canal Company they suggested that the canal hotel at Robertstown, where business was declining, might be suitable as a barrack, but the offer was declined by the military authorities.[86] The subsequent report of the board which surveyed the Curragh recommended that a site near the river Liffey at New Bridge would be more suitable, while the proximity [one and a half miles] of the exercise ground on the Curragh was seen as a bonus.[87]

However, less than half a century later even the details of the great musterings of the early nineteenth century and earlier on the plain seem to have been forgotten as, in 1855, when the exigencies of the service, due to the Crimean war, required the strengthening of the army, and the construction of a camp for 10,000 infantry on the Curragh of Kildare was ordered, the Royal Engineer officer assigned to the task could find no plans or records of the camps formed there even fifty years before.[88]

The status of the Curragh as a military station was to change over the years, as the British army was reformed and reorganised. From its conception as a temporary encampment for the immediate training of infantry recruits it was quickly designated a training camp, which status it was to retain. However, its organisational designation was to be that of the Curragh Military Division from 1856 until 1862 when it became headquarters of the Curragh and South East District. In 1865 it was known as the Division at Dublin and the Curragh (including Newbridge), in the Dublin or North District, and with the senior brigade commander in the camp in command there. In 1892 Kildare was included in the Curragh District, as was the Glen of Imaal in 1901. It became a Divisional Headquarters at the beginning of the century, and it was to remain such, with a major general in charge, until the evacuation in 1922.[89]

Apart from its association with soldiers and sheep, the Curragh has always been regarded as the home of horse racing, and the very name itself suggests that races of one sort or another have been held there since the earliest times.[90] The first surviving reference to organised racing there was in a seventeenth-century document of the eighteenth earl of Kildare which suggested that the earl might give 'a plate of about 40 pounds a year, which would bring vast concourse and expense among his tenants, the lands, on this account, will rise at least twelve pence an acre on that hopes, so that, tho' this seems a matter of pleasure, it will really be of great profit to many acres of my Lord's lands that lie adjacent to the Curragh of Kildare, and will improve the town and the rents of land contiguous to it'.[91] In 1696 the government gave two plates of £100 each to be run for annually at the Curragh races, and from that time onwards the winners of the government plates are documented.[92]

Lord Chief Baron Edward Willes spent a night in Kildare in 1760 and he found it to be 'a very pretty town by means of the gentlemen belonging to the Kildare Hunt. Their number sixty-one, and most of the gentlemen have built for themselves little pretty lodges in the town for the convenience of hunting. I came to the Curragh of Kildare, which is the New Market of Ireland, and I am afraid as much money betted there as at New Market. 'Tis a plain belonging to the king, of about fifteen miles in circumference. I never saw a finer turf; the sports say the sod exceeds that of New Market'.[93]

With the foundation of the Turf Club at Kildare in 1790 the status of the Curragh as the premier racing venue in Ireland was established. By 1815 it was the opinion of a clergyman that 'the Curragh . . . is too well known as an extensive race-course, or place of national amusement, to require description', but he was surprised that some of the inhabitants of the sporting neighbourhood contributed to the publication of his book on theology, 'a work so foreign in its objects to the views of the sportsmen'.[94]

The first of several visits by British royalty to the Curragh was that of King George IV who attended the race meeting on Friday 31 August 1821,[95] 'an event that will be long remembered with much pride and gratification, in the annals of the Irish turf', in the opinion of J.N. Brewer writing a few years afterwards.[96] Nevertheless, twenty years later, when Daniel O'Connell addressed a repeal meeting at the standhouse a journalist, commenting on the huge gathering, compared it with the attendance for the king's visit: 'True, there was, on the year King George IV dishonoured this country with his presence, a meeting on the Curragh, which, in point of numbers and respectability, has long been considered to bear away the palm from all Irish assemblies, but those who remember it are of opinion that it bore no resemblance to yesterday . . .'[97]

O'Connell had stayed overnight in Kildare town before the Curragh meeting on Sunday 7 May 1843. That morning, after mass, he came in procession to the standhouse where 'peasantry flocked in thousands . . . from all over the county, and further afield'. It was said that 'not a dozen [people] were left in Naas'.

Thomas Steele, in his introduction to the proceedings, referred to the 'baleful historical associations' of 'this lonely place, the Curragh was stained by the cold blooded massacre [of 1798]', but the Liberator himself confined his local comments to a mention of the members of the R.I.C. lying on the colloquial 'short grass' plain. 'There they were, poor fellows, lying on the short grass with the sky over them.'[98] The term 'the short grass' had long been applied to the sward, and in 1866 it was noted that 'the young men of Kildare are known as "the boys of the short grass"'.[99] O'Connell would have also known that a large force of military, including parties from the eleventh Hussars and infantry regiments, were located in Newbridge and elsewhere. That evening a dinner for 400 guests was held in the standhouse to close a landmark day in the annals of the Curragh.[100]

The ever increasing popularity in the mid-nineteenth century of the race meetings at the Curragh was largely due to the accessibility of the place since the opening of the Great Southern and Western Railway siding there in 1856 (Map 2, p. v). As the duke of Leinster, the largest landowner in the county, was on the provisional railway committee from 1843 there were no major objections to the making of the line, but when it was known that the railway would sever the plain the Turf Club and the Curragh Ranger registered strong disapproval. After months of discussions between the Turf Club and the railway company the former had a victory when they were asked to decide themselves on the route which the railway should follow. Subsequently the Turf Club was compensated with a railway siding laid to Kildare from the standhouse, and in 1853 a new standhouse, financed by the railway company, was opened.[101] The first excursion train ever to be run by the Great Southern and Western Railway was to the Curragh for the race meeting of 15 October 1846. Punters were charged two shillings and eleven pence in first class and two shillings and one penny in second class for the journey, and the train departed Kingsbridge at 11 a.m. In later years when third class carriages were added to the train even cheaper fares were available.[102]

The racecourses are depicted on Noble and Keenan's 1752 map of the county,[103] and it also has the earliest known engraving of a race there. It shows the contest between 'Black and All Black' and 'Bajazet' on 5 September 1751 for 1,000 Guineas. Taylor's map of 1783 depicts the standhouse for the first time.[104] The most detailed map of the race courses is that of Henry Walker, published in 1807. It shows that the four mile course crossed two roads, one of them the turnpike road from Dublin to Limerick, and came as far eastward as the Long Hill, which was to be chosen as the site of the military encampment almost half a century later. The only buildings shown on the plain are the standhouse and new stables at Round Hill, west of the turnpike road, and on the edges of the plain the various racing and farming properties are identified.[105]

The protection of the crown's rights on the Curragh was entrusted to the Commissioner of Her Majesty's Woods and Forests and Land Revenues, and to an officer titled The Ranger of the Curragh. His duties included the protection of the grazing rights and of the game, and to prevent encroachments. The first mention of a ranger was in 1687, and from that year onwards a list of those who held the office exists.[106] His salary was from the crown, and in the early years it was £20 a year and his livery; about the middle of the eighteenth century the salary was increased to £320, and £5.17.0 for livery. From 1717 the ranger was given responsibility for supervising the King's Plate Races, and the maintenance of the race-course.[107] His rewards for those tasks were a guinea for every horse entered for the royal or viceregal plate, and five guineas for a horse winning a race. The longest serving ranger (1818–1867) was Robert Browne, and it was he who had to cope with the major intrusions in his domain when the military settlement was being made. The Curragh of Kildare Act of 1868 drastically altered the terms of service of the ranger, and it ordained that 'there shall not be any salary, fees, or other pecuniary remuneration paid to or received by the ranger'.[108] The appointment of a Deputy Ranger was legislated for, and he was to have the status of a permanent civil servant.

Ranger Browne, a young militia captain from a landed Galway family, has been judged by Fergus D'Arcy to have made 'a decisive contribution to the preservation of the Curragh as a great open space'.[109] During his time of office he had to cope not alone with the establishment of the military, but also an expansion of racing and training, and 'a huge population increase of humans and sheep which respectively threatened encroachment and overgrazing'. His successor in office in 1868 was Henry Moore, third Marquis of Drogheda, senior steward of the Turf Club, and a prominent racing magnate who succeeded in his time to 'keep the soldiers and the graziers in check'.[110]

To judge by a report in *The Irish Times* in 1889 the glamour of the racing fraternity was then well established. The journalist who visited the Curragh in the February of that year found it to be 'a bleak and miserable spot in wintry weather'. Nevertheless he believed that 'life at the headquarters of the Irish turf is in jelly . . .' compared with that of the city dweller. There 'the racehorses, hunters and hounds share the divided attention of the civilian and the military population. We have heard more than once that there are two Irelands, but I am inclined to think that on and in close proximity to the land of the Turf Club exists and thrives a third and a cheerful Ireland'. He visited the training stables, and listened to local gossip, such as the fact that a veterinary surgeon had rejected 'Fethard', the animal which Maj. Doyne 4th Dragoon Guards was to purchase on behalf of Capt. J. Orr-Ewing.[111]

The reporter was also given some of the traditional lore of the plain. An old man told him about the Gibbet Rath and 'a brutal massacre of the Danes or other foreigners in the bad old times when we had neither the military nor the R.I.C. to protect us . . .' His jarvey spoke of the fairies that played music and danced in the rath, and of the hunchback who had his hump removed there. The story of Dan Donnelly's victory in the hollow subsequently named after him was known to the journalist, a gentleman whose impression of the Curragh, even on a bitter winter's day, was a happy one.[112] Donnelly had defeated the English champion George Cooper in a bout of eleven rounds, before an enormous gathering of spectators, in a natural hollow on the east side of the Long Hill in 1815.[113]

Apart from the little folk in the Gibbet Rath there were stories of other more ghostly spectres on the plain. Every seven years the Wizard Earl (Gerald eleventh earl of Kildare, who died in 1585), and his retinue of knights, were believed to rise from their enchanted sleep beneath the Rath of Mullaghmast and ride on their chargers to the Curragh. There they galloped around the plain before returning to Mullaghmast. This resurrection of the earl and his men was said to end only when the silver shoes of the earl's horse had worn out.[114]

The geographical integrity of the Curragh remained virtually intact until almost the middle of the nineteenth century. While the earthworks of Iron Age man and his successors, the roadways and the traces of eighteenth and early nineteenth century encampments, and the racecourses and standhouse of the Turf Club had altered the legendary sward, it was the coming of the railway which was to cause the first serious breach of the plain. The making of the Great Southern and Western Railway line, with the Curragh siding, in 1846 severed the north-west corner of the plain from the rest of it, and the separated portion became known as the Little Curragh.

Otherwise little had changed at least since 1777 when the agriculturist Arthur Young saw the plain as 'a sheep walk of above 4,000 English acres, forming a more beautiful lawn than the hand of art ever made'.[115] The French consul in Ireland, Charles Etienne Coquebert, in 1793 considered it 'a superb sward covered with sheep',[116] while almost forty years later Samuel Lewis expanded on the beauty of the place: 'The Curragh is a fine undulating down, six miles long and two broad; it lies in a direction from north-east to south-west, having the town of Kildare near its western extremity, and crossed by the great road from Dublin to Limerick; and is, in fact, an extensive sheepwalk of above 6,000 acres, [though he disagreed with Young's estimate of the area of the sward, Lewis repeated the Englishman's comment on the plain] . . . forming a more beautiful lawn that the hand of art ever made . . . it is depastured by numerous large flocks turned on it by the occupiers of the adjacent farms, who alone have the right of pasture, which greatly enhances the value of these farms. This plain has long been celebrated as the principal race-ground in Ireland, and is equal, if not superior, to that of Newmarket, in the requisites for this sport'. Lewis also commented on the 'numerous earthworks, most of which appear to have been sepulchral'.[117] Anna Maria Hall, collecting material for her three volumes on *Ireland, its scenery, characters, etc.* in 1840 agreed with Lewis' findings, describing 'the far-famed Curragh of Kildare', which she thought was unequalled 'perhaps, in the world, for the exceeding softness and elasticity of the turf, the verdure of which is "evergreen", and the occasional irregularities of which are very attractive to the eye'.[118]

Despite the intrusions of the eighteenth and nineteenth centuries the Curragh of Kildare in the mid-nineteenth century retained most of its geographical, archaeological, historical and mythical qualities. The earthworks gave testimony to the presence of pre-historic and early-christian man, the soldiers came and went, leaving little trace of their passing; the sites of their tents, field-kitchens and redoubts in time became confused with the ancient raths and trackways. And the flocks of sheep, of up to 30,000 in 1866,[119] continued to browse on the short grass as they had, in the belief of some of the shepherds, for over a thousand years. In more recent times Feehan and McHugh have concluded that 'the Curragh is perhaps the oldest, and certainly the most extensive tract of man-maintained semi-natural grassland in the country, having existed as such for two millennia and probably longer. Taking the cultural and natural facets together, it is very possibly the only landscape of its kind in the world'.[120]

That the Curragh of Kildare is a unique feature of the Irish landscape, providing abundant evidence of the presence of early man, has been acknowledged by visitors there for centuries. Since Cambrensis wrote his account of the plain in the twelfth century many other observant travellers have commented on the beauty of the place, its antiquarian remains, the folk traditions, the game, the luxurious sward, and the great flocks of sheep. The attraction of the space to the sportsmen, the racing men and the military was accepted by the seventeenth century. The people living in the neighbourhood of the plain became accustomed to the transitory musterings and presence of armies there. In the early nineteenth century, when as many as 13,000 men encamped for a month, the shopkeepers in the nearby towns and villages would have profited. The social life of the county also benefited from the military as the attractions of the Curragh expanded from field sports and racing to displays of military exercises, public ordinaries and balls. That period provided a foretaste of the years to come when equal numbers of soldiers were to inhabit the Long Hill intermittently, and their cavalry charges, manoeuvres and firing practices were to become regular occurrences. If the farmers, the cottagers and the traders were to benefit from the expanded population, there were also to be environmental, social and moral disadvantages. The character of the Curragh of Kildare was to change from that of Arcadia, of mythical and pastoral peace, to a place of Mars, of glamour, of sham battles and gun fire with, beneath the surface, squalor.

Chapter 2

THE BRITISH ARMY IN IRELAND AND THE BUILDING OF THE CURRAGH CAMP

The formation of the first standing army in Ireland has been dated to 1474 by the seventeenth-century English political writer John Trenchard.[1] It originated in a force of 63 spears and 160 mounted archers, drawn from the counties Dublin, Kildare, Meath and Louth by the lords of the Pale, under the leadership of the FitzGeralds of Kildare. It was the Guild of St George, which in 1473 became the Fraternity of Arms and was paid for out of public funds.[2] Trenchard described the brotherhood as consisting of 120 archers on horseback, 40 horsemen and 40 pages, and their task was the defence of the Pale. He also gave the strength of the army in Ireland in 1535 as 300, and in 1543 as 380 horse and 160 foot. In Queen Mary's time the standing force was 1,200, and during the reign of Elizabeth I it varied between 1,500 and 2,000.[3]

Commenting in recent years on Trenchard's findings, Kenneth Ferguson wrote that 'after 1535, the year "Silken" Thomas Fitzgerald's rebellion was quelled, there was constantly a standing army which fluctuated in size in the way Trenchard suggests'.[4] The army was billeted on the people, an arrangement which was always disliked by the populace; when Sir Henry Sidney addressed the matter in the Irish parliament in 1571 he explained that the army was necessary in Ireland, as the country was not 'quiet within itself', as England was.[5] England did not have a standing army until 1661.[6]

While in medieval times defensive castles had been built at strategic locations, such as on the limits of the Pale, and at ports and river crossings, they became more vulnerable when gunpowder was introduced here in 1361,[7] and the art of warfare was to change drastically. However, there is no reference to a gun being fired here until 1487, at the siege of the O'Rourke fortress of *Caisléan-an-Chairthe*.[8] When castles were found to be no longer impregnable, new defences were sought, and in 1549 the first forts were built for garrisons at Maryborough and Phillipstown.[9] In the early years of the seventeenth century the construction of the artillery fortifications commenced, and many examples still stand, in Cork Harbour, at Kinsale and Duncannon.[10] In County Kildare a substantial fort was constructed around an old Fitzgerald castle at Ballyshannon, south of Kilcullen. It was taken by Cromwellian Colonel John Hewson in December 1649.[11]

The strength of the Irish Establishment was laid down by the Parliament at Westminster, and the charge was borne by the Irish revenue. The English Act of 1699 (10 William III, c. I) ordained that 12,000 troops should be maintained in Ireland, with the country bearing the cost of them. That principle seems to have been maintained until the Act of Union.[12]

Though the forts provided accommodation for their gunners and sappers they were not barracks, and it was not until 1701 that the first purpose-built barracks in these islands were constructed, on the river Liffey close to Dublin city. The official policy of that period was reflected in a contemporary document: 'By the institution of barracks, these men [soldiers] are kept apart from the people, in the eye and obedience of their respective officers and thereby withheld from insulting or being insulted as is commonly the case in scattered quarters. By being active and powerful in the suppression of riot in others, they become also more formidable to the lovers of sedition, and peace is thereby preserved throughout the nations.'[13]

The Dublin barracks was designed by the county Kildare resident Surveyor – General Thomas Burgh.[14] In 1753 it was noted that *The Barracks* (as they were known until 1803 when they were termed the Royal Barracks) could accommodate three 3,000 foot and a 1,000 horse, and forty years later capacity was given as six troops of horse and six regiments of foot.[15]

In the establishment for Queen Anne's Army in 1704 thirty-six barracks for horse and 277 for foot were listed. It is unlikely that all of the barracks were custom-built as, for example, that in Munster, at Ross Castle near Killarney, which had a barrack block attached to the castle. Barracks were under the care of twenty-five barrack masters, and the entire annual cost of the establishments was £13,336.10.0. The only barracks then listed in county Kildare was that for horse at Athy.[16] It was described in 1729 as being in the area of Lieutenant General Napier's Regiment, and with its roof infirmary and stable in bad repair.[17]

Twenty years later two officers stationed in Athy barracks merited mention in the *Dublin Chronicle* of 13 December 1787 and 18 March of the following year. Lieutenant Mackenzie and his second, Cornet Gillespie 3rd Carrabineers (The Prince of Wales' Dragoon Guards) were 'honourably acquitted of the supposed murder' of twenty year old William Barrington in a duel which attracted a crowd of onlookers, and was fought outside the town on the banks of the Barrow. The combatants Mackenzie and Barrington, twice fired and missed but Gillespie objected to the proposed reconciliation between the duellists and he shot Barrington dead. Barrington was a younger brother of Sir Jonah Barrington, a judge and historian from near Abbeyleix.

Though not listed previously, there is a reference to a square fort at Naas in 1681, and it was later termed 'the south moat barracks'. It accommodated one Field Officer, two captains, four subalterns and 96 men. By 1824 it was found to be in a 'very dilapidated state . . . not worth repairs'.[18] A document titled *New Quarters of the Army* and dated May 1769 gave the stations throughout the country of horse, dragoon and foot regiments at 59 locations, including Athy and Castledermot in county Kildare.[19]

An annotated copy of Alexander Taylor's 1793 *New Map of Ireland*, believed to have been the property of Major General Loftus, shows the five military districts, as set up in 1796, and the locations of permanent infantry and cavalry barracks, and the numbers they could accommodate. There were 12 barracks in the northern district, with accommodation for 976 infantry and 196 cavalry. The western district had 15 barracks, accommodating 1,657 infantry and 374 cavalry. In the central district 216 infantry and 584 cavalry could be housed in nine barracks, including that at Athy which held 36 cavalry. There were eight barracks in the eastern district, with quarters for 2,302 infantry and 462 cavalry, and in the southern district 23 barracks provided billets for 3,120 infantry and 510 cavalry. The locations of barracks on the map were almost similar to those shown on earlier eighteenth-century maps.[20]

The revolutionary cataclysm in France and Ireland, and the war against France, allied to the war in America, during the last quarter of the eighteenth century, and especially the arrival of a French fleet in Bantry Bay in 1786 and at Killala Bay in 1798, had convinced England of the vulnerability of its traditional backdoor. To prevent any attempt of an invasion by Napoleon the construction of Martello towers around the Irish coast was commenced in 1804, and eventually over 40 were completed.[21] Barracks and military roads were constructed in the Wicklow mountains, to control the remnants of the 1798 rebels,[22] and a programme of barrack building in important locations around the country was undertaken. Simultaneously the size of the garrison increased from 41,000 in 1797 to 62,000 in 1801, but it was reduced to 45,000 in 1812 when troops were despatched to the Peninsular War.[23]

The commander of the forces in Ireland, on 5 January 1807, had expressed his views on the necessity for the maintenance of a large force in the country 'with a view to its defence as well as to the comfort and discipline of the army, it is to be his intention to propose that the worst of the ruinous habitations which are at present appropriated as barracks to the injury of the health, as well as the comfort of the soldiers, permanent barracks should be erected in certain situations and it being absolutely necessary sufficient ground for exercising should be hired'.[24] The duke of Wellington aptly described the attitude of the landowners to the building of barracks when he wrote: 'At all times the Irish gentlemen have been anxious to have barracks for troops established, each in his own neighbourhood, and they would sell land for the accommodation of government for that object. The establishment of a barrack affords a prospect of security, requires the outlay of a large sum of money immediately, and if occupied by troops, the outlay in the neighbourhood of their subsistence, maintenance, etc. . . .'[25]

In 1801 a barracks was commenced in Cork city, and over the following two decades the programme of building continued. A return of barracks in Ireland for 1811 distinguishes the permanent from the temporary billeting. Thirty-seven permanent cavalry barracks were listed, including Athy which accommodated four officers, 60 privates and 52 horses; 84 permanent infantry barracks included Naas with accommodation for 9 officers and 160 men. There were 17 temporary cavalry barracks and 154 for infantry, with two at Maynooth; that for cavalry accommodated one officer, 35 men and 35 horses, the infantry barracks accommodated three officers and 75 men. The barracks at Newcastle had accommodation for 25 infantrymen.[26]

The first new barracks to be built in county Kildare was that at Naas, to replace the old one on the South Moat. The construction of the infantry barracks to accommodate 18 officers and 300 men, or double that number in time of war, was commenced on a site near the town in August 1810, the architects being Messrs Bernell, Browning and Behan. It was almost completed three years later, at a cost of £17,900.[27] In November 1824 further work was contemplated there, and the figures for accommodation were then given as one field officer, four captains, seven subalterns and 600 men.[28]

Captain William St Leger Alcock 23rd Regiment Royal Welsh Fusiliers, who served in Naas in 1832, found Naas, then with a population of 4,777, 'the stupidest place imaginable, no one but Lord Mayo living in the neighbourhood, and he is soon going to England. However as a redeeming clause the barrack is excellent . . .' Later, when he was invited to Palmerstown and he found that the Mayo family 'lived in very good style, an excellent cuisine and if I mistake not a French artiste', and to Major Tandy's home at Milbankhouse, he cheered up. He enjoyed shooting snipe in the vicinity, and hunting with the stag hounds of the 5th Dragoons who were stationed at Newbridge. But when one of the deer kept by the dragoons was run to ground behind the barrack at Naas he pronounced the hunt 'a complete failure, the poor creature was not the size of a goat and made no play'. More simple pleasure, he found, was walking by the canal with the daughters of Dr Hall, even though they were accompanied by their governess.

Captain Alcock kept a diary in which he made several entries of the regiment's duties concerned with aid of the civil power. At Naas he 'had a good deal of unpleasant duty in attending the stipendiary magistrate to tithe meetings, tithe sales, driving cattle, etc.' He considered such duties as 'a very harassing and disagreeable duty for both officers and men', and the payment of tithes to the Protestant church as unjust. He welcomed the possibility of a reform taking place in the church. When his regiment moved to Baltinglass in January 1833 the march took them two days, 'a distance', he wrote, 'which though long is generally accomplished in one day'.[29]

The role of the military in aid of the civil power was also that of the 89th Regiment when it arrived at Naas barracks in January 1835. To maintain order during the forthcoming elections part of the 89th was sent to Carlow, Athy, Hacketstown and Rathvilly. At the end of August the regiment marched from Naas to Dublin 'where a much gayer year in garrison seemed to lie ahead'.[30] The quiet life of provincial towns was undoubtedly boring for men who had experience of garrison life in cities at home and abroad. In 1842 William Makepeace Thackeray remarked 'as for the town of Naas, what can be said of it but that it looks poor, mean, and yet somehow cheerful? There was little bustle in the small shops, a few cars were jingling along the broadest street of the town – some sort of dandies and military individuals were lolling about right and left'.[31]

In August 1855 the 240 men in the Depot Company of the 40th Regiment of Foot in quarters at Naas Barracks were replaced by militia units who were not always liked locally. The regular troops were generally popular in the town, and the local militia unit, the Kildare Rifles, also normally had good standing there.[32]

The second new barracks to be built in the county was the large cavalry barracks at Newbridge, on a site of 39 Irish acres, and costing £96,000. It was commenced in 1813 and completed six years later by the firm of Hargrave from Cork. Though the Curragh plain had been considered as a possible site for the cavalry barracks, it was decided that a location closer to the river Liffey would be more suitable, but the closeness of the Curragh for exercise purposes was seen as an advantage. The availability of forage and provisions locally was also appreciated. Accommodation in the barracks was provided for a cavalry force of 58 officers, 810 men and 980 horses, and for two infantry officers and 105 men.

An indication of the perceived isolation of the location of the new barracks was expressed by an officer of the 7th Queen's Own Hussars in 1822 when he wrote 'a very pretty part of the country . . . the only thing is want of society, for besides the barracks which is the largest in Ireland, there is not the vestige of a village or Gentleman's place for miles'. Gradually the town of Newbridge was to develop around the barracks which was described in 1837 as 'spacious and handsome, consisting of parallel ranges of buildings, connected by a central range at right angles; and are capable of accommodating two regiments, with apartments for their officers, and a hospital for 100 patients'.[33] The gradual development of the town beside the barracks in 1855 can be seen on the map of the Curragh bound with Lugard's narrative.[34]

In the first half of the nineteenth century the strength of the British army in Ireland varied from the lowest figure of 11,861 in 1802, to the highest figure of 24,918 in 1828. Between 1802 and 1845 the yearly average strength was 19,400.[35]

Three years was the usual rotation period for a battalion in and out of the country.[36] For administrative purposes the country was divided into six divisions, with headquarters for the north in Belfast, the east in Dublin, the south in Cork, the south-west in Limerick and the west in Athlone. Army headquarters was in the Royal Hospital, Kilmainham, Dublin. There the departments of the commander of the forces, the adjutant-general, the quarter-master general, the judge advocate-general and the medical director-general were based. For recruiting purposes there were three main centres, in Dublin, Cork and Newry.[37]

An indication that the military were already established socially in the county Kildare by 1849 was given by the wife of Lieut. Col. Henry Smith, landlord of Baltiboys, near Blessington, in her journal of that year. Commenting on a ball given by the Mansfields at Morristown Lattin, near Naas, she noted that 'the military' were there;[38] and a couple of years earlier she had encountered 'a decent little pedlar woman who was "long in the army"', and who had come to reside with a poor neighbour in west Wicklow.[39] As the various new barracks throughout the country received their garrisons the local residents adapted to the new social and economic opportunities, and very soon the military became an important element of the population.

The reason for the maintenance of such a large number of soldiers in Ireland was threefold: protection against invasion; internal security, that was keeping the peace and protecting property; and for the purposes of recruiting and training, 'a sort of strategic reserve for Great Britain and the Commonwealth'.[40] Official permission for the enlistment of protestant Irish had come in 1756, but it was not until 1799 that Roman Catholics could join the colours. However there is evidence that long before that date catholics had been recruited in some regiments.[41] Such was not the case in the 14th Regiment Light Dragoons (King's Hussars), at Carrick-on-Suir in 1794 when Private Hyland was court-martialled and sentenced to be given 200 lashes for declaring himself a catholic, and refusing to go to the protestant church. In Navan, in April of the following year, a soldier was confined for attending at chapel during holy week, but even then there was public disapproval of the impolicy of the military authorities.[42] Undoubtedly the tolerance or otherwise in a particular regiment was a reflection of the attitude of the commanding officer.

That both the internal and external roles of the army were exercised during the decades from the Act of Union to the outbreak of the Crimean War in 1854 is certain. In Ireland the aftermath of the 1798 rebellion, and the Emmet Rising of 1803 engaged many regiments in the internal security role. Social unrest was caused by a series of famines from 1816 and culminating in the great famine of 1845; the foundation of the Catholic Association, followed by the rise of Daniel O'Connell, the tithe war and the repeal movement proved that the benefits of the union were not being realised. For example, in county Kildare in August 1820 patrols were being sent 'into the neighbourhood of Kildare, Rathangan, Prosperous,

Robertstown and Kilcock and generally towards the line of the Grand Canal' where 'the existence of a spirit of disorder amongst the people and of a combination to raise the price of labour in an undue manner during the approaching harvest by forcibly preventing the farmers from employing any other than the resident peasantry in getting it in'.[43]

Captain Alcock, who served in Naas and Baltinglass during 1832 and 1833, was very critical of such duties, and during the following 90 years of British army presence in the county, as elsewhere in the country, the soldiers resented being called out in aid of the civil power, a task they considered more appropriate to the police.

War was the soldier's job, and England's wars with France and in America, and the Peninsular War, as well as the provision of troops for the empire, meant that there was a constant demand for enlistments and trained men. This latter requirement was to become a major one with the declaration of war against Russia on 28 March 1854.

Even though the Curragh of Kildare had hosted military gatherings frequently in the past the decision in 1855 to make there a camp 'of a more permanent nature' to accommodate 10,000 men[44] was to alter forever not alone the landscape of the plain, its environment, and its use, but it was also to cause major change in the lives of the local populations. The construction of the camp in itself was to attract to the neighbourhood of the plain a great number of soldiers and civilians, including hundreds of men employed by the building contractors, the craftsmen, labourers and tradesmen, and various other followers. The economy of the immediate area did benefit from the influx, with the demand for accommodation, food and recreational outlets. The public houses flourished. But the absorption of so many labourers from the county caused some landowners and farmers to complain of the difficulty of obtaining workers.[45]

The first official information of the government's plan to garrison a substantial number of men in the county came on 22 January 1855 when the quartermaster general, from the Horse Guards in London, enquired of the commander in chief in Ireland 'if, in the vicinity of Newbridge, there was any accommodation for 500 men and 200 horses?' A week later arrangements had been made for the erection of temporary accommodation for the 200 horses at Newbridge barracks, and for 100 horses at Portobello in Dublin, Dundalk and Cahir.[46] County Kildare was then in the Dublin Military District.[47]

The provision of accommodation for the militiamen embodied on the outbreak of the Crimean war was the principal reason for the building of the Curragh camp. The traditional practice of billeting the militia on the population was unpopular, and led to demands that either new barracks should be built, or that the expense of the men's lodgings should be paid out of the estimates.[48] In the middle of February 1855 Sir John Young, Chief Secretary for Ireland, replying to a question in the House of Commons, said that barracks were to be provided for the militia in Ireland,[49] and the provincial newspaper the *Leinster Express* carried a leader on the subject in its issue of 3 March. By then the enlargement of the army due to the war had forced the government to decide that progress on the matter of accommodation was necessary, and the military were taking action. A month later the same newspaper welcomed the establishment of camps at the Curragh and Aldershot as, not alone would they improve the quality of the soldiers lives, but they would also 'change a state of things irksome both to the men themselves, and those with whom they are compulsorily lodged'.[50]

Major Edmund Mansfield, who was a member of the 1866 Commission on the Curragh, told Lord Walter Fitzgerald in after years that 'when the Crimean War broke out in 1854, the military authorities established a camp of instruction on the Curragh without asking the permission of anyone, the idea being that it was only a temporary arrangement'.[51]

The initial surveying work on the proposed site, the Long Hill, was hampered by severe weather,[52] an indication of the severity of which was that the night train from Mallow to Dublin was snowed-in on the Curragh for twenty-four hours on 8 February. One passenger described the plain as 'the most inhospitable of all places in such weather . . . the snow driven by the hurricane over the Curragh is that of a sea of milk'.[53]

As no details could be found of the previous encampments on the plain the royal engineer charged with the task of making the camp, Major H.W. Lugard, decided to make a detailed record of the new undertaking. While the Ordnance Survey maps[54] showed a former camp site on the Long Hill, and which could be identified on the ground, enquiries to local residents and old officers produced little information, except for the remembered fact that distress was often caused to the men by a shortage of water.[55] Despite the fact that the ground was covered with snow Major Lugard initiated the task of finding a suitable site. Knowing that a plentiful supply of pure water was a vital requirement, and that in the past water had to be drawn over a distance to the encampments, the location of a source was of immediate concern.

It was realised that an enormous supply of water would be necessary for the needs of 10,000 infantry and their supporting establishments and camp followers, as well as for the fire hydrants required in a concentration of wooden hutments. The chairman of the Board of Public Works, Mr Richard Griffith, whose family lived at Millicent near Sallins, was of the opinion that an inexhaustible supply would be found deep beneath the gravel and sand of the plain. However, when experimental wells were sunk they proved to be unreliable. It was decided to invite tenders for the supply of water, and subsequently eight tenders were submitted. Amongst them were proposals to draw water over two miles from the river Liffey, but it was foreseen that such a project could infringe the rights of mill-owners on the river, as well as the possibility that the quality of the water might not always be acceptable. Finally the tender of Messrs A. and G. Holme of Liverpool, for the sinking of a well in a natural hollow near the site, was accepted. At 54 feet an inexhaustible flow of pure wholesome water was found, and there the pumping station was built.[56] It was a substantial brick building, with a tall chimney, and one of the few permanent structures planned for the encampment (Fig. 4).

The *Freeman's Journal* of 12 March 1855 reported that the foundations for the new barracks at Aldershot had been laid. This barracks was intended to replace an experimental camp on Aldershot Heath which had been initiated in 1851 with the intention of improving the efficiency and training of the army. The success of a major military exercise there in 1853 confirmed the suitability of the heath, and early in the following year it was decided to build a permanent barracks there. The site was described as 'salubrious and most eligible', adjectives which would also be applied to the Curragh of Kildare in due course.[57]

To facilitate the building of the Curragh camp the office of the Inspector-General of Fortifications in England forwarded plans to Dublin of the huts proposed for the encampment at Aldershot,[58] and specifications were prepared. When full authority for the works was received on 27 February 1855 the Commanding Royal Engineer placed advertisements for contracts on 1 March.

On 10 March 1855 the *Leinster Express* printed an account of the planned encampment in Kildare, giving details of the numbers and categories of huts which were to be completed by the end of May or early June. Tenders were to be invited for the building of one or more billets, or for the whole number, and which were to be completed in 60 days after the acceptance of a tender.

Messrs Courtney and Stephens of Black-hall Place, Dublin, submitted their tender on the 12 March to the office of the Commanding Royal Engineer at Dublin castle where Major Lugard was dealing with the project, and on the following day it was accepted. Sureties were given by William Dargan, the Carlow born railway contractor, and Henry Knox, an Iron Merchant of Bridgefoot street, Dublin. At a cost of £37,720 Courtney and Stephens were to construct 430 soldiers' huts, ten for staff sergeants, and ten for officers' stables.[59]

On 16 March a further advertisement was issued for the construction of 233 wooden buildings for officers' quarters, stores, guard-rooms, hospitals, etc.,[60] and on 29 March Messrs Holme were successful with their quotation of £25,907.[61] On 14 April they offered to erect ten more huts, for guard rooms and cells for £119 each and the same number for officers' servants' quarters at £117 each.[62] On 24 April a tender from the same firm to supply the water to the camp was finalised. It amounted to £7,739, with £396 additional for arrangements to render the water-works available for fire precautions; there was to be an annual payment of £582.16s for the working and maintenance of the fire system.[63]

Further tenders were arranged, including those with F. Ritchie and Sons, Belfast and Mr John Conolly. He contracted to build the permanent drainage system of brick, and covered with flagging, for £6,843. Mr A. Toole won the tender, at £2,946, for the laying of about ten miles of roads, but when he found that he had underestimated the cost he abandoned it. As winter was approaching and the necessity for good roadways was realised, the Royal Engineers took over the project, re-employing the contractor's labourers, horses, carts and gangsmen. This arrangement proved to be so economical and efficient, it was decided to retain it for the construction of the six rifle ranges and butts. The costs were £5,387.4s.6d, for 10 miles of roads, and £1,700 for the butts for rifle and artillery ranges, measuring from 300 to 2,000 yards. Amongst other items for which tenders were invited were officers' and soldiers' stoves, cooking stoves, privy trucks and other necessaries amounting in all to £8,486.11s.1d.[64]

The plan of the Curragh encampment was a series of ten squares (later designated A, B, C, D, E, F, G, H, I, K) extending from east to west, and fronting to the north, on the Long Hill which occupies a central position on the eastern side of the plain.[65] It was in two divisions with five barracks on either side of a more elevated area where were situated the staff officers' huts, brigade offices, two churches, post office, military police and clock and water tower. Each square of huts was designed to hold a regiment of 1,000 men, and the squares were separated from each other by eight feet high sod fences which were intended to facilitate discipline as well as being a precaution against fire. In the square subsequently to be designated G the staff sergeants' quarters hut was erected inside an ancient rath, and in the adjacent future F square a rath was similarly incorporated in the plan.[66]

The quarters for officers were built on a line advanced in front of each square, while the position of the headquarters and the accommodation for the staff of the general officer commanding was on a little hill some 300 yards from the troops, but commanding a view across nearly the whole line of the front.[67] At the southern side of the squares were the soldiers' cooking huts, ablutions, latrines, canteens, schools and accommodation for the families of men on the married establishment. There were barrack stores and an engineer park, coal and straw stores, stables for the military engineer train, and hospital accommodation for 432 patients in the regimental hospitals and 48 in the general hospital for contagious diseases. Quarters for chaplains and the commissariat officers were provided, and there were racket and ball courts, a market place, and a small depot for constabulary (for the observance of the civil law). The abattoir was some distance away, on low ground near Donnelly's Hollow.[68]

Most of the buildings in the camp were of American white fir timber, with iron being used for certain structures. Only permanent buildings were of brick. They were the water pumping station, the racket and ball courts and the clock tower (Fig. 3). The specifications for the various buildings were clearly detailed in Col. Lugard's narrative:

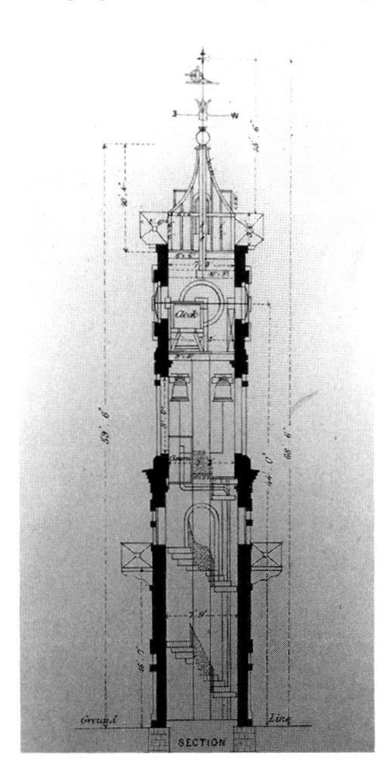

Fig. 3. *The Clock Tower in the Curragh Camp, from Lieut. Col. H.W. Lugard's* Narrative of operations in the arrangement and formation of a Camp for 10,000 Infantry on the Curragh of Kildare. *1858. It was to be the dominant building in the camp for half a century.*

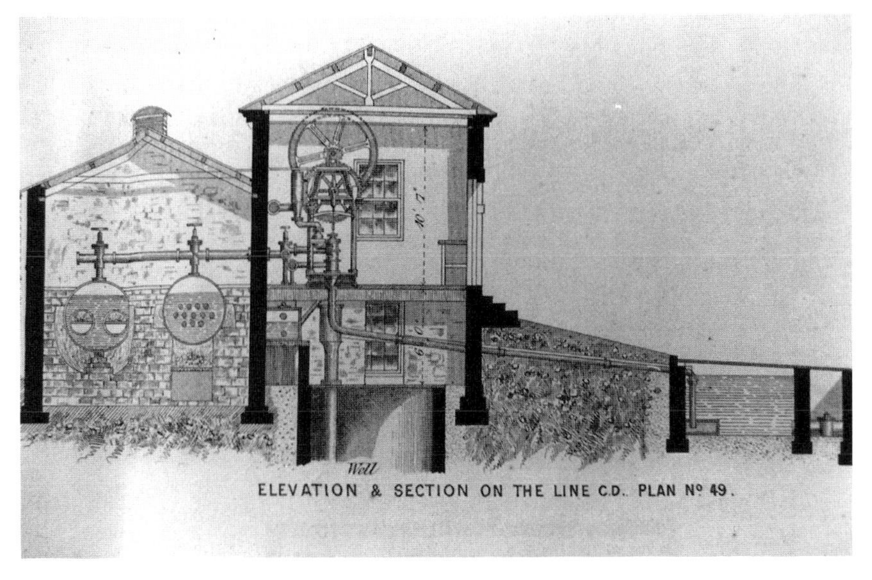

Fig. 4. *Detail of the Curragh Water Works from Lieut. Col. H.W. Lugard's* Narrative of operations in the arrangement and formation of a Camp for 10,000 Infantry on the Curragh of Kildare. *1858.*

> Officers' mess rooms were to be 50 feet by 20 feet, each having an anteroom, kitchen, pantry, scullery, mess-man's rooms, cellars for wine, beer and coal, and a verandah along the front and end. The soldiers' huts were to have openings in the ceilings, and ventilators in the ridges of the roofs, and extra windows were to be placed over the doors in each gable to afford additional light and ventilation. The walls were weather-boarded externally and internally, but the latter were reversed to provide better protection to the soldier from draughts, and to retain the lime-wash, 'and cause it run into laps, in case it should be necessary to resort to such a means of destroying vermin'.[69]

The health, hygiene and safety of the troops was a major consideration. Precautions were taken against fire, and Lieut. Col. Walker, Assistant Quarter Master General at the Curragh expressed satisfaction with the arrangements, but he recommended to the Barrack Department that the camp fire engines should be repaired.[70] Care was taken with the storage of ammunition and it was arranged that the ammunition carts were protected by earthen banks, surrounded by a *chevaux de frise*, and guarded by sentries. Proper drainage of the camp was important and extensive works were undertaken. The latrine system was also efficient. The latrines were 'divided into compartments, with small doors, so hung as to allow the feet of the occupant to be seen from the passage. Cast-iron trucks run under the seats on rails to receive the soil, which is kept deodorized by means of peat charcoal. They are emptied as required . . . by discharging the contents into covered carts, and there mixed with ashes, refuse straw, etc., and sold by the contractor as manure'.[71] During the construction of the encampment the commander in chief himself made regular visits to inspect the troops, and the progress of the work.[72]

At the end of March General Lord Seaton, who had been gazetted that year as commander of the forces in Ireland, came by special train to Newbridge to proceed on an inspection of the ground for the encampment.[73] Everyone seemed satisfied with the proposed site of the encampment, which was to be along the Long Hill from Hare Park. To amuse the general and his party Mr W. Disney J. P. had brought along a pair of greyhounds in slips and he got a hare sohoed in the park beforehand. The local paper reported that 'the run was short, but interesting, the dogs killing the hare in a short time. The works are going on for the encampment; a foundation of bricks and mortar has been laid'.[74]

The construction work commenced on 18 March, and by 9 July accommodation for 5,000 men was ready for occupation. The officers who supervised the work were Capt. F. H. Rich, R. E. and Lieutenant the Hon. W. Le Poer Trench, and the foremen were named Butler, Dobbin, Bergin, Farrell and McLoughlin.[75] Sometimes higher authority interfered in the work; on 24 March, for example, the general commanding directed that two churches, each accommodating 1,150 people, should be constructed in the rear of the centre of the camp, one for Anglican and Presbyterian soldiers, and one for Roman Catholics, and that barrack rooms and schools should be provided for the two chaplains of the army. It was agreed that the school and reading rooms shown on the plans of each barrack appeared to be adequate. As to the proportion of Protestant and Roman Catholic soldiers 'it was impossible to make any calculations as the regiments which are to assemble may be composed partly of English militia'.[76] The military secretary was also monitoring the progress of the building, and was satisfied that the plan prepared by Royal Engineer officer Col. Emmett for the 'temporary barracks to be erected on the Curragh of Kildare included provision for ovens, bakery shops, farrier's forge, etc.', which it is desirable should be provided that the young soldiers may not be dependant upon the contractors for matters of every day requirement'.[77]

The expressed intention of self-sufficiency by the military secretary certainly reflected army policy, but in reality the personal requirements of the soldiers would also be met by civilian entrepreneurs, as had been the case in military encampments on the Curragh since the seventeenth century when vendors of 'bread, wine, drink, flesh, fish, butter, cheese or the like' were encouraged there.[78]

At least one beneficial result to the county was expected from the building of the camp, that was 'the outlay of large sums of money in this county upon works of a more or less permanent, but invariably extensive character'. Thus welcoming the undertaking the *Leinster Express* noted that several contracts had already been entered into, and the proposal to take 70,000 gallons of water each day from the river Liffey and convey it three miles to the camp was mentioned. The employment of a large number of men in removing furze, clearing the ground and making mortar was remarked on.[79]

Then the newspaper gave a lyrical description of the environment into which the encampment was to be placed:

> The camp will be beautifully circumstanced, being placed on the side of the hill, extending from the walled deer park [sic] for more than two miles towards the end of the Curragh on the Wicklow side. In the rear it will be completely sheltered by the hill itself from north winds. The nature of the soil and inclination of the ground will make it completely dry in the wetter season. The view is magnificent, right to the mountains of King's and Queen's counties, left to the Wicklow mountains, in front the plain, with hills extending from Old Leighlin and Stradbally to Dunamase in bold relief.[80]

The building of the huts was divided between the two contracting firms, Messrs Courtney and Stephens, Blackhall Place, Dublin and Messrs A. and G. Holme,

Liverpool;[81] the former were building the quarters for the non commissioned officers and men, and the stables, the latter the officers' quarters, messes and offices. Both firms had built tramways from the edge of the plain to the building site for the transportation of materials which were brought in by railway to Newbridge, and by canal to Corbally Harbour. The scene was described: 'Trains of trucks traverse the train ways, drawn by powerful horses. Each firm has a large stud.' The Dublin firm employed 600 men, and had their own steam-driven saw mill on the site, managed by Mr Doolon. They also had a large work force in their city factory. Up to that date, 5 May 1855, they had imported 14 tons of nails to the Curragh, five tons of which had been used. Fifty-four tons of timber were worked up each day. At the end of their contract it was estimated that 3,800 tons of timber, 1,100,000 tons of bricks, 6,000 tons of stone, 7,000 tons of gravel, 70 tons of iron, and 13,000 superficial feet of glass would have been used. Courtney and Stephens workers on site included 46 carpenters, 40 assistants, 24 blacksmiths, 64 dray horses, and 365 labourers.[82]

The Liverpool contractor had 500 workmen in the camp, and 400 back in Liverpool where the hut sections were being made, mortised, joined and marked for fitting. Despite their apparent industry, by the month of June Messrs Holme were behind in their schedule and the Commanding Royal Engineer for Ireland was complaining. In their prompt reply to his letter the contractors excused the delay on the grounds that the Great Southern and Western Railway Company had not made the branch line which was intended to facilitate the building of the camp. Consequently they themselves had lost time by having the materials brought by canal to Corbally harbour, three miles from the Curragh. Transport by canal was found to be 'tedious, uncertain and expensive'.[83] Then they had to be moved three miles to the edge of the Curragh, from where they were taken to the building site by a two and a half mile length of tramway. Consequently the cost of materials had risen, and so had that of labour.

There had been trouble with the workforce, following a question in the House of Commons concerning the duration of the contract. Many of the workmen left the job, and the remainder demanded higher wages. The making of foundations for one of the battalion areas had caused difficulties, and a new site had to be chosen. Then the bad weather and the exposed locality had added to their problems. However, good progress had been made with the water supply by the sub-contractor; the well had been sunk and pipes were being laid by three teams of pipe layers. The chimney and flues of the pumping station were nearly completed, with two gangs of men working night and day. Messrs Holme assured Major Lugard that now 100 more men had been employed, and greater progress would be made.[84]

The *Leinster Express* remarked on the amount of food needed for the workforce, which in all totalled about 2,000, as well as their followers. To cater for their needs Mr Cleary from Kildare had opened a general store, and 'there was no exploitation of the workers as city prices were charged'. The progress of the making of the encampment attracted numerous groups of visitors from the locality and from Dublin. Later in the month the same paper was speculating on the possible military establishment and strengths for the camp, but could find no news as to those arrangements.[85]

The building of the encampment required such a large work force that soon the farmers in the county were complaining of a shortage of labourers, and a conversation overheard at Athy railway station in the August of 1855 reflected that concern; one man remarked on the difficulty of finding labour, and another suggested that the government should be asked to stop the works on the Curragh. A by-stander quipped 'you might as well expect they would stop the siege of Sebastopol', to which 'a gallant captain, and one of the Grand Jury, replied "Aye, a great deal better", which was greeted with laughter'.[86]

The large workforce employed by the contractors found accommodation in the neighbourhood of the Curragh, and canteens were opened by civilian caterers on the building site to cater for them. Correspondence in the files of the former British Army Headquarters at the Royal Hospital, Kilmainham, Dublin, contains applications from individuals from the city and more locally seeking jobs as canteen keepers, and by the end of May a fourth canteen was to be opened. Daniel Cleary of the hotel in Kildare, and who had already erected a hut for the sale of necessaries to the workmen at encampment C, sought permission to move it to a more suitable site, and he also applied for the job of canteen keeper there. It went to Mr Bobbett, who worked at the canteen in Newbridge barracks.[87] In June it was reported in the local press that Mrs Donald Michael had opened an hotel, but in September the same newspaper described the lady as 'providing refreshments for the workers, though in primitive surroundings', while Mrs Doyle from Queenstown had opened an hotel on the south side of the Curragh.[88]

Two of the men employed in the building of the camp were the brothers John and Bernard Denvir from Bushmills in county Antrim, but who had been reared in Liverpool. They came to the Curragh in the employment of contractors A. and G. Holme. John, who was a house joiner, later wrote an account of his time in county Kildare, and it reveals some aspects of the working conditions there. Many of the men slept in huts provided by the contractor, who also opened a canteen where they could dine and drink, 'too much drink, sometimes', Denvir thought. The Denvir brothers did not want to stay in the huts, and they found lodgings, with ten others, in the home of the Widow Walsh on the road to Suncroft from the Curragh. Early each morning the men left there, taking food for their breakfasts and dinners with them. On the building site they lit fires from scraps of wood to cook their chops or bacon on long sticks, or to boil eggs, and to make coffee or tea. Amongst the workers the Denvirs met a friend from Liverpool, an

Irishman named Tom Cassidy, who had a contract for planing the floors of the huts and he employed about a score of his fellow countrymen as 'floggers'. Such particularly laborious work was sub-contracted to a 'master-flogger', and his employees were known as 'floggers'.

Though both of the Denvirs were teetotallers, they were familiar with Igoe's public house which was situated at the corner of the Suncroft road on the Curragh. It did a great trade with the workmen, the soldiers and the militia, and one day John Denvir saw Capt. John O' Connell, a son of Daniel O'Connell, trying to get his men of the Dublin Militia out of the pub. Another morning he saw him receiving holy communion at the soldiers' mass which was celebrated on a temporary altar under a verandah in the officers' quarters. For recreation on a Sunday the Denvirs walked into Kildare to visit the round tower and the ruined cathedral of St Brigid. Newbridge they did not like as it was too like 'an ordinary English military station'.

John Denvir described how, as soon as they finished a hut, it was occupied by the military. 'I found among them,' he wrote, 'whether regulars or militia [they were] driven to wear the uniform by stress of circumstances, as good Irishmen as I ever met. Coming home from work one evening, I met on the road to the Curragh a party of them carrying, for want of a better banner, a big green bush and singing "The Green Flag" . . . as they came in sight of the famous plain itself a man struck up:

> Where will they have their camp?
> Says the Shan Van Voct,
> On the Curragh of Kildare,
> And the boys will all be there,
> With their pikes in good repair,
> Says the Shan Van Voct.

Denvir thought that some cynics might credit the mens' patriotism to Igoe's porter, but he himself believed it was a spontaneous outpouring from men whose hearts were burning for their homeland. John Denvir was to become an active Fenian when he later returned to Liverpool.[89]

By 28 April thirteen huts had been completed, but it would seem not without some problems. In its edition of that date the *Leinster Express* carried the short note that there was 'a mutiny of workmen employed in erecting huts on the Curragh. An advance of wages is one of the causes'. A week later it had a long report on the encampment, with no mention of the mutiny. Everything, it seemed, was perfect. It predicted that the Curragh of Kildare was 'likely to have a summer season of unusual gaiety and excitement', independent of the racing events. The exigencies of war had brought great numbers of young soldiers for training there for the battlefield. It was thought that the 'solidity of construction' of the huts suggested that the encampment was intended to be of some duration. Noting that the men were promptly and regularly paid by their employers, the journalist gave a summary of the statistics of the undertaking.[90]

While earlier the encampment had been described as 'a miniature town', in June it began 'to assume the aspect of military occupation, and the nondescript crowds of idlers and stragglers always found in the neighbourhood of the troops are beginning to muster strong in the vicinity. Mrs Donald Michael has opened an hotel, there are two post offices, duly organised and licensed with two posts out and two deliveries daily. There are spacious streets, cross streets and alleys . . . a supply of gas is being considered'.[91]

To coincide with the occupation of the first stage of the camp Lord Seaton decided to visit there himself, and on 2 June 1855 a special train from Kingsbridge, with horse boxes included, was arranged.[92] Representatives of the adjutant general, quartermaster general, Royal Artillery and the commissariat general were to accompany the general. One result of the visitation was a suggestion from Seaton that the streets and squares of the camp should be numbered and lettered, as 'such arrangement will be more convenient for the soldiers in the camp, and more easily recollected by them, than any system of names that may be devised'.[93]

Within a few weeks there was an even more important inspection of the encampment when the lord lieutenant came, accompanied by the general commanding in Ireland and his staff. They also travelled by a special train. The formalities of the visit included the notification to the commander on the Curragh that 'Lord Seaton will be in undress uniform, but if the day was unfavourable, he will not appear in uniform'.[94] Following the inspection Lord Seaton went to stay at Bert, Athy, the home of his son's father-in-law General Sir Ulysses de Burgh. Seaton's son, Lieut. Col. Hon. James Colborne, who was also his military secretary, had married Charlotte de Burgh, the second daughter of Gen. de Burgh, 2nd Baron Downes.[95] At Athy the commander in chief was honour-guarded by a troop of the Queen's Bays. On Sundays he attended divine service in the Athy parish church.[96]

On 13 July 1855 the general commanding again travelled by special train to Newbridge to inspect the Curragh hutments. Then he must have been appraised of the fact that the contractors working on the huts in C square had underestimated their time and had exceeded it by almost two months. The square was occupied by the 3rd Bn 60th Rifles and militia; two other corps were due there in the following week. Lord Seaton was soon asking for more plans of both C and G squares, and of copies of the plan of the entire encampment, as soon as they were engraved.[97]

There were other complaints also. The 2,000 men in occupation of the huts in July found that the rain came in torrents through the roofs which had split and rent in places during an earlier warm spell.[98] The unsuitability of the wooden buildings for the camp was becoming apparent, but the work proceeded.

The arrival of more men was reported in the *Freeman's Journal* of 15 August 1855. It said that the 8th Down Militia had left King's Bridge [Dublin] by special train at 3 p.m. for the encampment on the Curragh, accompanied by 'the gallant colonel', the Marquis of Downshire. When the Downs arrived at Newbridge station at 6 p.m. they were received by the bands of the 60th Rifles and the Westmeath Militia, who preceded them on the three mile march into the camp. That evening the officers of the Westmeath entertained the colonel and other officers to dinner in their mess-room, and the bands of the different regiments encamped performed during the evening.[99] But the fact that a week later some stores were still in short supply was the cause of a complaint from the marquis. He was not satisfied with the supply of medicine available in the camp for his unit.[100]

The man charged with the responsibility for the maintenance and equipping of barracks was the barrack master, and his duties were laid down in regulations published in 1824.[101] As the local representative of the secretary of state for war he had responsibility for all matters connected with the duties and expenditure of the barrack department. He might be in charge of more than one station, as in 1867, when Barrack Master Eveleigh, who was stationed at the Curragh, also administered Athy, Carlow and Maryboro' barracks. The barrack master and his staff attended to the 'comfort and convenience of the troops in barracks'; they issued barrack stores, and were especially busy when new troops arrived at a station. Bedding was a major issue as each man had to be supplied with straw for his palliasse (which was changed every two months), a pair of sheets (for which a charge was made of 2d for a single and 3d for a double bed. They were changed once a month, at the same cost). Roller towels were issued, and changed each week. One towel served twelve infantrymen or eight cavalryman for a week, and a penny a week was charged for the washing of the towel. The control of fuel and light was also the barrack master's responsibility; candles were issued by the pound weight, and coal by the bushel. The regulations allowed that: 'In Ireland where turf is issued in some places in lieu of coals, the turf kish is to be considered equal to a bushel of coals, the turf kish measures four feet long, two and a half feet deep and two feet wide in the clear, being twenty cubical feet'.[102]

The furnishing of the Curragh huts was the responsibility of the barrack master there, but within a month of the first occupation he had exhausted his stores, and he requested a further supply of soldiers' and officers' tables, bedsteads and boards and trestles.[103] By August the barrack master was again out of tables for the soldiers' huts, and as more troops were expected in the middle of the month he urgently requested a re-supply.[104]

Military arrangements for the smooth administration of the law at C barrack of the encampment were then being made. The appointment of provost sergeants there was being sought, and this was quickly approved, as was the sanctioning of an allowance of 17/6 per day to Mr Robert Browne, for his attendance daily at C

Square in his double capacity of Curragh ranger and magistrate. Lord Seaton ordered that Browne should live on the Curragh, but by the end of the year the military secretary was asking on whose authority Browne had been allotted a hut in the encampment.[105] By June a further part of the encampment was near completion. At that time, too, the proposal that the police should act as caretakers of the Curragh was rejected by their authorities. Subsequently the office of ordnance approved the employment of one head keeper, six local keepers and four pensioners for that function.[106]

The formalisation of military procedures at the camp continued and in July 1855 a warrant authorizing the general officer commanding, or the officer commanding the troops in temporary hutment barracks on C and K lines in the Curragh of Kildare, to assemble a district or garrison court-martial.[107] The military establishment was likewise being filled: the appointments of assistant adjutant general, assistant quartermaster general and brigade commanders were made.

Daily routine in the Curragh camp was governed by the regulations made by the deputy adjutant general in Dublin, on the orders of the general commanding in Ireland. A gun was to be fired by the artillery each morning at 6 a.m., at midday, and at 9 p.m. The last post was at 9.30 p.m., and by that time all men were required to be in their huts. The details of the duties of the barrack-guard, rear-guard, provost-sergeant, battalion picquets, field officer of the day in each brigade, and captain of the day for each battalion were specified. The movement of troops and civilians within the camp was to be controlled. Passage to or from the squares through the intervals between the huts was forbidden, and men could not go more than a mile from their regimental lines without a pass; if they were found out of bounds by the patrols they would be escorted to the guard room.[108]

The public was allowed to circulate along the front line and, during the day, through the streets of the camp, but only by the entrances specified, and not after tattoo. If they wished to enter the camp they had to have the permission of the officer at the guard. Sutlers trading in the lines were issued with passes, and carts bringing baggage or provisions into camp were to enter by the rear road only.

To confirm the status of the War Department property there was an annual perambulation of the camp. Usually held in the autumn, it included the 24 hour closure in one of two days to the public of the military roads through the camp. Barriers were set up, sentries posted, and only military personnel and H.M. mail carts were allowed to pass. Undoubtedly the arrangements caused inconvenience to traders and other habitués of the barracks, but as the closures were notified in advance, the situation had to be accepted by them.

There were fire regulations which prohibited smoking near the ammunition depots, the forage stores and the stables. As a fire gap the earthen walls between the lines were to be carefully preserved, and another precaution against the spread of fire was the placing of huts constructed of corrugated sheet iron at intervals between the wooden ones.[109] Lord Seaton, General Officer Commanding Ireland, recommended that as an additional safeguard in the large huts the stove pipe apertures should be enlarged, and distanced from felt and wood, and that large slabs should be put beneath the stoves.[110]

The matter of a uniform for the fire-brigade was also being given serious consideration by Lord Seaton. He suggested that the assistant fire-master and the sergeants should have tunics, the men blue serge shirts, but no pattern or the words 'Curragh Fire Brigade' was necessary (Fig. 5).[111] However, as the brigade was not permanent, and composed of men from various corps, Seaton did not propose any innovation in the dress of the army, and the use of serge was to protect the uniform 'as in the case of working parties in the Mediterranean', and it was not to be worn on parades.[112] It was finally decided that the crew's uniform was to be the same as that of the firemen at Aldershot.[113]

Five fire-engines were stationed in the camp, but they were not always effective when required. Despite the fire-precautions there were outbreaks of fire, such as that in November 1855 when a hut next to those occupied by the Limerick Militia was burnt. The men used wet blankets to try and quell the blaze as the fire engine was located some distance away. Commenting on the fire a local newspaper said 'the fire-engine of which we have heard so much was not available 'till the hut was almost consumed, though its after performance, flower-pot fashion, satisfied the

Fig. 5. *The fire-brigade outside the fire station at the Water Tower in the Curragh Camp. The royal arms are visible above the doors,* c. *1910.* Photo: *Eason Collection, National Library of Ireland.*

spectators that it was, like some other appurtenances of the camp, of a purely ornamental description'. If the wind had not been blowing away from the encampment it was thought that but 'a miracle alone would have saved it from destruction'.[114]

The efficiency of the fire-brigade was again being questioned in April 1857, this time by the military secretary at the Horse Guards in London, as there had been complaints of a bad performance at a recent fire there.[115] In September 1859 there was a fire in B square, following which officers of the 16th Foot made claims for compensation.[116] The fire-brigade also responded to calls from outside the camp, when requested.[117] In 1868 the status of the fire-brigade was raised when the general commanding in the camp ordered that the superintendent of the fire-brigade should not be the garrison sergeant-major. He would instead be of the rank of quartermaster, and he could additionally have the duties of looking after the camp market, and generally superintend in the camp.[118] A photograph taken in 1868 shows the thirteen-man fire crew of the 3rd Battalion Grenadier Guards posing before a fire screen. They are dressed in smocks, with pill-box hats.[119]

The self-contained nature of the camp is apparent from the existence of a postal system, and the preservation of a letter with the handstamp mark 'The Curragh Camp' and dated August 1855.[120] Other mentions of the early post office include another from August 1855 when the secretary at the G.P.O. in Dublin requested that the military assistant at the Curragh should not be removed before his successor arrived.[121] At the end of October the post master in the camp was given an allowance of fuel and light.[122]

In July 1856 the comptroller general of inland revenue stamps was considering the establishment of a post office at the Curragh to sell stamps. He was communicating on the subject with the distributor of stamps for the counties Carlow, Kildare and Kilkenny, and by the seventeenth of the month the sale of stamps on the Curragh was arranged.[123] However in October unknown difficulties had arisen at the office, and the post master was complaining, and it was reported that the post office was to be removed.[124] Subsequently the difficulties with the postal system were sorted out and an efficient service was established.

Changes were being made at the post office in 1859, and at the end of July the G.P.O. issued an order that the camp office would be kept open until 10.30 p.m., to allow fifty minutes for replying to letters. Orderlies were required to be at the office at 9.40 p.m., when the day mail arrived. Within a couple of weeks the G.P.O. was complaining that the post office was slow in issuing the letters for the camp.[125] A further advance in communications was also in hand at that time. The British and Irish Magnetic Telegraph Company had tendered for the extension of telegraph wires from Newbridge to the camp, and on 30 August 1859 their tender was accepted, a condition being that they were to keep the wires repaired. The company was able to provide telegraphic communication to all parts of the

United Kingdom.[126] Despatch bags came to the camp by train from Dublin to Newbridge, and they were in the custody of the train guards.[127]

The increase in military traffic between the railway at the Curragh siding and the camp demanded the making of a new roadway, to which the military contributed funds. But not sufficient, as the Curragh ranger was obliged to seek further monies to complete the road.[128] Early in September 1855 it was reported that the 'vast camp was all but finished'. The contractors were handing over the accommodation, which was inspected and approved, and the foundations for two churches, which were to be in the gothic style, had been laid.[129]

A visitor to the camp in the autumn of 1855 was Thomas Lacy from Wexford who was touring the country by train, to collect material for a book.[130] First he visited the town of Newbridge which he saw as 'comparatively new and thriving, consisting of one long and wide street; the military barracks, with their yards and offices, forming one side of it, and the new and fresh range of houses, which for the most part are licensed to sell spirituous liquors, being well frequented by the military, of which there was, at this time, an unusually large number in the town. Indeed, from the number of troops, and from the money necessarily spent by them, which falls as it does, into the hands of a comparatively small number of persons, the town could be scarcely anything but prosperous'. Then he went on to describe the 'evidence of taste and comparative refinement', the detached groups of three or four cottages . . . for the most part thatched, and with gardens, the ivy-covered Protestant church, and the demesne of Moorfield, the residence of Ponsonby Moore, Esq.

The importance of the military presence on the Curragh greatly impressed Lacy. Remarking that the outbreak of the Russian war prompted the making of the camp and that at Aldershot, 'in order that her majesty's forces might be rendered familiar with field exercises and also camp usage', he then went on to describe his first sight of the camp:

> . . . approaching the Curragh the visitor will perceive in the distance a long line of low habitations, which bear a resemblance to what might be supposed to be the city, or principal abode, of some king or chief of an uncivilised race, such as we have seen described by Mungo Park and other African travellers. These newly-constructed huts, if confined within a comparatively limited space, would strike the beholder as being of a very great extent, but from the vast size of the plain on which they have been erected, their appearance is by no means so imposing.
>
> Several regiments which were in the camp when I paid my visit happened to be out for exercise, and although some of them were pretty full, as for instance the County Limerick Militia, which musters 800 strong, they appeared from a distance to be a mere handful. There

> were at that time eight regiments, amounting to about 5,000 men, stationed here. The line of encampments may be said to extend a length of nearly two miles, in a direction partly east and west, occupying a gentle acclivity, which slopes downwards towards the north and south, whereby the water in rainy weather flows quickly off the ground, an advantage which must be considered of the greatest importance as regards the health and safety of the troops.[131]

However, another civilian visitor to the camp at that time thought that the soldier's view of the place might not always be a happy one: he found that 'a walk through the camp gives the impression of a very quiet, tame sort of place, with very little incident attached to it, and very little food for observation'. He described officers chatting in one place, soldiers in another and men and their wives relaxed in the mess room over a meal. The huts were found to be furnished in a most modest manner. When he asked a soldier how he liked the Curragh he answered, 'It's a goodish place sort of in dry weather, very healthy, but after 24 hours rain, why then, it was ankle deep in mud, like Sebastopol'. The visitor felt that the camp could be 'gloomy at night when nothing was happening, but that in the daytime there was a great stir with parades and field days. The drums and trumpets could be heard far and wide'.[132] That month gas pipes were being laid to provide lighting in the camp.[133]

Appreciating the extensive planning and work which had gone into the speedy creation of the county Kildare camp Lieut. Col. H.W. Lugard decided to set down all of the arrangements for its building, and his recommendations, so that if there was again such a requirement his compilation would be available as a blueprint. However, he did admit that 'the first reflection which naturally occurs is one of regret that this extensive amount of accommodation for troops was not constructed of more permanent materials; but it is necessary to bear in mind that the camps in England, as well as at the Curragh, were first called into existence by the exigencies of the service in a time of war, when the embodiment of the militia pressed for a large amount of barrack accommodation, and the necessity for training troops in large bodies became apparent. The first idea, probably, was to provide shelter for the soldier somewhat superior to that afforded by canvas tents in a humid climate, but with no view to more permanent occupation than was absolutely necessary for the training and requirements of the army during the continuance of the war – views that would appear, however, to have been enlarged upon during the early progress of the camps first undertaken. Hence the adoption of wood, and other but too temporary and perishable materials, in the construction of these military towns . . .'[134]

Some years later, during the sitting of the Commission of 1866, when the assistant quartermaster-general was asked if an extension was required to the camp, in which direction it should be built, he replied 'towards Kildare', and when Capt. Sim, R.E., was questioned on such a possibility he said that he thought the government would never again build huts: 'In my opinion the government will never construct wooden huts for a permanent encampment. It is not economical.'[135]

Before he had his project completed Lugard was attached to the Chinese Expedition, and he requested Capt. G.W. Leach, R.E. to complete it. Col. Lugard died at Hong Kong in 1857 and his *Narrative of operations in the arrangement and formation of a Camp for 10,000 infantry on the Curragh of Kildare* was published as an official document in 1858. It was dedicated to General Sir John F. Burgoyne, G.C.B., Inspector-General of Fortifications, and printed in Dublin by Alex. Thom and Sons. Lord Seaton gave permission for Capt. Rich's map of the Curragh Encampment to be lithographed to accompany the narrative.

The parliamentary paper *Return of gross sum expended in erection of Camps at Curragh and Aldershot 1857–8*, gave the costings for the building of the Curragh Camp as £184,081.10s.10d, and of Aldershot as £476,892.6s.8d. Also stated was the fact that the Curragh was to accommodate 10,344 men, and Aldershot 20,000.[136] However when three years later a further *Return* was published it included all the expenses of the undertaking on the Curragh, including barrack huts, water, drains, roads and ranges to 31 March 1861 as £192,821.14s.2d.[137]

The deputy quartermaster-general and the commanding royal engineer were complimented by the General commanding for their exertions in having 'temporary barracks provided. The new camp at the Curragh has been erected, and the necessity of resorting to billets has seldom occurred'.[138] Lord Seaton also complimented Col. Lugard for 'the rapid and efficient manner in which these operations [the formation of the camp] had been carried out'. The camp, in his opinion, fulfilled the objects for which it was designed: '1st: the separation of the regiments so as to assist in their regimental discipline, regimental police arrangements, and the interior economy peculiar to each corps. 2nd: to guard against the spread of fire into a general conflagration, without so widely spreading the camp as to militate against the comfort of the troops, and facility for assembling, both on ordinary occasions and for general parade'.[139]

The contractor Mr W.R. Stephens of Blackhall Place in Dublin, when he had closed his contract on the Curragh, wrote a letter of appreciation to Capt. Rich, Commanding Royal Engineer at the camp. Stephens described the contract as 'a heavy one . . . and a losing one, though not as bad as anticipated'. In closing the accounts the contractor wrote, 'I will try and end all unpleasant recollections attending it'.[140]

The final legal arrangements in the transference of part of the lands of the Curragh from the Crown to the military came when James Kenneth Howard, the commissioner of Her Majesty's Woods and Forests and Land Revenues, signed a memorandum with the Rt. Hon. Jonathan Peel, Her Majesty's Principal Secretary of State for the War Department, on 13 January 1859. The memorandum confirmed her majesty's right to 'a certain tract of land containing four thousand acres or thereabouts situate in the county of Kildare and subject to such rights of pasturage and common if any as are now legally exercisable thereon'. Reference was made to a request from late principal officers of her majesty's ordnance for permission to form an encampment for troops on the Curragh, and to erect certain buildings and other works, on the payment annually of the sum of £5 to the Commissioners of Her Majesty's Woods and Forests and Land Revenues. The commissioner gave 'full right, power and authority . . . for the making of temporary huts and other erections and buildings for the accommodation of ten thousand men' to the secretary of state for the war department, who was to bind himself and his successors to the payment of £5 yearly from 5 April 1855 as an acknowledgement of the crown's rights of pasturage or other common rights as a result of the making of the encampment, or from its occupation. On six months notice, if requested to do so, the military were to remove all structures and surrender possession of the occupied lands to the crown.[141] That year of 1859 also saw the opening of the constabulary barrack and the court house near the camp.[142]

The area of the Curragh plain to be available to the military was designated by the Act of 1868, and it was marked as Brown and Blue lands on a map which was lodged with the clerk of the peace for county Kildare (Map 2). The residue of the ground, marked as Green, was to be used for horse training and racing purposes, but it was also to be available to the military when required for reviews, drills and recreation. With the written permission of the lord lieutenant, in time of emergency, it might also be used for temporary encampment. Paragraph 18 of the Act confirmed that, subject to the provisions of the Act, all rights of common of pasture, rights of way, and other rights existing in, over, or affecting the Curragh should continue.[143]

Proposals made by Major General the Duke of Cambridge general commanding in chief of the army in 1857 concerning the new camps at Aldershot and the Curragh formed the basis for the use of the Curragh camp over the next 65 years. It was to be used for the concentration of large numbers of troops during the summer drill season, from May to September, after which the men would return to their permanent barracks for winter. A small detachment, for security purposes, would remain in the Curragh permanently. As the strategy of waging war evolved so did the tactics, and the expanse of the plain accommodated the innovations of the tacticians. From the mass movement of man and horse to the employment of mechanism armoured vehicles and the expansion of trench warfare the short grass was constantly fought over.

The camp on the Curragh of Kildare was to evolve into the largest military station in Ireland, and to continue to be an important training centre for the British Army until 1922. It was also to provide economic benefit to the tradesmen, farmers, labourers and smallholders, as well as the business and professional people, of the district and adjacent counties. The social lives of the people of all classes benefited from the presence of the garrison and its attendant civilian followers. However, at its inception, the encampment had caused concern on various grounds to many people, especially those living on the periphery of the plain. Their voices were to be heard at the inquiry which was convened at Newbridge in the autumn of 1866.

Chapter 3

The Curragh Commission 1866

Over the years as the presence of the military on the Curragh became established and the advantages and disadvantages of the encampment became obvious to the local people, certain issues remained unresolved. No legal basis had yet been put in place for the military occupation, and the public and individual rights of way, and rights of pasture and other rights had not been clarified. If the shepherds believed that their entitlements had been established for generations, the racing fraternity might also trace their traditions there for centuries. The often conflicting interests of the residents, the shepherds, the horse trainers, the Turf Club and the soldiers urgently required attention, and the first party to be thus facilitated was the army.

The military occupation of part of the Curragh for the past decade was to be legalised in 1864. A lease between the Commissioners of Woods and Forests and the War Office for a certain number of years, from seven to twenty-one, was suggested by Earl de Grey and Ripon, the Secretary of State for War.[1] It allowed for the continuation in use of the place by those people with rights of pasture, and of the Turf Club, under the office of the Curragh ranger. The military did not accept the proposals, and they believed that it would be more acceptable to them if the War Office assumed guardianship of the Curragh. Thus the matter rested for a while until Lord Naas, whose family home was at Palmerstown near that town, took it up in the House of Commons.

Lord Naas, in May 1865, raised in the House of Commons the subject of the status of the Curragh, and the apprehension in the county concerning the intentions of the crown and the military there. The ownership of the soil was uncertain, he said, causing the resident magistrate to dismiss cases of trespass against women. He asked 'what was the government's intention?' Robert Peel, the Chief Secretary for Ireland, replied that it was crown property, but this was greeted with cries of 'no, no'. 'It could be leased,' he said, 'to the War Department for 21 years for £170, and payment of £350 to the ranger, but it was proposed that that office should be left in abeyance as his duties conflicted with the rights of the War Department.' Colonel Dunne, M.P. for Queen's County, objected to the proposals, and was encouraged with 'hear, hear' from his supporters. Dunne held that the crown had no rights whatever on the Curragh, and the move was an atrocious invasion of public property. Having commented on the reduction in the numbers of sheep on the plain on account of the military presence, he said that in England the government had purchased Aldershot at a great cost, but in Ireland the land was occupied in an outrageous manner. The Curragh belonged to the Irish people, and the crown had no right to possess it. He hoped to raise a subscription to take a trial of the case. He was supported by Lord Dunkellin who said he had no idea the crown had such a scheme of spoilation. It was known for years as a matter of notoriety that the government had faulty title, and had no right to transfer it to the War Department. The camp, he thought, 'was a nuisance and an eyesore, and yet it was sought to perpetuate it. There was no knowing where it would all end. The Curragh was a place of national renown, and ought not to be cut up and destroyed. The bargain which the government was about to make was not only illegal, but would be most obnoxious to the whole county'. Col. French agreed that the crown had no right to the land, and should pause before entering an illegal course. Then William Smith proposed that the respective rights of the public and the Crown should be ascertained, but he regretted that Lord Naas had not proposed the appointment of a committee to enquire into the matter. Then the subject was allowed to drop.[2]

The speech of Lord Naas was the subject of a leading article in the *Leinster Express* on 5 May. It said that Lord Naas believed that it was clear to every observer that the military occupation of the plain was intended to oust every other interest, and while the advantage of such a large body of consumers would be good, the loss of such a great amount of grazing would be unacceptable. Already the race course had been occupied by military manoeuvres when it should have been clear for racing. The county could not consent to two such usurpations, and the title was being disregarded. Even the office of Ranger was in danger of being abolished.

Meanwhile, the writer of the leading article observed, the residents of Kildare were silent, their rights were being alienated without local remonstrance, but surely one should be got up? The parties should meet to discuss their grievances. The writer suggested that without the intervention of Lord Naas and Col. Dunne the transfer of the plain to the War Department would have taken place. But it was noted that there was no conclusion to the parliamentary discussion. A correspondent added a note to the leading article to state that if the Commissioners of Woods and Forests gave charge of the plain to the War Department the livelihood of the people living in the vicinity would suffer, such as the families in comfortable circumstances who earned their livings minding sheep or collecting droppings for sale. But not only the shepherds, but also the farmers would be at a loss. The members of parliament were exhorted to do all in their power to preserve the public rights on the Curragh.[3] A few days later the Irish members had a tea-room meeting in the House of Commons and they resolved to protect the interests of the public against aggression on the Curragh. It was proposed that a select committee should be formed to consider the matter.[4]

In mid-June the *Leinster Express*, quoting *The Irish Times* again returned to the question of the Curragh. The plain was described as 'impaired, untouched upon on any side', and no permanent squatting had taken place on it. There were public roads maintained by county *cess*, a form of rates collected by the Grand Jury of the county up until the end of the nineteenth century. The crown, it was believed, had let the 5,000 acres of the plain to the War Department for £530, a nominal sum as it was well worth £1 an acre. It was regretted that the House of Commons had been called out early on Friday night, before Lord Naas could bring forward a motion requesting that a Select Commission should be appointed to inquire into the claims made by the crown on the Curragh. *The Irish Times* writer opined that 'the Curragh is well worth a struggle, were it only because it is the finest race-course and training ground in the United Kingdom'.[5] In the issue of the local paper of 1 July, was a letter from Lord Naas in which he stated that 'at present there were no proceedings to lease the Curragh to the War Department'.[6]

The same paper on 11 November, and in a few subsequent issues, carried an official notice to the effect that commissioners were to be appointed to ascertain the boundaries, rights of common, rights of way, etc. . . . on the Curragh.[7] A map, coloured to indicate the areas concerned, was to be available for scrutiny by the public. Authority was to be sought for the stopping or alteration of roads and the halting of rights of way, as necessary. Lands coloured brown on the map were to be vested in the Secretary of State for War. There were to be provisions for awards and compensation, and the making of rules and regulations, as well as arrangements for the racing and training of horses. The map might be inspected at the office of the clerk of the Poor Law Union in Naas, or at the office of the clerk of peace after 30 November 1865.[8]

Commenting on the situation in that year, Lord Walter Fitzgerald later wrote that 'the military authorities caused a great deal of discontent in the neighbourhood by stopping, or threatening to stop, certain roads, and by informing the public that they were permitted to cross the Curragh through the Camp on sufferance only. Indignation meetings were held, and it was decided to have the question settled as to the ownership of the Curragh; for the purpose Parliament granted a Commission . . .'[9]

A Bill to make better provision for the Management and Use of Curragh of Kildare was circulated in the House of Commons in May 1866. It acknowledged that 'divers persons' claimed rights of pasture, rights of way and other rights there, and that public rights of way were also alleged to exist. Then it referred to the agreement between the Secretary of State for War and the Commissioner of Her Majesty's Woods, Forests and Land Revenues under which part of the Curragh was occupied by an encampment of some of her majesty's forces, and to the possibility that the property in the Curragh should be vested in the Secretary of State for War. It was proposed that a commission should be appointed to ascertain what rights existed, and for the purchasing and extinguishing of any rights that might conflict with the interests of the military. Before the second reading of the bill the financial secretary to the Board of Treasury announced that it was the intention of the government to appoint a commission to make inquiry into the question on the spot, with a view to legislation in the next session.[10]

The lords commissioners of H. M. Treasury, by a minute dated 10 August 1866, appointed a Commission of Inquiry and it sat at the Court House, Newbridge on Friday 14 September 1866; the hearings lasting for eight days.[11] Opening the proceedings the chairman summarised the purpose for which they were assembled, to ascertain rights of common and public rights of way,[12] and then he invited Mr Hallowes to make a statement on behalf of the crown. Hallowes said that the crown claimed fee simple to the Curragh; that the interests of the crown and of the public were identical, and that the commission should endeavour to find the rights of both and have them protected by legislation, if necessary. Then he referred to the office of the Curragh ranger, but the first such appointment by the crown which he could trace had been in made in 1710. The ranger's salary was paid by the crown, and there were caretakers to prevent trespass; in 1856 six cases of trespass had been brought before the magistrates in Kildare, and fines were given. Hallowes made it clear that the sole purpose of the commission was to hear claims, and to report to the government, who might consider introducing an act of parliament. The commission could not compel attendance of any witness, nor could they examine under oath. Nor did they have the power to determine the rights of any party making a claim.

The first response from the floor came when the chairman remarked that 'the camp had, to a considerable extent, benefited the farmers and others residing in the neighbourhood'. It drew cries of 'no, no' from the claimants who were waiting to demand their rights to the plain.[13]

The majority of the witnesses were laying claims to rights of pasturage on the plain. One man said his family had grazed sheep there since the time of St Brigid,[14] but Thomas Keenan of Ballymanny could only go back 300 years with his family's entitlement![15] J.F. Meekings of Ballysax Manor also claimed that his wife's family had possession 'for nearly 300 years, on a grant from James II'.[16] The complaints against the military encampment were for loss of the grazing ground for 2,000 sheep on the 700 to 800 acres on the strip of about two miles by half a mile, which it was claimed was taken up by the site of the camp.[17]

There were allegations that sheep had been shot near the ranges, though William Belford, who lived behind the butts, alleged that he 'had several of his sheep shot close to his own door'.[18] It was said that dogs from the camp worried the sheep, sometimes at the behest of the soldiers, and it was charged that they had also stabbed a farmer's horse with their bayonets. One claimant protested that 'some of the sheep came home dead, and more came home wounded. I have some pensioners in my flock now'.[19] If the sheep wandered near the butts the mounted sentries, according to a claimant, 'always run them with great celerity. The sheep were injured, not being mountaineers, they could not stand that, and I had to take them away altogether'.[20] Other military activities made large areas of the plain dangerous on account of weapons practice, or useless for grazing due to the traffic of military trains and such activities.

From the cases made to the commission it became apparent that the grass suffered from persons other than the army; the sward was being damaged by overgrazing, and there were thousands of unauthorised sheep on it. Sheep left overnight were herded into hollows, and their droppings, which should be scattered by the shepherd, were being allowed to lie there, and the grass was destroyed. Elsewhere manure was being removed, to the detriment of the pasture. Gravel, sand and sods of turf were being taken, and one local farmer regretfully remarked that he 'could remember the Curragh when it was breast high in furze'. Matthew Lawlor of Frenchfurze expressed annoyance that the soldiers had taken sods from his frontage [pasture before his house] which would take years to again grow.[21]

There were many complaints to the commission concerning the loss of, or interference with, rights of way across the Curragh. The Grand Jury [the body responsible for the administration of the county] expressed disapproval of the loss of any rights to the public, including that of passage on 'the grass road from Athgarvan to Kildare, which they used every day, and which road is the basis or foundation of the present road through the camp'. While the military would have preferred if no public road went through the camp, it was accepted that that by the clock tower would be kept open to the public. One resident of the perimeter of the plain said that 'on the 12th March 1854 the first sod of the Curragh was turned and the passage from my place to the Curragh was very much interfered with by the parties during the building of the camp'. Another resident held that travelling on the road through the butts at Ballysax was dangerous when firing was in progress. He added that his servant was frightened by the noise and 'several times heard the bullets flying past'.[22]

Commissioner Wetherell sought to distinguish between the different classes of road which were on the plain. Referring to a map, (Map 2, p. v) he explained that the double red line represented county roads, double black lines were roads which had been presented for, but were not now used, a single black line was the symbol for a track or grass path. The public, he said, had no right to pass over them, except in such cases as Knockaulin were there were a great many houses. Fourth class military roads were marked on the map with a single red line. The road which passed through the camp was not shown as a public road made by the county, or a public road presented. The complex question of roads, tracks and rights of way was debated, with the chairman admitting that there were people who claimed a public right of way over every part of the grass. Even though the public was then denied the use of the road across the camp by the water tower, Gen. Gordon announced that 'as far as the military are concerned, in order not to inconvenience the public, we propose to make the public a present of it . . . to please the people in the neighbourhood, and ourselves at the same time if we can.'[23]

The firing of artillery was amongst the objections raised by George Knox who lived near Flagstaff Hill, on the eastern fringe of the plain. He said that the reverberations had shattered his windows, knocked the gate pier, damaged the stables, broken his pump and cracked the wall of his drawing room 'so that, at this moment, you could put your hand into the wall. I brought down some of the military men who were quite astonished, and Capt. Rich R.E. could not understand it, but he happened to be standing in my bed room when the 12 o'clock gun went off and away went two or three panes of glass'. Asked by the chairman if the glass was broken every time the gun was fired, Knox replied 'no, but I get them to fire down at your side and it did not have the same effect at all'. General Gordon quipped 'you want them to break my windows I suppose', but Knox was not amused, replying 'I say my house and stables have been injured'.[24]

William Belford, who lived at the back of the butts, and whose flock grazed between Knockaulin and Ballysax, protested that not only had several of his sheep been shot, but that also his house 'was shot, and sixty feet of stabling thrown down by a 12 pounder that hit his house during artillery practice'. There had been firing on 29 June and 27 September 1865, and damage had been done on both days. Nine balls, between nine and twelve pounds each, had landed in his land, and he brought the twelve pounder that penetrated his house to the brigade office. An engineer had been sent to estimate the damage and Belford subsequently sought compensation, but had not received it, though he had a letter from the government saying that they would take steps to secure him, and there had been no firing since. The complainant must have been relieved to learn from Capt. Edward Sim R.E., acting District Commanding Engineer, that there had been no artillery practice since he had come there, and that as far as he knew, there was no intention of ever having it there again.[25]

In evidence the Turf Club agreed with the sheep owners that the grass, including the race-course and gallops, had been damaged by the military trains, cavalry charges, field days and reviews. When it was suggested that the turf of the plain was torn up by the cavalry practising their charges, Gen. Gordon remarked that he had often heard it said that 'sweeter grass grows where the cavalry pass'. To this a farmer replied 'certainly there does'.[26] However, the insensitivity of the military to the racing folk was illustrated by a story that 'the late General Browne galloped over the whole of the Rutland course, and charged over where we were going to race'.[27]

When the evidence of the Turf Club was about to be heard commissioner Major Mansfield said that 'with regard to the Turf Club, I am able to state that the Earl of Charlemont will appear here on behalf of the club on any day that may be appointed for the purpose'. In fact, the earl did not appear during the hearing, but a letter from him was read out instead. Evidence on behalf of the club showed that the building of the camp had caused some inconvenience. The rath gallop had been cut off, and other gallops had been damaged by the laying of sewers and the cutting of trenches. Both race-courses and gallops were regularly cut up by the artillery, military vehicles and the charging of cavalry, and one gallop near the butts could only be used before firing commenced as bullets regularly struck the area. By far the greatest damage was done on field days, or during reviews, when great crowds gathered to enjoy the activity. The passage of carriages, horses and pedestrians across the turf was harmful, and neither the courses nor the gallops were respected by the visitors.[28]

One proposal made to correct matters was that military activity should be entirely confined to the south side of the Limerick road, though Lord Charlemont in his letter did not think this necessary. He wrote 'in cases of grand reviews in the presence of royalty or distinguished strangers there would be no objection to the entire plain being used, but it would be expected that some assistance would be given by the military to repair the damage'. He also recommended that a new course should be made in substitute for the four mile course (part of which was on the south of the road) so that the entire distance could be run on the north side. All of the plain, he thought, should be open for horse training, and the course and gallops excepted from military use.[29]

Michael Clancy, from Brownstown Lodge, official starter for the Turf Club, said that a regular staff would need to be hired to keep the damage by the military repaired. At that moment compensation for damage amounting to £50 a year was paid to the club by the military. It was the opinion of Robert Hunter, deputy Curragh ranger, that this payment should continue, and also that 'the arrangements of cavalry, general officers and officers in command of regiments as heretofore laid down for not interfering or working on their training ground on ordinary field days be continued, a custom which has been found very advantageous to the Turf Club'. From his evidence it would seem that if the £50 continued to be paid yearly that the Turf Club would tolerate the holding of review days when it was known that the plain was 'cut up by the troops within a circle of six miles', and the public damaged the four mile course, the mile and the

Peel courses. Clancy was also the agent for keeping the courses and gallops in order, for which his annual pay was £80. His duties included preventing trespass, seeing that the damage done by the military was repaired, keeping the posts whitened, putting chains and stones on and off courses as necessary, and for working on the courses and gallops. He revealed that the average number of horses in training on the plain was about 170, and he explained how, when they trained in a single line on the centre of the gallop it developed a rut, as from wind and sun, it became as dry as powder, and in wet weather 'like a sewer'. Then he stated that the military horses and vehicles crossing the gallops destroyed the run, and made them dangerous to both horses and riders.[30]

The presence of the abandoned women on the plain was a cause of serious concern to many parties, including the military themselves. The provost sergeant spoke of the difficulty in maintaining order within the camp when, during the past six months, no less than 556 cases of trespass by prostitutes into the camp, and about 150 cases of drunkenness, had occurred. Trespass, he said, occurred at all times, but mainly between 7 p.m. and gun-fire in the morning. There were many additional cases against the girls who left the lines if their names were taken for nuisance. When brought before the magistrate at Newbridge only one conviction would be given against a girl, even if she had trespassed ten times in a week, and the punishment was ten shillings or a week in jail. The same person was fined over and over again, and there was one girl who always paid at once.[31]

In summer, the sergeant said, there were about 100 women 'hanging about the Curragh, if not more, and in winter thirty or forty at Frenchfurze, at Ryan's farm, or in the ditch at Furry Lane'. They all lived in the open throughout the year, except for some in a house at Athgarvan, and others in the only brothel in the neighbourhood. When the military police were ordered to turn the women out of a 'nest' near the camp, most of them went to Newbridge, but some were harboured in Ryan's field. 'Now, if I turn them out one end of the camp, in ten minutes they are at the other,' he complained.

William Ryan, from Ballymanny, was indignant at the suggestion that he harboured the women. They came there, he claimed, because his land was on the way to the town from the Camp Inn. When he had them summoned they got only four or five days, and then were back again. On another occasion the magistrate, Col. Shaw, released them. To this statement the chairman exclaimed that Shaw was acting in a civilian capacity only. Ryan continued his evidence telling that one day when he went and destroyed the women's pots and delf that on that same night one of his sheep was killed; he added that it would be more than his farm was worth to turn the women off. Another nuisance to him was having the soldiers walking on his land when coming to visit the women. An officer, he claimed, had killed two of his sheep, but when he went to the camp to try and find him he was unsuccessful. Other witness objected to the women as they stole hay and turf, and they gave bad example to their families. Property, it was believed, was morally damaged.[32]

Ryan also alleged that the drinking houses which had opened on the fringes of the plain had brought disorder and petty crime, and the increase in the admissions to the jail and workhouse at Naas, which caused an upswing in the expenses of those establishments, was blamed on the immoral characters from amongst the camp followers. Major Borrowes, who was vice-chairman of the Naas Union's Board of Guardians, and a member of the Board of Superintendents of Naas jail, told the commission that, as a result of the establishment of the camp, there was an additional expense to the Naas workhouse of about £500 a year, as there was an average of 30 camp followers there. Of about 90 prisoners in the town jail, about 60 of them would be brought there for crimes connected with the camp, amounting to an expense of £410 annually. He expressed the wish that the government might give some pecuniary assistance to the institutions. However, his estimate that 60 out of the 65 committals to the jail during the past month of August were of camp followers was challenged, and it was held that 15 committals had no connection with the camp.[33]

That the commission, even before it sat, had been fully appraised of the position of the military in relation to their requirements on the Curragh was to be expected. The chairman, General Gordon, as general officer commanding the camp, was fully conscious of the situation there, and of the training grounds and welfare facilities necessary for all arms. Nevertheless, to ensure that the military case was fully presented to the commission the Under Secretary of State for War forwarded a series of letters relevant to the encampment, and which were later published in the press. The covering letter, addressed to Gordon from the Under Secretary of State for War at the Horse Guards, was dated 11 September 1866. It conveyed the instructions of Field Marshal the duke of Cambridge commander in chief as to which officers should attend the commission to give evidence as to the requirements of the troops; they included the general the Right Hon. Lord Strathnairn, G.C.B., commander in chief of the forces in Ireland, Maj. Gen. Key, Cavalry Brigade, the deputy adjutant-general, the deputy quartermaster-general, the assistant quartermaster-general, and the officers commanding the 4th Dragoon Guards and C Battery, Royal Horse Artillery.[34]

Of special concern to the military were the boundaries of the land reserved for drill purposes. Cambridge, it was emphasised, was strongly opposed to any curtailment of the ground coloured blue and green on the plan of the area which was placed before the commission. Any reduction, it was held, would be most injurious to the military advantages possessed by the Curragh as an admirable position for the concentration of large bodies of troops. It was also conveyed that the commander in chief was strongly opposed to the suggested enclosure of the ground which, he believed, would answer no good purpose and would be a great disadvantage to the free action of the troops, and it would be objected to by the local residents. This part of the scheme should be abandoned, and it would save much undesirable expense. That the public should be fully appraised of the rights of the crown was also recommended.[35]

The objections to the enclosure of land surrounding the camp of instruction were based on a submission made by Lord Strathnairn on 4 June. Three reasons were given:

1. That it would interfere with and embarrass the field instruction and movements of the troops.
2. That the great extent of the enclosure will render it difficult to be guarded; that consequently, it will not prevent the soldiers from breaking out of barracks and bounds at undue hours.
3. That the Curragh is a sort of Irish heirloom in which the whole country takes a pride and interest; that there are various claims to right of way, pasture, etc. . . . across the Curragh, which would be affected by the enclosure. These rights are probably not well defined, but they have more or less been legitimized by immemorial usage and sufferance, and are of considerable benefit to the poorer classes who enjoy them, and a great convenience to high and low.[36]

Strathnairn concluded his submission by stating that as the enclosure would be a very unpopular measure it would probably involve the government in constant lawsuits, 'which would be tried by persons with Curragh sympathies'. The discipline of the troops, he believed, could be furthered by the making of a 'well conceived obstacle round the huts of each regiment, to which no local interests would offer the slightest objection, and which could be well guarded'. He concluded with the information that 'the country people and others connected with the Curragh had never shown any opposition to, or put any obstacle in the way of the field exercise or movement of troops of all arms across the whole extent of the Curragh, and on both sides of it. They have always displayed the best feeling in this respect, and none more than the duke of Leinster'.[37]

While Lord Strathnairn's letter was sensible and conveyed some knowledge of the actual feelings in Kildare, it is obvious that his information came from a source not close to either the racing men or the shepherds, both of whom had long standing grievances with the military, as expressed in the evidence afterwards given to the commission.

During the summer months of the drill season General Lord Strathnairn was in the Curragh to inform himself of the requirements on the ground. On 3 September he sent his recommendations to the War Office. Having listed which officers he thought (including himself) should advise on the amount of land necessary for drill and instruction, he replied to the request that he should give his opinion on the area of ground required by concurring with the opinion of the commander in chief 'that the whole of the land coloured blue and green (on the plan of the Curragh) should be reserved for military purposes and vested in the name of the secretary of state for war'.[38]

Interesting information about the use of the plain emerged from the letter. The ground on both sides of the railway line was then being used for strategic movements and the requisite instruction in them of the whole of the troops of the three arms stationed at the Curragh during the drill season. Strathnairn wrote that he had never seen ground so well suited, on account of its undulations and other features, as the Curragh, to the masking of bodies of troops and to the instruction of the officers and men of all arms; and on the other hand, of delivering their own fire under cover. The extent of the turf, he believed, qualified the plain especially for the exercise of troops, and 'to curtail in any way such a valuable training ground would deprive it of advantages which render it unequalled as a camp of instruction'.[39]

Strathnairn thought that the commission should be appraised of his experience as commander of the forces during his stay on the Curragh during the drill season, and as already expressed in his letter to Lord Hartington Secretary for War, on 4 June last. He welcomed the understanding that the proposed enclosure of part of the plain was abandoned, as it would have been disadvantageous to the troops, and unpopular locally. It was always his wish, and that of the officers under him, to be as agreeable as possible to local interests, including the Turf Club. He wrote that when he or his officers were notified of a race or trial in any part of the Curragh care was taken that the presence of troops did not disturb the event.

The general emphasised that the constant good feelings from all gentlemen connected with the Curragh towards the military was appreciated, and that no complaints had been received of the interference of the troops with the legitimate or even the traditional rights of the Curragh, and that 'as regards the position of military and civil interests in the Curragh, no change or improvement is desired'. He did admit, however, that he had heard of 'one or two persons of extreme opinions, with a mistaken view of gaining popularity, declare (in the House of Commons) that the crown have no right whatever to the Curragh, and that the troops must quit'. It was the general's belief that those agitators had mistaken the popular local feeling. It was superfluous to observe, in his opinion, that the presence of a large body of troops must be of the greatest benefit to the neighbouring landed proprietors, and cause an acquisition and circulation of money, which but for their presence there would not be.

When the people realised, he continued, that there was no intention on the part of the crown to interfere with the rights which had existed from time immemorial on the Curragh, such as pasture or a night's rest to herds going to market, etc. they would be satisfied and 'the presence of so large a body of men appreciated'. His lordship recommended that the rights of proprietorship of the crown over the Curragh, which had existed since 1200, and those of ancient custom, should be clearly defined and published in the *Dublin Gazette*. Further, as a popular measure, the crown should declare that no wish whatever existed on their part to interfere with the ancient rights, but that they would give them the fullest effect.[40]

The suggestion of Lord Strathnairn that the advent of a large force of military would benefit the neighbourhood was reflected in some of the evidence. The property of Mr Steele at Rathbride had, according to his agent Richard Kelly, been increased in value by £5,000 by reason of the camp being established. Colonel Grey, a retired officer, who had taken a property at Great Connell, said that the one inducement to taking the lease was the nearness to the Curragh, and the existence of the camp. He saw the availability of society on the camp as an advantage, and when the chairman asked him if he had paid £3 an acre in consideration of the society of the camp, Grey replied: 'for keeping up my military acquaintance, that was a portion of the advantage'.[41]

Nevertheless, everyone did not appreciate the proximity of the military in the same way as did the colonel. John Cullen, Ballysax, expressed the wish that 'we never saw the camp. I like the people in it, but it brought a curse [in the persons of the prostitutes] on the Curragh since it came here, and the curse of God will fall on it yet'.[42] Later, when Mr Hallowes, solicitor to the Woods and Forests Commissioners stated that 'it was obvious to all that the establishment of the military camp on the Curragh cannot be otherwise than beneficial, not only to the public service, but also to the surrounding country. I believe it is an admitted fact that the camp has to a considerable extent benefited the farmers and others residing in the neighbourhood', he was greeted with a chorus of 'no, no' from the unhappy claimants in the body of the hall.[43] However, other claimants agreed with Hallowes. Richard Kelly, agent on the Steele estate at Rathbride, thought that 'the camp did us no harm. The camp market, and the want they have for everything grown in the county make a great improvement in the county'.[44]

Simon Ryan, whose lands were at Frenchfurze, when asked if he thought his farm had increased or decreased in value since the camp was formed, replied that 'I do not think it has decreased in value . . . it has increased in value by my own application to care and improve it'. But when he was pressed as to whether the camp was a good market for him, he rejoined: 'we farmers don't use it much in that way. We get manure out of it, and it is all we expect'.[45] Major Borrowes had a more positive attitude to the camp. He said that his 'personal residence in the county [Gilltown, near Kilcullen] enabled him to say that within a radius of two miles round the camp the inhabitants are greatly benefited by it. The camp draws its milk, butter, eggs, fowl and vegetables from the people. I know people well off, who had nothing before the camp was established, and who have since used their one or two acres for spade husbandry. The pecuniary advantage of the camp does not extend beyond that radius'. When the chairman remarked that the farmers sold hay and oats to the camp, Borrowes reminded him that Kildare was not a tillage county. 'I think', he said, 'the supplies for the camp in the way of forage are obtained from outside the county, King's county, the county Wicklow and other counties. These supplies are chiefly brought by rail.'[46]

On the last two days of the hearings two senior military officers were examined. On Friday, 21 September, the seventh day, General Key, commanding the Cavalry Regiment in Ireland, appeared, and on the eighth day, Monday 24th, General Lord Strathnairn, commander in chief of the Forces in Ireland came before the Commissioners.

General Key, who had been in command of the Curragh at different periods, gave his views on the proposal that, at some future time, if the capacity of the camp was to be increased from 10,000 to 15,000 men the area designated for the expansion, to the north of the encampment, was unsuitable as there was a slope there, and a swamp which was damp in wet weather. The south side, he thought, was 'generally occupied by buildings of one sort or another, while to the east there was a very steep slope that would have to be scarped out if buildings were to be put there. Beyond that slope, eastwards, were the cavalry lines, which, although suitable for horses in the summer, would not be suitable for permanent buildings'. The area of the rifle ranges, he recommended, should not be disturbed 'as in a camp where 10,000 men can now be accommodated and which is a camp of instruction, full-distance rifle ranges are most necessary'.[47]

The area to the west of the camp, which was used for drilling and instruction of the three arms, and temporary encampments, should, in the general's opinion, be preserved for those purposes. 'I consider,' he said 'that the exercise of the troops [in that area] should be paramount to any other rights which may be supposed to exist over it.' Then he referred to a proposed road in front [north] of the camp which, he thought, 'might be of some small benefit to the public, but would be of great inconvenience to the camp'.[48]

General Lord Strathnairn was equally determined in his opinions. After the chairman, Major General Gordon, had invited him to give his views on the proposed area of the Curragh over which the crown would have absolute control, and indicated on a map the three areas to be used, respectively for the buildings, the ranges, and for drill instructions and temporary encampments, his lordship rejected the map as it did not have the degrees of elevation. Another map was produced, and on it he outlined his proposals. The high ground towards Gibbet Rath he recommended as excellent for further building sites as it was above all miasma; the low ground would not be healthy for the troops. However, if a different area on the plain were selected, it could also be suitable; about 600 acres could be needed in all.[49]

While the general accepted that an act might be necessary to give the crown full control over a designated area, he questioned the necessity for legislation. In his view legislation respecting the Curragh should be avoided as much as possible:

'things go on very well as they are. We are never obstructed in performing the evolutions necessary for the proper instruction of the troops stationed here.' Then he repeated the fact that from his own experience, and he had been on the Curragh for some time, there was a good relationship with the Turf Club.[50]

When asked if he was satisfied with the advantages enjoyed by the military, Strathnairn replied that he was, with the exception of the police. The number of low public houses on the confines of the Curragh, he said, 'were the resort of bad characters from different parts of Ireland, and there are also those unfortunate women who traverse the Curragh more prominently than is necessary, particularly on Sundays'. He wanted them to be controlled while admitting that 'of course these women cannot be removed altogether, but they need not be allowed to appear so prominently', and that they should be kept out of the huts. If the proposed right of way through the camp was to be allowed, the general thought that measures would have to be taken to prevent a right of way through 'the heart of the camp degenerating, as it would, into a nuisance and an abuse'.[51] This way had been a green track until metalled by the War Department.[52]

The general held that the police given responsibility for the area should be under the crown, rather than military police, and that they would patrol the plain, clearing disorderly houses and taking up people frequenting unlicensed premises. They would also prevent vagrant sheep from being grazed.[53]

Commissioner and barrister Nathan Wetherell expressed the view that the Curragh might be placed under the control of a civil officer, who would be appointed by the Ranger, and the general agreed, but he felt there might be some objection as such an officer 'might not concur in the views of the military authorities as to the removal of vagrants and prostitutes'. His wish was that a 'discreet [military] officer, well thought of in his own service, but not on active service, would be the best qualified'; such a man's military experience, especially if he was also familiar with the Curragh, would be ideal. A condition of his appointment should be that he was essentially connected to the military interests, and be in constant communication with the general officer commanding.[54]

General Gordon specified that he was speaking of a government officer, appointed by the crown, who would control all of the various parties who frequented the camp, and Strathnairn was in agreement, providing he had definite instructions. The officer, in his understanding, would have the power to grant and take away licences, of removing the keepers of low or disreputable drinking, gambling or other houses on the Curragh or its borders. The troops had excellent canteens, and did not need spirit and beer shops. But he emphasised that he did not want to 'stop proper public houses, as they are the custom, both in this country and England'. He thought that if the officer appointed was a magistrate that it would be very suitable.

Alarmed at the mention of the possibility of the use of military police, commissioner Mansfield, a local landowner, told the attendance that 'on a former occasion where there was a question of transferring the Curragh to the War Department there was a serious apprehension among the people in and about the Curragh that they were to be handed over to the military. I think the appointment of a military police would be likely to create that feeling, they would think there was something arbitrary to be carried out, whereas at the present time I am personally aware that the best feeling exists between the parties'. Lord Strathnairn was sure that the police 'could rid the Curragh from one of its chief annoyances, the improper women who frequented the neighbourhood of the camp and made themselves prominent on occasions of reviews, field days and such. If there were police they would remove these women as they do in London. They were not only an annoyance, but were also hurtful to the troops', he added. Major Mansfield agreed that it would be proper, not only for the camp, but the county, if the police removed the women.[55]

Lord Strathnairn showed concern that there should be as little enclosure of the encampment as possible, as it would not then interfere with the manoeuvres of troops, or with the pasturage or rights of way. The road which was to be kept open through the camp, however, would be subject to closure by the magistrate or the military if there was a breach of the peace. Mr Wetherell explained that the intention was that an area of 600 acres would be taken exclusively under the crown, and leased to the War Department, and that all rights of commonage would be discharged, and compensation given for the loss of those rights. The ranges area was to be open for grazing. He asked the general if any of the 600 acres, except that actually occupied by the huts, should be purchased by the crown and the rights of common suppressed, or that only a certain portion should be discharged. An area of about 50 yards around the camp should be discharged of commonage, and it would be patrolled by sentries who would challenge trespassers.[56]

As to the road behind the butts, the general believed that the public passage there of footmen, carriages and horses was a nuisance. The troops had to cease firing if a person passed, and cavalry videttes had to be employed to watch the road and to escort persons and carriages clear of the ranges. He suggested the road should be re-routed, or that during the season of rifle practice the area should be reserved exclusively for the troops. When the season finished, grazing could then be resumed.[57]

The north-east area of the camp, which was occupied by the headquarters and the slaughterhouses, was a busy place, with continual traffic to and from the camp, and as a result was a poor pasturage and might have to be compensated for. Lord Strathnairn, who billeted in that area, remarked that the sheep frequently came there, especially at night when they passed close to the hut occupied by Col. Mackenzie, Deputy Adjutant General. Strathnairn informed the commission that it was important that the north-east area should be controlled by the crown as otherwise 'gypsies, tramps, etc. might pitch their encampments close to the military huts'. He accepted that the constant passing of soldiers to and fro had much damaged the good qualities of the grass there, and that he had heard that since the camp was established the number of sheep on the plain had diminished considerably.[58]

When he was finished his evidence the commander in chief of the forces retired, and the clerks set about the transcription of the shorthand notes of the proceedings. They were to fill two volumes of manuscript, totalling 614 pages.

During the eighth day hearing 71 tenants, or their representatives, gave evidence. So did eight army officers, the commander in chief of the forces in Ireland; General Key, commanding the Cavalry Regiment in Ireland; the deputy adjutant general, the quartermaster general and the assistant quartermaster general, Capt. E. Sim, R.E., Col. Shute, and Maj. Sarsfield Greene R.H.A., and the Provost Sergeant James Savage. There were representatives from the County Grand Jury and the Naas Board of Guardians. The County Kildare Clerk of the Crown gave evidence, as did the rector of Ballysax and the dean of Kildare. The former represented some of his flock, while the dean made a claim for grazing rights on behalf of the cathedral, which he could not substantiate as 'the deed had been lost. It had been stolen by the Roman Catholic Bishop, and was now traced to Maynooth College Library' [where there is no knowledge of it now].[59] The Turf Club was represented by Mr Christopher St George, and the Curragh Ranger by his deputy; evidence was heard from three other officials associated with the turf. Twenty-three solicitors or agents appeared on behalf of the claimants.

The commissioners submitted their findings in a report to the Lords Commissioners of H.M. Treasury, and they were subsequently published.[60]

The report acknowledged the fact that 'a considerable extent of the Curragh [was] appropriated by the troops for ball practice, and the whole of the Curragh is used by them for drilling and exercises', and that the plain was 'exceedingly well adapted for all the purposes for which it is now used, being undulating in various parts, it affords special advantage for the instruction and evolutions of troops'.[61] The clear definition of the external boundaries of the Curragh was recognised, as well as the fact that leases of rights of pasture existed since the time of Queen Elizabeth. Of the maximum number of 15,000 sheep then on the sward it was estimated that about 4,000 were 'foreigners'.[62]

The Grand Jury's request that the public right of way through and upon the Curragh should be entirely permitted to the county was noted by the commissioners, as was the proposal that if the Ballysax to Athgarvan road should be closed that another might be made as near to it as possible. Recognising that the upkeep of the existing roads on the plain fell on the barony of East Offaly, it was suggested that the contractor for the upkeep of the roads should be allowed to take gravel from the Curragh, and that if any income was derived from grazing or otherwise that it should be used for road maintenance. Also noted was the proposal of the Commissioners of Woods and Forests that, apart from the Ballysax to Athgarvan road and that which ran from east to west through the camp might have to be diverted and replaced with new roads, that all other roads, including that which went from north to south through the camp, should be permitted to the public.[63]

It was accepted that legislation would be necessary to secure to the crown the power to procure absolute control over the marked areas as specified on the accompanying map, and to discharge all public and private rights of way, except the public right of way over the north-south road through the camp.[64] It was proposed that the management of the plain should be transferred to the Commissioners of Public Works in Ireland, giving them power to appoint bailiffs and make byelaws. That a more efficient mode of dealing with offences against the soil of the plain should be initiated, and that more police should be provided to control the prostitutes, was also recommended.[65] Finally, the appointment of another commission was recommended to assess the rights of common pasture and of other rights, and the value of same was to be ascertained. The amount of compensation, if any, which was to be made was also to be fixed by the new commission.[66]

In all, the fact that the commission had achieved a reasonable balance between the different interests seeking rights on the Curragh was illustrated in the response of the military and racing gentlemen. While the latter were pleased that the existing privileges of the Turf Club on the sward would remain, they were disappointed that the military had not been restricted in their activities to an area south of the main road. The military regretted that they did not procure as large an exercise area as they wanted in Blue Lands, the lands to be used for weapons practice; nor were they pleased with the retention of the office of Ranger as it had been proposed by the commander in chief in Ireland that 'a discreet retired military officer' would be a preferable official.

A Select Committee, chaired by the Earl of Mayo, sat for three days in June 1868 to consider the Curragh Bill of 1866. Amongst those who gave evidence to committee were two members of the commission of 1866, Maj. Mansfield and Mr Wetherell B.L. Much of the material covered by that commission was discussed, such as that in the evidence of Keith Hallowes, solicitor to the Woods and Forests Commissioners in Ireland. From his description of the deterioration of the sward due to the increased traffic since the making of the camp he revealed that local enterprise was profiting: 'The eggs and butter and milk and so on are brought from different farms around the Curragh, and parties used to drive in every direction in a straight line', [thus damaging the turf].[67]

A new designation for the plain was given by Hallowes when he said 'we consider the Curragh to be a Royal Chase. Other than neighbours, the crown had the power to exclude pedestrians and horsemen'.[68] It had been given as an assurance from General Browne that no parades or exercises by the cavalry or the artillery would be held south of the Limerick road, but that the Crown might have troops in all parts of the Curragh if necessary.[69]

Major E.A. Mansfield, speaking as a local resident, emphasised that the public had always enjoyed uninterrupted right of way across the plain. But now the local people had found it necessary to make representation to the authorities about the improper characters who had located themselves on the Curragh; he described them as 'the greatest public scandal, and of the greatest indecency'. He also spoke of the huge increase in traffic from Newbridge to the camp,[70] a fact with which Fitzjames Clancy, the head caretaker on the Curragh, agreed. As to the improper women, Clancy said that while they had been banished from the plain some years before, they then went on the roads 'and were a greater nuisance there than they are now on the common'. Asked if he had any problems with people squatting or erecting huts he said no, as they were summoned, and the huts removed.[71] The banishment of 'the wrens' [as the prostitutes were known] by the resident magistrate was also mentioned by Michael Cahill, land agent to Lord Clifden. They were a nuisance to the farmers, he said, as every ditch was crowded with them, or they put up huts or occupied farmers' outhouses in summer. They also lived all the year round in the furze.[72]

Lord Mayo questioned Maj. Arthur Leahy, Assistant Director of Works in the War Department, about the reservation of a further 100 acres of ground on the hill, for a possible extension to the camp. The major replied that there were no immediate proposals to build.[73] At the end of the hearing Mr Watson, solicitor acting on behalf of the Commissioners of Woods and Forests in England, handed in a schedule of claims to pasture rights on the commonage.

The Report from the Select Committee on the Curragh of Kildare Bill, which included the proceedings of the committee and the minutes of evidence, as well as proposed amendments to the bill, was published by order of The House of Commons on 25 June 1868.[74]

On 16 July 1868 *An Act to make better Provision for the management and Use of the Curragh of Kildare* was passed.[75] It confirmed much of the content of the 1866 bill in relation to the various rights claimed, and ordered that three commissioners should be appointed to hear any claims that would be made, and whose decisions might be appealed in a court of law. The award of the commissioners was to be made before 31 December 1869.[76] The act detailed the duties of the Curragh ranger (who was not to receive any payment) and his deputy (to be paid),[77] and it designated the three divisions of the plain which would be controlled by the Secretary of State for War, and it empowered the war department to take up to 100 extra acres into the camp area, if required.[78] The secretary also had the powers to make sewers and drains, and to take gravel from the plain, and to divert, if necessary, the road from Ballysax to Athgarvan and to replace it with a new one.[79] Authority was given to the Lord Lieutenant to make bye-laws for the Curragh.[80] If it was ever found that the Curragh was no longer required for military purposes, the act specified that the plain was to become subject to the authority and control of the ranger.[81]

The passing of the act had important consequences for the three principal users of the plain. Over a century later it was described by Fergus D'Arcy in his history of the Turf Club as having 'constituted a decisive check to the march of the war lords'.[82] By it the status of the ranger was confirmed; he was to care, manage and preserve the plain, and to be appointed by the lord lieutenant and not by the Secretary of State for War. This did not please the generals, and Lord Strathnairn was unhappy when he realised that henceforth the racing fraternity would have 'a direct line to the highest level of the Irish government'.[83]

Even though the army had exclusive use of the site of the camp, with all rights of common of pasture, rights of way and other rights suspended there until such time as the site might be vacated, and the making and use of the ranges was covered, and another area of the plain was identified for review, drilling and recreational purposes, the soldiers could no longer manoeuvre over the entire Curragh without the written permission of the lord lieutenant.[84]

The status of the Turf Club was confirmed, and the area over which horse racing and the training of horses might be undertaken was defined. Subject to the provisions of the act, all rights of common pasture, rights of way and other rights were to continue as before.[85] The determination of those rights was to be a matter for the commission which the act had legislated for.

The three commissioners appointed under the act to hear the claimants were Henry H. Joy, one of Her Majesty's counsel, Alexander Stewart, former solicitor to the Board of Works, and Major E.A. Mansfield, one of the land-owning family from Morristown Lattin, Newbridge, and a major in the county Dublin militia. They had a notice posted on 14 October 1868 which informed the public that they would sit at the petty sessions court house in Newbridge on 5 January 1869 to consider any claims which had been lodged before 30 November with the clerk of the peace for the county of Kildare at Naas. Over 400 claims were considered by the commissioners, of which 58 were withdrawn by the applicants, and 216 allowed by the commissioners. They published their award on 30 June 1869. There were 117 claims for rights of way, and 37 claims of loss of grazing ground due to the building of the camp. The majority of claims were for right of pasture on the common.[86]

Bye-laws for the Curragh, as provided for in the act of 1868, were promulgated on 4 December 1868. They included provisions to cover:

1. Unauthorised grazing on the plain.
2. The removal of sod, turf, sand, gravel or furze.
3. The dumping of rubbish there.
4. Interference with notices, or the damaging of posts, fences or chains.
5. Trespass.

It was decreed that each morning the shepherds were to scatter the manure of their flocks which had been herded together overnight. Bye-law number eight was that which was most likely to affect the military establishment. It specified that: 'It shall not be lawful to encroach or trespass on, or injure, or commit any nuisance on the said Curragh, or otherwise to make any unauthorised use of the said Curragh, or to interfere with or obstruct the authorised user thereof.'[87] How effective the bye-laws were in respect of the intention to eliminate nuisance will be considered below.

The final nineteenth-century act concerned with the Curragh was that to confirm the award under the Curragh of Kildare Act, 1868, and which became law on 9 August 1870.[88]

The problem of the Donnelly's Hollow/Ballysax road remained a contentious one, and bye-laws were made on 17 May 1894 by the Secretary of State for War for regulating the Curragh range ground. The bye-laws provided for the closing of the road when firing was in progress, and the passage of man or beast was prohibited. The limits between which the road would be closed were marked with posts on which copies of the bye-laws were posted. During firing practices red flags were to be flown on the Clock Tower in the Curragh camp, a point known as Moteenanon, and Flagstaff Hill. Military personnel, designated by the general officer commanding the Curragh District, were authorised to take into custody any persons found contravening the bye-laws, and to remove any vehicle or animal in the area of the road. On conviction, a fine not exceeding £5, might be awarded.[89]

The County Grand Jury passed a presentiment consenting to the bye-laws at the summer assizes of 1895, and in February of the following year an arbitrator was appointed to hear claims from *bona fide* property owners in the immediate district whose right of property were injuriously affected by the temporary closing of the road. Mr Edmund Murphy, 81 Pembroke Road, Dublin, the arbitrator, published his draft award on 3 July 1896. It named 66 applicants seeking compensation, and awards were made to 23 of them. The highest award of £750 was given to landowners William Pallin of Ballysax and Athgarvan Lodge, and Mr and Mrs Gordon, Ballysax Hills, who received £225. The remainder of the awards ranged from £40 to £5, with the majority being under £15.[90]

The 400 people who submitted claims to rights of pasture, rights of way or other rights in 1868, and the 66 people who sought compensation for the intrusion of the military activity in their everyday lives in 1896, were an indication of the social and material consequences of the advent of the military to the Long Hill on the plain in 1855. A list of 15 existing public roads over the Curragh was given in the second schedule to the Award document, and the commissioners determined that there were 'no other rights in, or over the Curragh save those by this our award'.

Lord Strathnairn's opinion, given at the Curragh Commission in 1866, that the presence of the army would benefit the neighbourhood, was undoubtedly true in some ways, but in other ways the local residents suffered. Even the length of time taken by the War Department to finally acknowledge the inconvenience caused to the people by the range practices must have been a constant annoyance, and of the 66 persons who felt aggrieved by the road closure only 23 were rewarded, and that was after enduring the bother for over 40 years.

The publication of the Award of the Comissioners in the month of June 1869 was the conclusion of a series of legislative measures intended to improve the management of the Crown lands of the Curragh of Kildare. Since the promulgation of the Act of Edward I in 1299 which sought to protect the 'common pasture in the soil of the lord King' from the depredations of swine, the ownership of the sward was claimed by the crown. Almost five and a half centuries were to pass before the next official document was specifically concerned with the state of the Curragh appeared, the Report of Ranger Robert Browne in 1833 which found that the palin had been well preserved.

There can be little doubt that the legislation governing the Curragh did protect it in many ways. Apart from the further development of the camp itself, and the ancillary road systems, there was no further encroachment on the plain during the time of British administration. But, despite the efforts of successive rangers, the military continued to damage the sward, and the vagrants to infest it. If the Curragh was to serve as an arena for major racing and military occasions, attracting thousands of spectators, it was also to be a temporary home to thousands of soldiers who experienced the less glamorous side of life there. Initially poor barrack accommodation and unsurfaced roads made their lives miserable, while the strenuous duties and training, often in inclement weather, and the isolation of the place, did not endear the short grass to the men.

Chapter 4

MILITARY ROUTINE AND TRAINING

There had been speculation in the locality of the Curragh in the autumn of 1855 that all the military would evacuate the camp for the winter, but in mid-November it was known that they would stay, and that arrangements were being made to make the living conditions more comfortable. The roads were being repaired, and supplies of fuel were being delivered. The conditions of the roads were so bad that the social life of the district was adjusted to allow the officers to return to camp early in the evening. As the local paper revealed 'when balls or entertainments are given in the vicinity [of the camp] they commence at such an hour as to enable the officers to return to quarters at an early hour'. An official reason for keeping the camp open during winter was that there was a shortage of barrack accommodation, and that the camp would accustom new levies to actual experience of campaign life.[1]

Winter time in the camp was less pleasant than earlier in the year. Many of the men occupied the long evenings by attending the regimental schools[2], but their daily duties were adversely affected by the elements. The ground easily turned to mud in wet weather, and for sentries who spent long periods outside it could be very uncomfortable. Early in November it was arranged that 'for the health of the troops' wooden platforms would be made for them, two planks wide and four inches above the mud and water'.[3] The laying of roads or paths to facilitate the passage of reliefs from the main guard to sentries posted over the magazine and the coal stores was also recommended as 'a comfort to the officers and men on that guard'. After Christmas drilling had to be cancelled, on account of the rain, and a marching-out parade was also abandoned. By January 1856 no money had become available for the paths, nor would it be available until the estimates for the year 1856–57 were voted. Lord Seaton suggested to the officers commanding the brigades and regiments that they should 'make the best of the existing means, and lessen as far as practicable the present discomforts until fresh funds are available.' Capt. Rich R.E. was directed to assist any fatigue parties as far as he had in his power.[4] However, some relief for the men came when approval was given for the issuing of an extra pair of boots to each man doing duties during the winter.[5] Nor were the horses forgotten; heavy cart, wagon and dray horses were to be given four pounds extra hay.[6]

That winter a journalist wrote, 'the severity of the weather prevented field days [on the plain] until the opening of spring, when the field works now in progress of completion will be ready for the practice of artillery and minie rifle. In the camp both officers and men are beginning to feel the effects of winter. Except as on the main line of road there is mud everywhere . . . as on the heights before Sebastopol. It appears entire huts may not be waterproof. The militia regiments here, however, have better quarters than at Aldershot'.[7] Early in 1856 there had been complaints about the condition of some of the stables. Those intended for use by Lord Seaton's staff were in bad condition after use by the cavalry and artillery during the winter, and the swinging bars of the stalls were dangerous. It was directed that instead wooden partitions should be erected.[8]

The achievement of the workforce at the camp within such a short time was considered remarkable, especially as it commenced during a period of adverse weather, and when the military themselves were preoccupied with the embodiment of the militia. The fact that the Curragh itself was considered to be an isolated and distant place to which to transport materials and to attract and accommodate workmen did not hamper the successful completion of the work.[9]

A young officer who served on the plain in 1856 left his impression of the station.[10] Seventeen year old Lieutenant Alexander Bruce Tulloch of the Royals (later Royal Scots), returned from the Crimea, and was posted to the Curragh. He did not like the place: 'A more dreary quarter for a lot of young fellows than the Curragh in those days could hardly be imagined, the only excitement being the races. Those who could afford it went to Dublin, but those who could not had no amusement but country walks and occasional games at cricket with the other regiments. We sometimes had a divisional or brigade drill, and of course the usual daily parade, with the march past in slow and quick time. The new institution, musketry instruction, was now regularly carried out. In the dreary wet winter our regimental theatrical company was decidedly useful in giving occupation, but I for one wished the regiment was back in the Crimea again, and I fancy many others also did.' Eventually the subaltern got a long leave period and went to London, but 'all too soon my leave was over, and then back again to the objectionable Curragh. By this time our theatrical company had quite made a name for itself. On one occasion Hawley Smart was playing "Cool as a Cucumber" before the lord lieutenant in one of the Dublin theatres taken for the night, whilst Charles Mathews played the same piece in the other. On the last day of the year we had some special piece on at our Curragh theatre, and as the clock struck twelve all our actors appeared with a sash inscribed 1857. Little did they know what that year [of the Indian Mutiny] meant for the British army, and that for so many of the audience then present that was their last New Year's Day'.[11]

Even if Lieut. Tulloch found the new camp to be a dull place he must have appreciated the endeavours of the Turf Club to facilitate the social activities of the officers. In July 1856 the marquis of Waterford, on behalf of the Turf Club, sought permission from the Department of Woods and Forests to permit the military to hold an archery meeting, breakfast and ball at the Stand House, as the lease of the Stand House specified that it could only be used for racing purposes. Of course at that early date the extensive field exercises of the army had not yet caused friction between them and the racing men, and it was understandable that Waterford was 'anxious to make the living at the Curragh as agreeable as possible to the officers quartered there'.[12]

In the Spring of that year, as part of the post war reforms, the five military districts in Ireland were abolished and three new divisions substituted, at Dublin, Cork and the Curragh. The new organisation was intended to create a more unified and disciplined army, with, in the case of the Curragh, two infantry brigades and a cavalry brigade commanded by a lieutenant general. From April 1856 the headquarters of the Curragh Division were fixed at the camp with a staff consisting of one lieutenant general in command, three major generals as brigadiers, one being for the cavalry brigade quartered in Newbridge, one as assistant adjutant general and one as assistant quartermaster-general. There were three majors of brigade and five aides-de-camp.[13] Three years later, at the end of January 1859, it was decided that the Curragh Division would be broken up, with the reduction of a lieutenant general and of one major general.[14]

The building of the camp was almost completed when peace was made with Russia, and the militia was disembodied.[15] To celebrate the end of the Crimean War a great national banquet was given to the victorious soldiers returned from the battlefield in the great bonded stores at the Custom House in Dublin. Amongst the regiments represented in the 4,000 officers and men at the banquet were soldiers from Newbridge, Naas and the Curragh camp.[16] With the ending of hostilities the purpose for which the camp had been created no longer existed, and it was decided to make it a camp of instruction, and as a place for the accommodation of a large force of infantry which could be trained and manoeuvred with the artillery and cavalry on the 5,000 acres of the Curragh. Its proximity to the large cavalry barrack at Newbridge was a decided advantage.[17] As part of the post-Crimean war reform in the British Army camps of instruction for all arms were also formed at Aldershot, Shorncliffe and Colchester in England, all of which were established by the year 1862.[18]

In March 1857 Major General H.R.H. the duke of Cambridge (Fig. 6), who had commanded the Dublin District from 1847 to 1852, and was general commander in chief of the army in 1856, expressed his views on the retention of the new

Fig. 6. *Field Marshal H.R.H. the duke of Cambridge, K.G., Colonel in Chief 17th Lancers 1876. In March he had ordained that the Curragh Camp should become a Camp of Instruction.* Photo from The Military Life of the H.R.H. the duke of Cambridge, *by Col. W. Verner, 1905.*

camps at Aldershot and the Curragh. He was of the opinion that troops should merely be concentrated in the various camps of exercise during the summer months; and he held that it was desirable, in the non-training season, that they should mix freely with the civilian population, though the brigade and divisional organisation would still be maintained. His biographer, Col. Willoughby Verner, late Rifle Brigade, writing in 1905, believed that 'in view of recent history, there can be no doubt that he was right; and his memorandum, replete with sound common sense and pertinent comment, applies as well to-day as it did then'.[19]

This is the duke's memorandum, dated March 1857.

> The question as to the advantages or disadvantages to be derived by the troops from stations such as Aldershot and the Curragh have been much canvassed and various opinions have been expressed, coming from quarters deserving of the highest consideration. It may be well, therefore, to consider the objects for which these stations were formed, and at the same time lay down certain principles upon which they ought to be conducted for the future.
>
> At the commencement of the late war it was felt that, however good our troops were, they laboured under the great disadvantage of being wholly unaccustomed to act in masses, and that neither the superior

officers nor the staff ever had opportunities of studying their profession practically in field operations, even on a limited scale. The troops in general, too, were in want of that constant supervision so essential to the efficiency of an army, however small in numbers, and the various branches of the service, being isolated and detached, were never brought together for unlimited operations, save in one locality (Dublin), and even in that garrison their sphere of action was extremely limited.

Hence the selection of Aldershot and the Curragh as stations affording space and every possible advantage in point of ground to correct these deficiencies.

During the war it was found necessary and desirable to concentrate as much as possible the militia at these stations with a view to drill, and also in order to obviate the great inconvenience of billeting the militia regiments in their various localities. On the return of the army and the disembodying of the militia the troops took the place of the militia, and during the present winter [1856–7] both stations have been fully occupied by them, much having to be done to reduce the army from war to peace establishment, which could be better effected when the troops were concentrated than when scattered about in small detachments all over the country. It was also desirable in this manner thoroughly to establish the brigade and divisional system which had been decided on as essential for the efficiency of the army for the reasons stated.

As soon, however, as these arrangements are completed, and that new system is practically introduced, it becomes a matter for consideration how these stations can be made most useful for the public, and whether they should be constantly occupied by the troops or only filled for drill and exercise at certain stated periods of the year. After giving the subject the fullest consideration, it certainly would appear that the interest of the service and the advantage to the troops would be best consulted by sending the greater portion of them into permanent barracks for winter quarters during a considerable portion of the year, whilst concentrating them there for drill and exercise in large masses during the summer months only, say May till the end of August or September. At other times Aldershot should only be occupied to the extent of the permanent barrack accommodation which is now being erected; and the Curragh, with reference to the protection of the huts, which will always require a small force for their security.

During the summer months specified above, as large a number of troops should be collected as can be spared from garrison duty, every species of field exercise should be practised, officers and men should be

> thoroughly trained and prepared for war, and uniformity and combination be established and enforced.
>
> Such is the practice which obtains in all armies, whether great or small. Why should the British Army form the solitary exception?
>
> On the other hand, during the winter months, the troops are better in barracks in various parts of the country, and the inhabitants of the several large towns and districts where barracks exist are gratified by their presence. The friendly intercourse that is carried on between the armed force in the country and its population in general is of great benefit to the army and of advantage to the country, and feeling favourable to the army is thereby encouraged and kept up.
>
> But even during this period of dispersion the divisional and brigade system must be continued, and this can easily be effected if due regard is paid to the mode in which the various portions of the troops are quartered and the proper localities are selected for the staff that accompany them.
>
> The arrangements by which these views can best be carried out are appended to this memo., from which it will be seen that though the barrack accommodation throughout the country is not altogether as convenient as could be wished, still it is sufficiently so not to present any great difficulties to the attainment of the important objects which ought never to be lost sight of.[20]

That the general commanding in chief's thoughts and instructions were apposite to the Curragh could be seen in the suitability of the plain for mass military exercises, and of the hutments for billeting the militia during their annual training periods. As a summer training ground for the army in Ireland it was considered ideal, while out of season a small garrison provided security for the huts.

General Lord Seaton would have been aware of the commanding general's directions, and no doubt aware of their implementation as he continued to make regular visits to the camp. When he inspected the 8th Hussars, who were about to embark for India, at Newbridge on 29 September 1857, he afterwards went to the camp, and in the following March he again travelled by special train from Dublin to visit the same two stations. In March 1859 he undertook the same inspections, and afterwards he went to stay with his son and daughter-in-law at Bert.[21] That summer he stayed at the camp in quarters which had been repaired and wall-papered before he came. Then he went again to Bert, but he remained in touch with his headquarters. Each day he received a letter-bag which was sent by train from the military secretary's office in Dublin.[22] He would have been aware of the fact that that year, in India, a Victoria Cross (the first of three to be won by Kildare-born soldiers) was awarded to a man from Kilcullen. He was

Lance Corporal Abraham Boulger, 84th Regiment who distinguished himself at Lucknow. Later, for further bravery, he was given the honorary rank of Lieutenant Colonel.[23] A Victoria Cross found on the Curragh plain in recent years is also believed to have been awarded to a soldier in India at that time. As the bar of the cross is missing the recipient cannot be positively identified, but he is believed to have been a member of the 8th Royal Irish Hussars, a regiment which distinguished itself at the battle of Gwalior in 1858; the regiment was back in the Curragh from 1869 to 1875, during which time it is assumed that the holder of the V.C. lost his medal on the plain.[24]

A sketch of the camp from about that time by Lady Fremantle compares well with the 1861 painting of the camp by Surgeon Lamprey. Lady Fremantle sketched from the high ground to the east of the camp, looking towards Kildare cathedral on the horizon. She clearly shows the dominance of the clock tower over the churches and lesser buildings in the camp.[25]

The clock tower also impressed another observer of the camp who wrote this description in 1867: 'Coming from Newbridge at a sudden sweep of the road, close by a neat Wesleyan church built of corrugated iron, you see at once the immense line of the camp far above on the topmost ridge of that Long Hill. A tall clock tower, whence every portion of the Curragh can be seen, shoots up in the centre. Close in front six pieces of cannon guard the flag of England. On the right of the tower is the Catholic chapel, on the left the Protestant church, each capable of containing eighteen-hundred worshippers. Here are the schools, marvels of neatness and efficiency, here also the post-office, conducted with true military precision and regularity, the savings-bank, the telegraph station, and the fire-engine depot. If you could see through the hill, you would discover on its further side a considerable market where traders bring their goods, and the country people their produce. A busy stirring scene it is, and a gay one too, when the trim, neatly dressed, and comely wives of the soldiers come forth to cheapen and purchase what they can.'[26]

The dominant clock tower has an internal stairs which permits access to a balcony that goes around the tower. From it, according to local tradition, the senior officers could follow the training exercises on the plain. The tower is now the last important building of the original encampment which survives in the camp.

Following the dictum of Major General H.R.H. the duke of Cambridge, general commander in chief of the army in 1857, the encampments at Aldershot and the Curragh were to be the principal summer training grounds of the regular and reserve army in England and Ireland. On the plains the various services would be brought together to exercise *en masse*, thus gaining vital training experience, while the officers would have the opportunity of studying their profession practically in field operations. Using the organisation of the brigade and divisional system (a division is the largest formation which is permanent in composition; it usually consists of three brigades, each of three battalions), and under constant supervision, the duke decreed that 'every species of field exercise should be practised, officers and men should be thoroughly trained and prepared for war, and uniformity and combination be established and enforced'.[27]

Preparation for war was the ultimate aim of the training carried out each season on the plains of the Curragh; it was the main purpose of the encampment, and the annual rifle practices and competitions, the inspections, the drill periods and field exercises there were the major events of the year.

During the months from April to September each year an average of 5–10,000 men, with up to 12,000 (during the royal visit) in 1861, were accommodated in the huts or under canvas on the plain.[28] In addition to the regular troops, units of the militia were usually there. The Irish militia constituted a reserve force of 32 infantry regiments and 11 artillery regiments in 1879, the strength of which was about 20,000.[29] Enlistment in the militia from 1874 was for a six year period, and the bounty paid for that period was £6. In 1874 £1 was paid for each training year, but by 1894 it could be either £7 or £6.10s. The £7 was paid as follows: 10/– on drill enlistment, £1.10s. for first training period, and £1 each for the subsequent five years. The bounty of £6.10s. was given in a payment of £1.10s., for preliminary drill and £1 for each of the following years.[30] The numbers of militia reporting for training varied. For example, in 1855 over 3,000 militia paraded there before Lord Seaton.[31] In June 1896 it was reported that, included in the 10,000 men encamped on the Curragh, were 2,000 militiamen, and three years later 1,500 militia were there, with 7,500 regulars.[32]

Preparations for the training period commenced well in advance of the formation of the camp, with such details as additional pay for the staff which had been detailed to oversee the camp being resolved. When the rifle practices were held special payments were made to the field officer in charge of the detachments of troops assembled there, providing they did not average less than 500 men; the non-commissioned officer employed similarly was given pay which brought his remuneration up to that of a sergeant major. His Royal Highness, the commander in chief, was interested in these arrangements in November 1858, and he suggested that the duty might be carried out by a field officer detached periodically from the garrison of Dublin. He asked if Lord Seaton, general commanding in Ireland, thought a sergeant from one of the regiments could be sergeant major, with pay up to that rank, or if a permanent sergeant major should be appointed, and not borne on the strength of any regiment?[33] Whatever was the outcome of those proposals is not clear, but in the spring of 1860 a Lance Corporal was given 6d a day while acting as orderly room clerk for the detachments assembled there for rifle practices.[34]

During those practices Capt. John Collins 26th Cameronians was killed by a stray bullet when returning from the range on 18 May 1860. Later his comrades erected a monument on the spot.[35] That accident would have aggravated the fears of the local residents that their lives were endangered by the shooting, fears which were to be expressed at the Commission of Inquiry to be held in Newbridge six years later.[36] Thirty years afterwards, in 1896, the introduction of a rifle with increased power necessitated the closure during firing of the road from Ballysax to Donnelly's Hollow on which it was considered dangerous to travel. Public meetings of protest were held and a compromise was reached.[37]

On occasions the grazing sheep might interfere with the ball-practice. Early in January 1863 the Curragh Ranger was complaining of sheep being killed by dogs from the camp, and by ball-practice. Major General Ridley promised an enquiry on the matter. However, when claims were made they were not entertained as it was thought that 'owners of sheep should take the necessary steps for the protection of their flocks'. The Secretary of State for War directed the Officer Commanding Royal Engineers to order the Ranger 'to have the sheep driven out of the way of [rifle] practice to prevent recurrence of such incidents'. To protect their flocks, it was suggested that the farmers should put them in at night, but it was found that that suggestion could not be acted on as the commoners had no right to erect hurdles or break the surface of the plain to make sheep folds.[38]

Local residents found that they could profit from the ball-practice by burrowing in the butts for the lead shot which they could then sell. As the burrowing damaged the butts the military sought to prevent the practice, and the caretakers of the plain were charged with its prevention. Caretaker James Ryan brought brother and sister Mary and William Doolan from Brownstown before Newbridge Petty Sessions in June 1865 on a charge of 'rooting and injuring the butts'. They had been out at 3 a.m., he claimed, but Mary swore she was then in bed, and William said he was looking for his two asses, and collecting a little manure. Pleading for her children Mrs Doolan said that within the last six years she had paid over £100 in fines, to which the sessions' chairman remarked 'it certainly must be profitable business when you can afford such fines'. But fined they were, 2/– each, with an additional 1/– for damage and 2/6 for costs. A note in the court report in the local paper was that 'the public cannot conceive the amount of lead that is daily consumed at ball practice, but when we state upon authority that over 16,000 ounces is lodged in the butts every day, we do not wonder at the temptation. Owing to the targets being generally pitched in the one place the difficulty of collecting it cannot be very great'.[39]

Drilling was an important part of the training of both officers and other ranks, and it did not always go as well as it should. Colonel Masters, O.C. 1st Bn Royal Northumberland Fusiliers, was asked by the commander of the forces in September 1865 to explain why the second line of the unit of which he was in charge, moved forward about 20 paces after deployment, and which, according to the assistant adjutant general, threw the first line out of position?[40] A couple of years later it was the turn of the Royal Artillery to be reprimanded; it had been requested that three field batteries of the Royal Artillery should assemble at the Curragh in mid-July 1867, and those from Dublin, Athlone and Kilkenny were selected. Detachments of cavalry from Longford, Carlow, Fethard, Bandon and Skibbereen were also to be there for the Drill Season.[41] During the field day which was held on 8 October Chesnut Field Battery was not manoeuvred properly, and the officer commanding Royal Artillery got a 'please explain' on the matter.[42]

If the spectacle of a large number of troops of all arms, cavalry, infantry and artillery, manoeuvering together in their colourful trappings (Figs. 7, 8, and 9), with the music of the bands, the rolling of the drums and the braying of bugles, greatly entertained the spectators who gathered on the plain to see the military activity, the exercises were not enjoyed by the troops. As a man from the ranks wrote 'as a rule, after the novelty has worn off a little [the men] are apt to loathe field-days with heart and soul, designating such exercises as "a . . . nuisance", and often not entirely without reason. A field-day invariably means a hurried breakfast, a long march to the scene of the day's manoeuvres, many hours spent under a hot sun, or exposed to wind and rain without shelter, during which the troops are bound to considerably damage their uniforms, which have to be replaced at their own expense. Added to this, the energy of even the keenest of them evaporates somewhat when returning tired and dusty from a long day's exercise, he has to turn out spic-and-span in a couple of hours' time and mount guard'.[43]

Fig. 7. *The Curragh Camp Headquarters* c. *1860. The engraving shows a group of officers and their ladies watching a section of infantrymen drilling while other sections practice musketry. Courtesy Daniel Gillman.*

Fig. 8. *Infantrymen in an advance skirmishing at the Camp Curragh during the training period there of the Prince of Wales in 1861. Such exercises were a normal part of the drill season, but the numbers of troops engaged would not always be as large as in that year. Copyright: Royal Archives Windsor Castle.*

Fig. 9. *The lord lieutenant the Earl of Dudley, General Lord Grenfell, Commanding in Chief the Forces in Ireland, and Maj. General. G. de Courcy Morton, General Officer Commanding at the Curragh, at a review on the Curragh. Following the Royal Field Artillery on the sward were the 6th Dragoons, 11th Hussars and 19th Hussars. General de Courcy Morton died while serving on the Curragh in 1906.* Photo: Curragh Camp and District. Illustrated and Described. *(Dublin)* c. *1908.*

Nor were the members of the Turf Club enthusiastic about the great mock battles. The manoeuvering of the cavalry and the artillery cut up the sward, and after a series of complaints to the military it was agreed in 1861 that exercises would not in future be undertaken on the Peel Course, though one officer insisted that on division field-days it might be necessary to do so.[44]

Inspections were a constant and routine part of army life, but the frequent visits of the commander in chief of the Forces in Ireland from Kilmainham meant that the troops on the Curragh had to demonstrate their standards of turn-out and training regularly. General Lord Seaton, for the drill season of 1855, decided to move his establishment to the Curragh, and he took Martinstown House, which had been built as a shooting box by the duke of Leinster on the south-eastern side of the plain,[45] for three months from August.

On the first Monday morning of the general's visit all of the troops were under arms to be inspected in line; then the general ordered that evolutions should be made, and finally there was a march past. Within a few weeks the lord lieutenant came on a visit, with a retinue which included Lord Otho Fitzgerald, an officer in the royal horse guards, from Carton. Greeted at Newbridge railway station by a guard of honour from the Scots Greys, they mounted and rode into the camp. There they were met by Lord Seaton, but the regiments drawn up a short distance from the huts were commanded by Col. Webber. There were between 3,000 and 3,500 men from the North Cork, Clare, South Downshire, Longford and Co. Dublin Light Infantry militias, and the 60th Rifles, on parade. Accompanied by General Seaton, the lord lieutenant rode around the line and inspected the men. Then the regiments deployed in line and proceeded to manoeuvre, but the evolutions of a field day were abandoned when it began to rain. Nor were there many 'fashionables' in attendance as 'the weather was threatening'.[46]

October 1855 saw a continuance of important occasions. A large attendance from Dublin and the surrounding counties saw 10,000 troops performing evolutions for several hours under General Seaton, but the highlight of the day seems to have been the playing of the band of the Clare Militia, of which the chief performers were Hungarian. It was reported that the 'band [had] acquired more respect than any other band in the camp'.[47] A week later the local paper again noticed that on the previous Sunday (7 October), the earl of Cardigan was present at a Field Day when the 10,000 men were again exercised. The earl was already familiar with county Kildare as he had commanded the 11th Hussars at Newbridge in 1843. Subsequently famous for his leading of the charge of the Light Brigade at Balaclava in 1854, he was now back in Ireland as inspector-general of cavalry.[48]

In mid-October the Wexford Militia was the centre of attention when it was presented with new colours by Mrs Carew from the home county, with General Seaton presiding.[49] Another exercise, in which the troops were ordered to pitch tents by companies before Lord Seaton, was soon mentioned in the same newspaper, after which 'each regiment was put through field evolutions by Lord Seaton'. Just after Christmas 1855 the commander in chief decided to leave his quarters on the Curragh and return to the Royal Hospital for the winter. The departure of the general must have been a relief to all ranks in the camp as it was known that he was 'unremitting in his inspections, and was often seen riding across the Curragh at dusk [to Martinstown], after having spent the day at the camp'.[50]

Captain A. Montgomery Moore, A.D.C. to Lord Seaton, and Lieut. Col. Sir T. Alexander, K.C.B., occupied some of their leisure time on the Curragh by making archaeological excavations into 'about a dozen of the Curragh tumuli', including the Gibbet Rath. Digging down to a depth of eight feet the gentlemen uncovered a cist burial (the stones from which were taken for display at the headquarters garden in the camp), pottery, human and animal bones, a tenth century English coin and what was thought to be an insurgent's pike from 1798. A paper from Capt. Moore on the excavations was read to the Royal Society of Antiquaries at a meeting in November 1859, and reported in the Society's journal for that year.[51]

Sometimes the weather disrupted military exercises; from a travel claim submitted by Staff Captain F.D. Maclean 13th Light Dragoons for 22 August 1860 it is apparent that on that day while General Sir George Brown (who that year had succeeded Lord Seaton as commander in chief, but who never seemed to have the same affection as his predecessor had for the plain) was visiting Newbridge, and about to proceed to the Curragh when the weather disimproved. The general sent his aide, Capt. Maclean, by car to the camp 'to order some of the cavalry to quit the camp forthwith, without waiting for the post'. He wrote a note to that effect on the aide's claim for 3/– for the hire of the car.[52] Again in the summer of 1861 Sir George Brown, sent a message to the Curragh that he would not go to the Field Day as it was too wet. Nevertheless the hussars and artillery were to march to Dublin after they had dined, leaving their tents standing on the Curragh. Exactly a year later Sir John again showed reluctance to travel to Kildare when he sent a telegram to General Ridley at the Curragh Camp asking 'what he thought of the weather and the state of the ground for to-morrow?' Another telegraph followed that one ordering the whole of the 15th Hussars to march back to Dublin on the morrow, and that none of the 11th were to come.[53] Detachments reporting to the camp usually marched there, as did a detachment of the 2nd Battalion Lancashire Fusiliers from Ballinrobe in December 1860.[54]

But the train to Newbridge was also used for the conveyance of troops and stores, and in the autumn of 1857 the 8th Hussars came by rail from Dublin for the inspector general's autumn inspection at the camp. That year Lieut. Col. Lefroy was the inspecting officer for the various military schools in the camp.[55] Such annual inspections were intended to ensure the maintenance of standards in a unit and they were much anticipated by all ranks.

In addition to the normal military instruction imparted in the camp there was also voluntary training in classes which might be organised, as in 1866 when Maj. Gen. Napier of the Royal Barracks in Dublin proposed that a voluntary class for instructors in military sketching and reconnaissance should be held in the Curragh.[56] There were also visits from officers concerned with troops stationed in the camp and if part of a unit was detached to the camp the officer commanding could seek permission to go there to visit them. Such a request, made by the O.C. 5th Battalion Military Train in Dublin, in January 1859 was recommended, as he had a troop there.[57] New accommodation for the Military Train in the camp had been provided for in the estimates in 1857. That the activities of the military were being scrutinised, other than on official inspections, was evident in the winter of the following year when there was a complaint from Army Headquarters in Kilmainham to the officer commanding the 5th Battalion Military Train that the horses and men were being overworked at the Curragh and in Dublin.[58]

There were 11,000 men either in huts or under canvas on the Curragh in 1860, a time when the facilities of the encampment must have been stretched to the full, the efficacy of the regulations tested, and the impact on the plain and its environs of such a great number of soldiers and their followers have been considerable.[59] The firing of the artillery practices there was having a particular impact on the property of one resident, as revealed at the Commission of Inquiry some years later.

George Knox, who lived at Flagstaff Hill behind the ranges, was especially annoyed when he complained to the Commission in 1866.[60] His house had been damaged by the canon, and he had protested to the authorities in the camp. Reaction to his complaints was reflected in a letter dated January 1860 from Lord Seaton who asked the officer commanding Royal Engineers 'what measures had been adopted to make more secure the artillery practices at the Curragh Camp next summer, and if steps had been taken to lengthen the screen on top of the hill to prevent the shot from striking off at an angle and entering the grounds contiguous?' By 10 February the artillery butts screen had been lengthened, and in the summer the firing again took place.[61] That consideration which was also given to the nature of the duties of the men working with the artillery and the cavalry horses, and for the wear and tear of their clothing and boots, was shown when they were given a special allowance of four pence a day.[62]

Troops from the camp might be called out in aid of the civil power. Requests for the presence of the military on election days were usual, and in April 1857 Major Browne, commanding the 4th Bn Military Train, was asking headquarters for copies of the printed instructions for parties in aid of the civil power.[63] In May 1859 a troop of cavalry and 100 infantry were sought for elections at Maryborough.[64] Troops might also be sought for the maintenance of order amongst troops socialising in a town, as when the resident magistrates, Mr Browne and Mr Powell, requested the officer commanding 14th Light Dragoons at

Newbridge in June 1860 that a picquet might be available on Sundays to preserve tranquillity on the streets.[65] In July 1865 troops were requested for the preservation of order during elections at Portarlington and Mountmellick.[66]

A study by D.N. Haire of the army's role in aid of the civil power at local or parliamentary elections, when the troops were required to escort conservative voters to the polling places, found that not alone did the duty frequently bring the soldiers into conflict with the people, but, it could 'in certain circumstances be called unconstitutional and an undue interference with democratic rights . . . it was a task thoroughly disliked by both the army and the civilian population'. To quote General Lord Strathnairn, the troops were placed 'in a false and as disadvantageous position as it does freedom of opinion and right of constitutional election of members of parliament'. The passing of the ballot act in 1872, with the introduction of the secret ballot and of more polling places, meant that the army was no longer employed at election times.[67]

The resident magistrates might also ask for a military presence at race meetings, as Lieut. Col. Shaw did for the Curragh meeting in June 1860.[68] By 1865 the military were proposing that the constabulary might be used for the races at Punchestown, and that troops should not be given unless urgently required.[69] However, in May of the following year, a picquet of two officers and 42 men went to the Curragh race-course to maintain order there, if required.[70] Fear of disturbance by Fenians may have necessitated the strong picquet, and intelligence reports could have alerted the authorities to the intention of John Devoy and his associates of being present there.[71]

The reception by the people of the picquets called out in aid of the civil power was not always favourable, as at Callan, also in July 1865. Expressions of disloyalty by the townspeople there caused Maj. Macpherson 1st Bn South Wales Borderers to write a report of the incident to his superiors.[72] As the century progressed and civil unrest increased due to agrarian agitation, the prospect of Home Rule, economic crises and the Land War, the involvement of the military with the police in endeavouring to maintain order lowered the status of the soldiers in the estimation of many of the people. It also caused some of the officers to question their role in support of the civil power, a task generally resented by the military.

From the first season of military training based in the new encampment on the Curragh in 1855 the local civilian residents were to find that the annual spring to autumn activities of the army were to considerably effect their lives. Thousands of regular and reserve soldiers from England and all parts of Ireland would spend their days exercising on the plain, and their leisure time recreating, possibly in the hostelries within reach of the camp. Financial gain to the area came in the increase in business in the neighbouring towns, and property owners could rent accommodation to the officers. The influx of visitors on the occassion of major field exercises further enhanced the local trade, especially in Newbridge where travellers from Dublin arrived by rail. The inclusion of county gentlemen, such as Lord Otho Fitzgerald, the son of the duke of Leinster, in the military entourages, would not have passed unnoticed.

The negative side of the training, apart from the disruption to the natural peace of the plain, was the noise from the weapons ranges, the thundering of the cavalry, and the damage to the sward, which concerned both the shepherds and the racing men. The accidental shooting of an officer and of sheep near the ranges alarmed the local people, which the gain from scavenging for lead in the butts did little to relieve.

For the troops the training period was the principal event of the year. If they anticipated the time in camp or under canvas, and the renewal of old friendships, they also endured the tedium of preparation for the drills, the ranges, and field exercises: the barrack economy, the cleaning of equipment and weapons, the constant inspections and the endless hours in all weathers on the barrack squares or the plain. If engaged in aid to the civil power at race meetings or elections the men might enjoy the sport, or have an easy day in a country town, but they might also find themselves faced with a hostile mob and an unpleasant close encounter with the natives. But that was army life, and veterans of the Crimea, or men returned from service in India, were accustomed to boredom and tension, while newcomers to the ranks accepted the training routine and other details as a normal part of their duties.

With the regularisation of the status of the British army on the Curragh of Kildare, as defined in the Curragh Act of 1868,[73] and protected by the byelaws of the same year,[74] the military had settled into a routine of occupation of the camp which was to continue for the next 54 years. It was the main training ground for the British garrison in Ireland, operating on the basis of the instructions laid down in 1857. If the Curragh was to be at its busiest during the summer drill period when the regular and reserve forces assembled for training, it was increasingly used at other times for the firing of range practices and recruit training. It was to be the location for the All-Ireland shooting and sporting competitions, when the regiments stationed throughout the country met on the ranges or sports grounds.

While the plain continued to provide space for regular infantry, cavalry, artillery or combined arms movements, it also became the rallying place for the troops participating in the manoeuvres which extended into neighbouring counties from 1894. They were amongst the many innovations in training introduced by General Viscount Wolseley, general commanding in Ireland from 1890 to 1895. Wolseley also encouraged night training, and route marches for both seasoned men and recruits along the roads in the Curragh area. The former participated in continuous marching, which could last for up to six days and cover well over 100 miles. The object of all training was to prepare the men for service at home, in the colonies, and for war; for many that meant service in the Afghan War in 1878, the Zulu War in the following year, the Egyptian campaign of 1882 or the Boer War from 1899 to 1902. In addition to displaying their war skills on field days and manoeuvres on the Curragh the men showed their mettle on ceremonial parades marking royal or battle anniversaries, and in the military tournaments held in Dublin.

As the training of the troops changed over the years to encompass the introduction of new weapons, tactics or strategies, so did the soldiers change. The introduction of a short service system as part of the reforms introduced by E.C. Cardwell, secretary for war, in 1870 was to alter radically the age of the soldier. Enlistment for twelve years, six of which were to be on the reserve, was introduced; gradually this brought about a significant reduction in the age group of the soldiers, a factor which made them more agile on the battlefield, and more socially acceptable in garrison areas.[75] The Curragh camp was also changing; when the nineteenth century was in its final decades a programme of rebuilding was commenced on the Curragh as the original wooden and tin hutments became unfit for habitation.

It was the policy of the War Office, but one which was not always strictly adhered to, that regiments should not remain more than a year in a barracks. Thus it was intended to discourage the men from marrying locally, no doubt in deference to the army adage that 'a soldier married is a soldier lost'.[76] In Ireland an exception was made for the cavalry as the country was so well suited to the horse, with a plentiful supply of forage, and splendid hunting country. From the recorded duration of service in stations on the Curragh and Newbridge a sample of 333 postings in the Curragh between 1803 and 1922 shows that there were 131 for the three month drill periods, and 141 for less than one year. There were 58 stays of from one to two years, and but three of three years. At Newbridge a sample of 107 postings of cavalry regiments shows that 69 were for periods of less than a year, and 38 for between one and two years. The regiments which stayed for the longer periods, or which returned frequently to a station, established contacts with their civilian neighbours through sport or social meetings, and inevitably some connubial unions were made.[77]

Glimpses of the military training conditions in the Curragh Camp can be gleaned from material in the contemporary Kilmainham Papers and military autobiographies. In the autumn of 1871 and the summer of 1872 the weather was so bad for the soldier, especially those men under canvas, that approval was given for the issue of waterproof sheets to the men in September 1871,[78] and in June of the following year it was found necessary to provide wooden bottoms for the tents of the 84th and 97th Foot.[79] That year the 8th Hussars were also there. The Surgeon Major of the Regiment was Irish-born John Smith Chartres and he was accompanied by his family, including his ten year old son John. Half a century later John junior, having served in British Intelligence, was converted to nationalism and in 1921 he was secretary of the *Sinn Féin* Delegation to London, and he is credited with drafting the Irish section of the Treaty in the following year. He died in 1927.

Factors other than the weather could also create problems in the camp. When the North Cork Regiment of Militia reported to the camp for training in the last week of July 1873 they were billeted in lines next to those of the Queen's County Rifles. There a row started between the two units and stones and sticks were used. When the Queen's County men retreated into their quarters the Corkmen attacked the huts and Col. Freer of the 27th Infantry Regiment had to intervene before order was restored. To prevent further encounters the regiments were put under canvas at opposite ends of the camp, the Queen's at Donnelly's Hollow, and the North Cork at Frenchfurze. An inquiry was held into the incident on 2 August, and a few days later the regiments were disbanded. Such affrays between militia units were not uncommon.[80]

The ordinary infantry training programmes in the camp consisted of foot and arms drill, bayonet practice, musketry and range practices, culminating in participation in the reviews or the evolutions and sham battles of the field days. An infantryman left a description of his training in the camp: initial musketry training consisted of preliminary drill, and lectures on the 'theoretical principles of musketry', which many of the men found beyond them. When they were on the range, he noticed, the recruits walked around with loaded rifles, or one might be seen 'sitting on an ammunition-box with a lighted pipe in his mouth . . . perhaps the providence who watches over drunkards and fools keeps a watchful eye on recruits at musketry'.[81]

Cavalry training involved mastering the skills of horsemanship with the use of the lance, and battle charges, while artillery training on the plain, after 1866, concentrated on the movement of the guns into firing positions and preparation for engagement. But all arms shared basic instruction in barrack duties, sentry and guard drills and in the maintenance of their weapons and equipment. The mounted regiments had the additional and unending task of caring for their mounts and the stables.

As well as fulfilling their normal training camp duties the men were regularly called on for special details, such as the maintenance of order during elections or at race meetings, guarding of the judges at the assizes, or sentry duty outside their lodgings.[82] But the relevance of some of the normal training to the benefit of the troops was not always obvious. For example, Lord Sandhurst, who was commander of the forces in Ireland from 1870 to 1875, on one occasion during a field day 'put his troops through a representation of the battle of Aliwal', in which he had participated during the war in India in 1846.[83] Such an eccentric approach to field-training was but another indication of the necessity for the long overdue re-organisation of the army which was introduced by Edward Cardwell when he was secretary of state for War from 1868–1874. He managed to cope with a dominating commander in chief, the duke of Cambridge, who resented civilian interference, and Cardwell brought unified control at the War Office, abolished the purchase of commissions, and linked together the regular, militia and volunteer units.[84]

In 1876 the reporting of the advent of regiments for training in the camp showed that the Kildare Rifles, under Lieut. Col. the Marquis of Drogheda, arrived in mid-May; the Carlow Rifles came a week later by a special train of 12 carriages to spend 27 days under canvas, and the Dublin Light Infantry were there in July.[85] While in camp the regiments were inspected by Col. J. B. Spurgin, commanding 66th Brigade Depot at Naas, after which they were disembodied.[86] In September the permanent staff of the Kildare Rifles returned to the Curragh to fire their own rifle practices. Naas barracks was also a recruiting depot, and that month it was noticed that almost twice as many men had joined up there as in other places during the previous three years.[87] In the same month a draft of 50 non-commissioned officers and men left the depot for the 102nd Regiment Madras Fusiliers at Gibraltar.[88]

During the summer drill periods, when the regular and militia regiments gathered for training on the plain the numbers present could vary from year to year. In 1879, for example, the secretary at the War Office sent a message of regret to the military commanders to the effect that that year the annual assembly of militia regiments at the Curragh for drill would be confined to those located within a distance of 50 miles from the camp, and the number of regiments was not to exceed eight.[89] At the end of each drill season a report was submitted to the adjutant general at the Horse Guards.[90]

Sometimes military duty might be combined with pleasure. Such a happy conjunction arose for the duke of Connaught and Strathearn on Wednesday 8 November 1876 when, in his capacity as officer commanding 1st Battalion Rifle Brigade, he came to Naas barracks to preside at a courtmartial. When the business of the day was over the duke lunched with Col. Spurgin and the other officers, after which the duke enjoyed a game of tennis on the barrack square. Later he left by train from Sallins to return to the Royal Barracks in Dublin. Such a royal visit to the barracks must have been a major social occasion for the officers, and source of considerable interest to the people of the town.[91]

As a training camp the Curragh became familiar to almost all of the regiments of the army serving in Ireland as each summer they assembled there for drill and rifle training. The population of the camp averaged some 4,000 throughout the year, but in the summer training months that number was often doubled (Fig.10). The local press followed the arrivals and departures of the various units from and to their stations, and commented on any special happenings. According to the *Kildare Observer* of 23 May 1885 'the elite of the county' were invited to

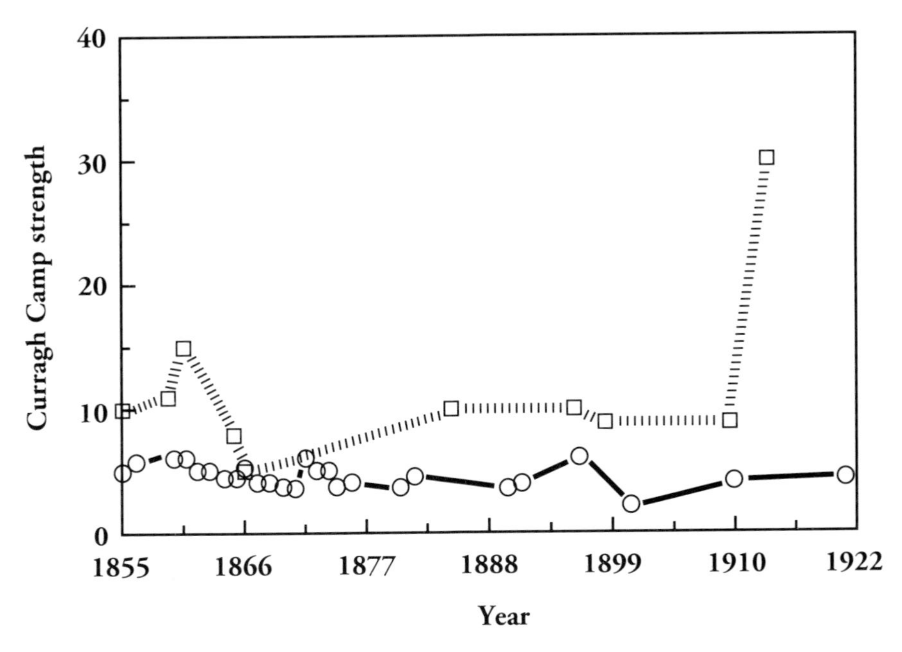

Fig. 10. *The average annual (circles) and maximum (squares) population in thousands of the Curragh camp from 1855 to 1922.* Sources: *census reports; army medical department reports; provincial newspapers*

a ceremony in the camp at which the Marchioness of Kildare, on behalf of the duke of Leinster, presented colours to the 3rd Battalion Royal Dublin Fusiliers (formerly known as the Kildare Militia). For the parade the battalion was formed in a square under the command of Col. the Hon. E. Lawless, and the colours were blessed by the Rev. Ritchie. After the ceremony the visitors were entertained to lunch by Col. Lawless, who was a brother of Lord Cloncurry of Lyons.[92]

In September of that year the same newspaper reported on the annual army rifle meeting in Ireland which was held on the Curragh ranges in the second week of the month; despite unsettled weather there were good scores. Major Gen. the Hon. C.W. Thesiger, commanding the Curragh Brigade, presided at the competitions, and the 5th Dragoon Guards came over from Newbridge. In the evenings the competitors were entertained by the Curragh Amateur Dramatic Club.[93] A week later the newspaper reported that during a Field-Day practice two men of the 5th Dragoons were injured when their horses ran away and fell on them.[94] With the participation of such large numbers of men and horses in the frequent exercises accidents were unavoidable, and the camp medical staff and the two station hospitals were prepared for all emergencies.

During Maj. Gen. Thesiger's command in the Curragh he also held the appointment of Inspector General of Cavalry in Ireland. He lived at Crotanstown Lodge, on the edge of the plain, while his aide de camp, a married captain in the

Inniskilling Dragoons, resided in the Head Quarter Block.[95] Lieutenant Colonel Hon. G.H. Gough, 14th Hussars, who was Deputy Adjutant General, Curragh Brigade and Brigade Major Cavalry in Ireland, was quartered in the Flag Staff Hut. He was Hon. Secretary of the Officers' Recreation Club for the camp and Newbridge, the facilities of which included a reading room, library, daily telegrams, billiard room, whist room, lawn tennis, football, athletics, war games, polo, rackets and cricket. The library contained 400 volumes belonging to Morrow's Circulating Library, which were exchanged monthly, and military works belonging to the Garrison Instructor's Reference Library. Members also had the use of the camp gymnasium where foils, singlesticks, boxing-gloves, footballs, quoits, hammers, shot, etc. were kept for their use. Non-commissioned officers and privates also had access to the facilities at the gymnasium.[96]

General Sir Alexander Godley, recalling his time as a subaltern on the Curragh in 1886, wrote 'the training we did, though not conducted on the principles of today [1939], was sufficient to keep men fit and well accustomed to the handling of arms. Staff tours and tactical exercises only existed in the minds of a few officers rather in advance of their times. Adjutant's drill, at which he and the Regimental Sergeant-Major reigned supreme, were frequent, and physical drill to music, at which the feature of most importance was the keeping of time to the band, was also the province of the adjutant'.[97] Godley described the field days which were held periodically. The battalions lined up 'in full rig, red tunics, busbies etc., with scouts out, would advance in a steady "thin red line" upon Camp Inn Hill, which was perhaps 1,000 yards, or less, away. On nearing it we would fix bayonets and charge, and so home in good time for dinners'.[98]

At that time the participation of the captain company commanders in the exercises was being introduced, and one such exercise, according to Godley, a captain led his company to Donnelly's Hollow where he gave them 'a most spirited account of the great prize fight [of 1815] after whom it was named, and of the many great fights which had taken place in it'. Godley also recalled the financial difficulties of living on five shillings and three pence a day when the mess was run by a contractor 'whose daily charge was seven and sixpence, so that one was considerably out of pocket before beginning to live outside the mess at all. A bit of cake or an apple was extra to the daily charge, and one was expected to drink a glass of port every evening, and champagne on the weekly guest-night'. With the additional expense of providing one's uniform, laundry, and furniture (officers then travelled with their own effects, camp bed, chest of drawers, wash stand, table, bath, etc.) gentlemen often found it difficult to manage.[99]

As a training ground for recruits the camp was a more austere station. The daily routine for the soldier was as follows: Reveille 6 or 6.15 a.m., according to season. Then there were ablutions and bed-making before check parade, followed by breakfast at 8; then orderly-room (for the disposal of prisoners and official business) from 10–11; dinner at 1 o'clock, and tea at 4 p.m.; tattoo roll-call was at 9.30, and lights out at 10.15. Parades were usually from 7 or 7.15 a.m. to 7.45, and again from 11–12. Recruits had a more demanding programme with additional drill from 9–10 in the morning, and in the afternoon from 2–3. Field-days and route marches frequently took the place of the ordinary parades.[100]

While the trained soldiers and recruits underwent their instruction on the barrack squares or on the plain, their comrades in unit headquarters continued with their daily tasks in the barracks. A rare glimpse of one such man at work is a drawing from August 1888 which shows the shoemaker of the 1st Bn Highland Light Infantry at work in his shop in the camp. Surrounded by boots and accoutrements for man and beast the cobbler works at his last. That the hut was lit by oil lamps, and heated from a free-standing stove is obvious from the drawing.[101] That battalion had arrived from Belfast in the camp on 29 August 1887,[102] and remained there in I and K squares until mid-May 1890.[103]

Another battalion that arrived in the camp in 1887 was the 5th Battalion of the Royal Dublin Fusiliers (late Dublin County Militia), which reported on 30 May with a strength of 26 sergeants, 8 drummers and 102 rank and file, for 56 days training. Subsequently the inspection of the battalion by Col. Frankland, O.C. 102nd Regimental District, based at Naas,[104] and the presentation of new colours by the Marchioness of Londonderry to Col. Vernon in the camp, was reported in the press.[105] The Royal Dublin Fusiliers had been associated with Naas since 1873 when the barracks had become the Brigade Depot of the 102nd Royal Madras and 103rd Royal Bombay Foot Regiments. When the territorial system was introduced in 1881 the Madras Fusiliers became the 1st Battalion and the Bombay Fusiliers the 2nd Battalion of the Royal Dublin Fusiliers. With those two regular battalions were linked the Kildare, Dublin City and Dublin County Militia Regiments as the 3rd, 4th and 5th Battalions of the R.D.F., with the regimental depot at Naas.[106] Men who returned to the barracks from service in India called the stream in which they bathed behind Naas Barracks 'the Madras river'.

At Naas barracks in the month of June 1887 a *feu de joie* was fired to mark the anniversary of the queen's accession to the throne, and in the Curragh there was 'immense attendance' at a review 'on the short grass' to mark the jubilee. It was reported that 'the evolutions were well done, and there were three cheers for the queen'. During the three day holiday the 1st Battalion Cameroons held their sports, which again attracted many ladies and gentlemen from the neighbourhood.[107]

In Newbridge barracks N Battery 1st Brigade Royal Artillery (the Blazers) raised a cheer to celebrate the queen's birthday in July 1887, and afterwards they entertained their invited guests to a collation.[108] On display there, in the mess room of the 11th Hussars, was the Balaclava trumpet which had been sounded at the famous charge in 1854. It was described as 'perfectly black in colour, battered and bent'.[109] The preservation of the trumpet formed part of the tradition of the regiment, and when the hussars were posted to another station the trumpet, with the regimental silver, glass, china, pictures, trophies and other memorabilia went with them. In Newbridge the instrument had added significance for the regiment as Lord Brudenell 11th Light Dragoons, afterwards Lord Cardigan of Balaclava fame, had been stationed in the barracks in 1843.[110]

Fusilier Horace Wyndham in 1891 described the Curragh camp as being modelled after 'the better-known ones at Aldershot and Colchester, and as the great school of practical instruction for all the troops in Ireland, cavalry, artillery and infantry, and with detachments of the various departmental corps from the garrison', he saw that it was constantly subjected to changes. As a military station he believed that the Curragh was not popular, 'for battalions are sent there for work, and a drill season there, with its ceaseless round of route-marches and field-days, is a severe purgatory after the gaieties of garrison towns like Dublin, Belfast, and Cork'.[111]

Wyndham described service conditions and outlined the daily duties and chores for the private soldier. The early morning parade lasted about three-quarters of an hour, and it included doubling half-a-dozen times around the square and fifteen minutes 'physical drill'. From 9–10 a.m. the men cleaned their billets, scrubbing the tables and blackleading the trestles and form legs, and whitewashing the hearth. Then they cleaned their accoutrements in preparation for the 11 a.m. parade. This parade might be taken by the sergeant-major, the adjutant or the commanding-officer, the latter usually doing so once a week. On that parade every available man, including the 'employed men', those were the orderlies, tailors, shoemakers etc., had to appear.

After the hour long parade the men returned to their rooms and arranged their accoutrements and bed-spaces for inspection by the orderly-officer of the day. At 1.20 p.m. the bugler sounded the dinner-call, and the orderly-men of the day went to the cook house to collect the meals for their respective messes. The men ate in their billets, and when the meal was over the table was again scrubbed down, and the eating utensils washed. In the afternoon the old [trained] soldiers, if they were not on guard, picket or fatigue duty, might play football or simply lie on their beds until tea time at four. In those hours also men who wished to attend the regimental school might do so, to study for a third-class certificate, which was necessary for promotion. In the evening the soldiers visited their friends in the other barracks, went to the canteen or to the nearest town. The bugler sounded 'the first post of tattoo' at 9.30 p.m., and at ten o'clock the orderly-sergeants went around the rooms to call the roll of their companies. A quarter of an hour later it was 'lights out', and so the soldier's day ended.[112]

Alexander Godley, who served as an officer in the Naas depot in the early 1890s, admitted that 'of barracks, I really saw very little, but we sometimes had cheery-dinner parties there, and after one of them, for a bet, Butler Brooke (son of Sir Victor Brooke of Colebrooke, county Fermanagh) and I undertook to walk to

Dublin. The distance was not very great, only about twenty miles, but after a hard day's hunting or polo (I cannot remember if it was winter or summer, probably the latter), and a very good dinner, I have a vivid recollection of sitting down on a heap of stones at Blackchurch, a little more than half-way, and wishing I was comfortably in my bed! However we plodded on, arrived at Kingsbridge soon after sunrise, and won our bet'.[113]

He also recalled a day when he was detailed to meet at Sallins station Lord Wolseley, commander in chief in Ireland, who was coming to inspect the depot [Wolseley's grandfather came from Tullow, county Carlow].[114] Godley was driving his colonel's dog-cart, while the commander's staff followed in a hack-car. 'I was much alarmed at the prospect, but my apprehension proved to be quite unfounded. He was most kind and talked most interestingly during all this and the return journey. The only remark which might have been construed as one of dissatisfaction, was that the old dog-cart rattled a good bit, which it certainly did!'[115]

That Wolseley was not always so easily pleased is apparent from his correspondence with H.R.H. the duke of Cambridge commander in chief. Writing from the Royal Hospital in Dublin on 18 August 1892 he lamented that 'our manoeuvres here at the Curragh have been somewhat depressing to me as yet, for they displayed a sad want of technical knowledge and military instinct on the part of all the commanding officers of battalions and majors of batteries. At all the field days of importance that we have had as yet at the Curragh the two cavalry brigadiers have invariably broken up their brigades . . . ,' a tactic of which Wolseley did not approve, and he also found that the artillery was poorly trained, while 'the infantry struggle, [their] attacks are absurd in conception and futile in execution. After two seasons here now I am convinced that our senior officers require to be well grounded in tactics'. He suggested that during the winter a board of examiners should test the tactical knowledge of the majors and lieutenant-colonels.[116]

A couple of years later the 1st Battalion Royal Munster Fusiliers at the Curragh was chosen as the first unit to test a new system of training ordered by the commander of the forces in Ireland. The battalion was to undertake a weeks continuous marching over a distance estimated at 96 statute miles, and subsequently all troops within the command were to be similarly exercised. Early in February 1895 the Assistant Adjutant General congratulated the battalion on its achievement of covering a distance of 123 miles in six days without a single man falling out. The march was especially difficult due to the 'trying state of the roads, and deep snow for two marching days'.[117] In 1896 the battalion, with a strength of 20 officers and 616 other ranks, again distinguished itself in another forced march of six days duration. No man fell-out during the 156½ mile marathon, and the battalion could boast a high percentage of young soldiers in the ranks and their endurance was compared to that of old campaigners. In August 1896, after over two years on the Curragh, the 1st Battalion Munsters moved to Fermoy.[118]

The precision and bearing of the troops formed during the long hours of training were displayed to the public on ceremonial days such as the royal birthdays. In 1897 the observation of the queen's birthday at the Curragh was reported in the *Navy and Army Illustrated.* Photographs of the marching past show the infantry and the engineers passing before the reviewing stand where Major-General Combe, commanding the Curragh District, and his staff had taken up position. The order of review was, first the Royal Horse Artillery, followed by the cavalry, Royal Artillery and Royal Engineers. The mounted troops first passed at the walk, then at the trot, and lastly at the gallop, during which the bands played the tunes which belonged to each regiment. The infantry first passed at the slope, then at the trail, and the officers gave the order 'eyes right' to the men as they passed the saluting base. In the Curragh District at that time there were one regiment of cavalry, two batteries Royal Horse Artillery, two field companies of the Army Service Corps, a company of the Medical Staff Corps and a detachment of the Army Ordnance Corps.

The firing of the *feu de joie* was explained as having required much practice in advance by the infantry. The men were drawn up in two ranks, and on the order to fire the right hand man of the first rank immediately fired, and then each man fired a round in turn, to be immediately followed by the second rank; both ranks fired three times. Then bayonets were fixed and arms presented, after which three cheers were given for the queen, with each man waving his head-dress above his head.[119]

That the month of June was a high point in each training season was indicated by the two major events of the 1895 season. From 8 June to 14 June the Royal Irish Military Tournament, for the benefit of Irish military and other local charities, was held at Ballsbridge, Dublin. Two performances were given each day and the participating troops were drawn from stations throughout the country for the tournament. It was held under the patronage of the queen, the lord lieutenant, H.R.H. the Field Marshal commander in chief, and Field Marshal Viscount Wolseley, commanding the forces in Ireland.[120]

The lord lieutenant was also patron, and Field Marshal Wolseley president, of the All Ireland Army Rifle Meeting which took place on the Curragh from 26 to 29 June. The major generals commanding the four Irish districts were vice-presidents, and there were two committees to manage the meeting. A detailed poster gave the details of the 28 competitions to be held, and for which many of the prizes had been donated by contractors to the forces. These included a military tailor, a jeweller and an arms manufacturer from Dublin, five breweries including the Phoenix Brewery in Dublin which gave three prizes, and the army contractors Sir R. Dickeson and Co., and three local firms. The local firms were Mr Charleton's Garrison Studio in Newbridge, which gave a silver cup value £7, and three cash prizes; Mr. G. Searight of the Curragh camp who presented the

Searight Prize of £5; Mr J. Boyle, Market Square, Curragh camp, who gave two prizes, one of £5 for a 500 yards rifle competition, and a cup valued at three guineas to be awarded to the warrant or non-commissioned officer of the Curragh District making the best aggregate during the meeting.[121] The excitement and business generated in the locality during the shooting competitions was a bonus to the traders, and the donation of prizes by some of them was a token of appreciation for the military trade.

During the quieter months of the year the regiments garrisoned in the camp underwent regular training; war games for the officers were played in the class rooms during the January of 1896, and on the 27th of that month a course of continual route marching was commenced. Every possible man was ordered to take part, including those employed at regular jobs who were to be relieved by special duty battalions. The 1st and 2nd Yorks, 2nd East Yorks, 1st Battalion Munsters were each given special marching dates, and they marched up to 18 miles a day. Daily returns of strengths were to be submitted.[122] The 1st Yorks completed their week of continuous marching having totalled 104 miles, during which only two men fell-out and had to be admitted to hospital.[123] The Munsters, 700 strong, marched from the Curragh via Kilcullen to Naas where they were greeted by the band of the Dubs and led to the depot. There they were given refreshments, with the sergeants of the Munsters being entertained by the sergeants of the Dubs.[124]

Routine training continued in the camp, and in February 1896 a draft of 90 men was being prepared by the 1st Battalion Royal Irish regiment for the 2nd Battalion which was in India. At the camp gymnasium a course in the new sword drill was being held for gymnastic staff sergeant instructors in Ireland. Before they returned to their stations they were entertained to a smoker in the gymnasium.[125] At the end of that month the musketry practice for the 3rd Battalion Militia (Dubs) commenced. By 7 March the route marching had ceased in the district, and the programme for drill, physical training, etc. was announced.[126] Regiments were relieved of all other duties to enable them to practice their regimental drill.[127] The programme for the year allowed one week's musketry to each unit, and two weeks to recruits. Recruits from the 4th and 5th battalions of the Dublin Fusiliers were also to undergo their preliminary drill training in the camp in May, while the 'old hands' of three battalions were to assemble there in June for 29 days.[128]

By the middle of March 1896 the 1st Battalion of the Hampshire Regiment, the 2nd Battalion 1st Yorks Light Infantry from Mullingar, and the 1st Battalion Lancashire Fusiliers from Athlone had arrived in the camp for musketry practice.[129] On the 21st of the month it was announced that a compliment of just over 500 Dublin Fusiliers would be quartered in I and K lines in the camp for preliminary drill.[130] Advance planning was also in progress, and towards the end of March a meeting was held in the officers' club in the camp to discuss the All-Ireland Army Rifle Meeting which was to be held there from 16 to 19 June.[131]

The administration of the camp establishment continued as always, and in March the stock-taking of the ordnance stores was completed and regiments were being notified of the receipt of necessaries, such as soldiers boxes.[132] Inspections, such as that of the fire engines by Sir Eyre Shaw (late captain of London Fire Brigade) in January,[133] and that by the Inspector General of Cavalry of the 10th Hussars and the 18/5th Army Service Corps,[134], and by the Director General of the Army Veterinary Department of the horses in the Army Service Corps and the Royal Engineers, were held later in the year.[135]

A special train from Kingsbridge in Dublin brought the 4th and 5th Battalions of the Dublin Fusiliers to the Curragh for their annual training in April 1896, and at the same time 'the whole of the adjutants and sergeants major of the regimental depots in Ireland arrived [in the camp] for physical training'; they were attached [for administrative purposes] to the 10th Prince of Wales' York Regiment.[136] The A.S.C. was then running a course in transport duties for the 1st Battalion Munsters, but it, and all training, was as usual, suspended for the two days of the Punchestown races to enable everyone to attend (Fig. 11).[137]

By the month of May 1896 the strength of the Curragh District was 5,896, and it consisted of one cavalry regiment, two batteries of artillery, three infantry battalions, two field companies engineers, and four companies of the Army Service Corps.[138] That month also the headquarters block was being prepared for Field Marshal Lord F. S. Roberts, commander of the forces in Ireland, who with Lady Roberts and their family, was to reside there during the summer months.[139]

The queen's birthday was again celebrated in the camp on 20 May when 'the troops in Ireland were under arms at noon'.[140] There was a *feu de joie*, and three cheers led by the major-general. Then the 'whole of the troops in the camp and Newbridge' marched past twice, before many spectators.[141] Some of the troops from Newbridge were absent as A Battery R.H.A. had left for Glenbeigh for gun practice.[142] *En route* to county Kerry the artillery men were entertained 'very hospitably' at Cassidy's Brewery in Monasterevan; Cassidys had the contract to supply both the Curragh Camp and Glenbeigh camp with 'the best stout, porter and ales'.[143]

In the first week of June 1896 the troops began assembling for training in the camp where some of the new barracks being built there were ready for occupation; amongst those arriving were 428 members of the Hampshire Regiment from Birr, who were based on number 8 camping ground, and where they were to be joined after a couple of weeks by seven more officers and 212 men. The 4th Battalion Royal Dublin Fusiliers marched in, and the mounted and dismounted elements of the 28th Field Battery R.A. and the 78th Field Battery R.A., who were to be located south of the new cavalry barracks.[144] The 1st Royal Dragoons were also there, having marched from Dublin where they left their women and children in Island Bridge Barracks, and the 1st Squadron of the 15th Hussars camped at Donnelly's Hollow during their musketry training period. There was

Fig.11. *Engraving depicting the Royal visit to Punchestown in April 1868. The Prince of Wales is mounted on the white horse on the left; behind him, the Prince Teck, the Duke of Manchester and the Duke of Cambridge. Of the 132 gentlemen named in the picture 46 have military rank, and 37 have titles. The Princess of Wales and the Viceregal party are in the covered grandstand, above the prince.* Photo: National Library of Ireland.

then also a brigade of 2,000 militia in the camp.[145] When the drill season was over the establishment of the camp was to be increased with the posting of a cavalry regiment to the new cavalry barracks there.[146] The headquarters of the officer commanding Royal Artillery, Curragh District, was moved from the castle yard in Dublin to the Curragh in July 1896.[147]

The All-Ireland Rifle Meeting was held in the week commencing 14 June in beautiful weather. The newly appointed Officer Commanding the Forces in Ireland, Field Marshal the Rt. Hon. Lord Roberts presided, and he presented his own cup to the winning team. Even though there were more entries than usual the competitions were not prolonged as the closing of the Donnelly's Hollow road enabled the firing to be carried on continuously all day.[148] At the Military Tournament in Ballsbridge non-commissioned officers from the Curragh gymnasium were the winners of some of the fencing competitions. A sergeant and a gunner from the Royal Horse Artillery based in Newbridge were injured during the tournament, and a subscription was raised for their welfare. To encourage soldiers from the county Kildare to attend the tournament special fares were offered by the Great Southern and Western Railway Company.[149]

The glorious weather that summer ensured that by the end of June when the militia of the Royal Dublin Fusiliers had finished training, which was assessed as 'exceptional', they were seen to be 'quite bronzed by exposure to the sun'.[150] Early in July Field Marshal Lord Roberts came to Naas to inspect the 3rd Battalion Royal Dublin Fusiliers at the Depot.[151] No doubt but the men were glad to finish the strenuous training, though sorry that the comradeship of the camp was over.

While the various military schools in the Curragh camp closed for the summer leave period,[152] the visiting units got on with their work. Field operations were held between the Hill of Allen and Timolin, and the cavalry brigade and Q Battery R.H.A. formed up in a brigade mass near Ryan's farm on the Curragh for brigade drill.[153]

When manoeuvres were held a special financial grant was made by the ministry; it amounted to £5,000 for the Curragh in 1896, an increase of £1,000 on the previous year.[154] Manoeuvres had been held for the first time in Ireland in 1894 when Lieut. Col. Sir Anthony Weldon Bart. from near Athy, had accompanied the general officer commanding in Ireland, Viscount Wolseley, as his aide-de-camp during the exercises, a function which he also filled during the manoeuvres of 1895.[155] Wolseley had improved training standards during his time in Ireland (1890–95) by introducing long marches, battlefield manoeuvres and night training. His successor at Kilmainham was another Irish born general, Lord Roberts, who was to hold the appointment until he was sent to command the British forces in the Boer War in December 1899. He was also concerned with improving the training in the army, and he expressed disapproval with the Curragh as a training ground. There, and at Aldershot, he considered 'the ground to be too restricted to admit of anything more than ordinary drill instruction. Manoeuvres on an extended scale are out of the question'.[156] Perhaps that was why ground chosen for the July manoeuvres in 1896 was 'on the other side of Kilkenny, in the direction of Cashel'.

The troops marched from the Curragh to the selected area, with fatigue camps at Youngstown and Pike, five miles from Athy.[157] All of the troops participating in the exercises were reviewed by General Lord Roberts at the Curragh: three regiments of cavalry, four batteries of artillery, nine regiments of infantry. A total of 5,578 men paraded for his inspection.[158] Subsequently he sent his compliments to them all.[159] A news item in the *Leinster Leader* of 1 August 1896 gave a figure of 10,000 soldiers as having passed through Athy en route to the Curragh and Dublin after the manoeuvres[160] (and it was later noticed that one soldier died after the exercises[161]). An official report on the conduct of the manoeuvres was submitted to the War Office by Major General Combes.[162] So pleased were the people in the manoeuvres area, they petitioned Lord Roberts 'praying him to exercise all the influence he can to insure the manoeuvres of 1897 being held over the same ground as this year, in order to show the soldiery how much they are appreciated'.[163] It would be safe to presume that the relative affluence of the soldiers, rather than the company of the men themselves, was what was appreciated by the civilians concerned.

Back in the Curragh the summer training continued with, in the first week of August, the 'men composing the Dublin column' pitching their tents at Hare Park; they had started their training by firing their range practices at Kilbride, county Wicklow, at 9 a.m. on 6 August, and at 4 p.m. they had struck their camp there to march 14 miles 'over very rough country' to Ballymore Eustace, where they arrived at 9 p.m. Having tented there over-night they marched on the next morning to the Curragh where they arrived at 3 p.m. looking 'very smart and soldierly like'.[164]

Early in September it was reported that in another exercise on the Curragh outposts on a line of observation from Knocknagarm to Croatanstown were manned by the infantry brigade at 8 a.m. on the morning of the 4th 'to protect a western force bivouacking on the plain', and earlier that week the general officer commanding had inspected the 1st Battalion Sherwood Foresters on the polo ground prior to their departure for Malta.[165]

The forty-first anniversary of the victory at Alma in the Crimea was celebrated on Saturday 19 September. The 1st Battalion Prince of Wales' Own Regiment (Yorks) trooped the colours on the lower polo ground near the camp, and an open invitation to attend was issued to everyone, including mounted officers and officers of the garrison, who were to be in undress uniform.[166]

Late in the month of October the drafts prepared by Q and R batteries of the Royal Horse Artillery and the 2nd East Yorkshire Regiment departed from the Stand House siding for Queenstown, en route to India.[167] Then the 1st Battalion

Sherwood Foresters in the Curragh was ordered to prepare a draft of 138 other ranks for Bengal, and which was to be ready for departure on 23 November.[168]

The intensive winter training programme continued with night training by the brigade on 2 October, and by the end of the month route marching had commenced for the infantry in the camp. The ranges were also in use as detachments from 2nd Scots Guards, 2nd Hampshire, 1st Oxford, 2nd Royal West Kent and 1st Durham regiments arrived in the camp and were billeted in J and K squares.[169] Training for the officers of the garrison continued with war games, played in the garrison instruction office in F square.[170]

As that outline of training on the Curragh in 1896 has shown, it was a bustling place for the greater part of the year. In the 40 years which had passed since the designation of the Curragh as a camp of instruction, and as a means of accommodating a large force of infantry for the purpose of being trained and manoeuvred in conjunction with cavalry and artillery, the Curragh of Kildare had hosted numerous great musterings of the three arms, sometimes in weather which made conditions for the men almost intolerable.

Route marches on the roads in the neighbourhood of the plain brought the troops into formal contact with the civilian population, while the grand manoeuvres which commenced in 1894 introduced people at a considerable distance from the camp into acquaintance with great numbers of soldiers. The extensive rifle practices held on the Curragh during the shooting season necessitated the disruption of passage on the Donnelly's Hollow road, and culminated in compensation being given to those persons most aggravated by the shooting in 1896. But the local graziers would, in the opinion of Major General Combe, Commanding the Curragh District in 1899, benefit from his plan to clear the furze on the south side of the camp. More grazing land for the sheep would be available if he was given permission to clear the furze to create a greater area for cavalry training, necessary as there was then a large increase in the strength of the cavalry quartered at the Curragh. The Quit Rent Office, which administered the interests of the Crown, had no objection to the proposed clearance being done for the convenience of the War Department, and at that department's expense. The deputy ranger was to supervise the work.[171]

Over the decades thousands of soldiers had served in the Curragh and Newbridge districts when, in addition to their professional experiences, they had encountered the local residents. They had met socially with some of the town and country folk, and there had been pleasant and unpleasant encounters. Unlawful behaviour occurred between both communities, but there were also conjugal unions made. The commissioned officers who left impressions of their time in county Kildare generally enthused about the enjoyment of the social and sporting life there. The quality of the hunting and racing available was especially appreciated.

If the training facilities of the Curragh had been found to be good, the living quarters were not so described. As early as 1858 Lieut. Col. Lugard, the architect of the hutments, had regretted the fact that they were constructed of timber. Over the years they weathered badly, and were especially uncomfortable in inclement weather and in winter time. The concrete billets which replaced some of them from 1871 proved to be equally unpleasant, but in the early 1890s a comprehensive rebuilding programme was commenced and over the next 30 years well designed brick barracks were to transform the camp into a large modern military town.

Chapter 5

A Royal Visitor and a Fenian Interlude

The regular training programme in the Curragh camp, on the ranges and plains, and in the surrounding countryside, were always thoroughly planned in advance, and subsequently took place during the appointed periods and in the designated locations. Occasionally there were minor interruptions to the schedules due to such factors as the weather or the exigencies of the service which might necessitate an immediate movement of some troops. However, there might also be extraordinary disruptions to the normal training programmes, such as that experienced in 1861 when the heir to the British throne was sent to the camp for military training, or when the Fenian activity from 1865 onwards demanded an intensifying of security and operational duties.

The training period of the year 1861 was to become a memorable one in the military annals of the Curragh camp as it was chosen as the season during which the heir apparent to the crown of the United Kingdom of Great Britain and Ireland was to undergo a period of military instruction on the plain. The regiments which rotated there that year were the 4th Queen's Own Hussars, 3rd King's Own Hussars, 11th Hussars (Prince Albert's Own), 15th King's Hussars, 1st and 3rd Battalions Grenadier Guards, Devonshire Regiment, East Yorkshire Regiment, Green Howards, Lancashire Fusiliers, Worcestershire Regiment, King's Royal Rifle Corps, Royal Irish Rifles, Manchester Regiment, and the Rifle Brigade. The census for the year, taken in the month of April, gave the military population in the county as 5,103,[1] but during the drill period on the Curragh alone it was estimated variously as from 10,000 to 12,000, and even as high as 15,100.[2]

The strain on the administration of the camp that season must have been considerable following the announcement from London that Prince Albert had decided that his nineteen year old son, Albert Edward, Prince of Wales (Fig.12), 'should be subjected to ten weeks course of infantry training, under the strictest discipline which could be devised, at the Curragh Camp near Dublin'.[3] It was

Fig. 12. *The Prince of Wales aged 19. His training period on the Curragh in the year after this photograph was taken proved to be a memorable one for himself and his parents, as well as for the people of Kildare. From* Royalty in the World. *(London), 1860.*

further planned that, after some weeks, Queen Victoria and her consort would come over to visit the prince, and to review the troops.

Prince Edward was happy with the arrangements. The *Freeman's Journal* remarked that this, his fourth visit to Ireland, was by his own choice. He was welcomed in Dublin 'with banners, flags, bells and fashionable ladies'. The journal suggested that if he 'could divest himself of his princedom, and go about like the Arabian Caliph for the three months he will be fighting sham battles in Kildare, he would acquire far more fruitful knowledge and Ireland would command a deep place in his affections. The campaign on the Curragh will however be varied with leave and furlough'.[4]

The Prince of Wales' sojourn in Kildare was also intended to educate him in the social requirements of his station. It was ordained how he was to occupy each evening of the week: on two evenings he was to entertain senior officers to dinner; he was to dine with the regiments in strict rotation on three days, and on the remaining two evenings (one of which should be Sunday) he was to dine privately in his quarters, after which he might read or write. His relationship with other officers was to be 'placed upon a becoming and satisfactory footing, having regard to his position both as a prince as well as a field officer in the army'.[5]

On his arrival at the camp on 29 June the prince was met by General Sir George Brown, the General Officer Commanding in Ireland, and greeted by a royal salute from a field battery of the Horse Artillery. He was to be accommodated in the headquarter hut which had formerly been used by Lord Seaton. It was surrounded by 'tastefully laid out grounds' where two small tents had been erected, and it was guarded by men from the Grenadier Guards, to the 1st battalion of which regiment the prince was attached for training. The battalion was brigaded with the 36th Regiment. The *Illustrated London News* of 13 July 1861 carried a report on the Curragh, with pictures of the royal quarters, and of a grand review (Fig.13).

Unfortunately, Edward's progress on the barrack square was poor, and prior to the visit of his parents the prince had been given a bad report from his superior officer who had 'abandoned all hope that he might be fit to command a battalion by the end of the month'. As a colonel the prince felt that he should not have to perform the duties of a subaltern, and when he asked that he might be allowed to act as a company commander his colonel replied 'you are too imperfect in your drill, sir. Your word of command is indistinct. I will not try to make the Duke of Cambridge (the general officer commanding in chief) think you are more advanced than you are'.[6] When the queen heard of this rebuke she praised the colonel 'for treating Bertie just as other officers', but before leaving the Curragh it was arranged that the prince would manoeuvre a brigade. The Prince of Wales was learning the practical duties of a regimental officer, and he was attached 'for drill' to the 1st Battalion of the Grenadier Guards, acting as captain of No. 9 Company. It had been decided that there was to be no distinction whatever between the prince and the other officers and, wet or dry, he was to turnout for all parades, drill and exercises. In the afternoons he was 'to play some games of racket'.[7]

The royal party, consisting of her majesty, the prince consort, and several of their children, were aboard the royal yacht which anchored at Kingstown on 21 August. The first day of the visit was spent in Dublin, and on the 23rd the prince consort made a quiet visit to the Curragh Camp. The prince was accommodated in the hut which was normally that of the general officer commanding. There he entertained his father to lunch, even though the visit was unexpected and he had been about a mile from the camp with his regiment when the prince consort arrived. He had ridden over to the exercise ground and for about an hour watched his son, with his company, going through battalion drill.[8]

The following day the prince consort accompanied the queen back to the Curragh when there was a grand review of all the troops in the camp. The queen, standing in her carriage and protected from the rain with a parasol, saw her son 'perform manfully the subordinate duties of a company officer in a force of some 10,000 men, [another account said 12,000 men] a very complete little army, consisting of a brigade of cavalry, 3,500 strong, a brigade of horse artillery, with 18 guns, three brigades of infantry, and five field batteries with 30 guns. Unhappily, the day proved so wet and stormy that the queen, after a vain effort

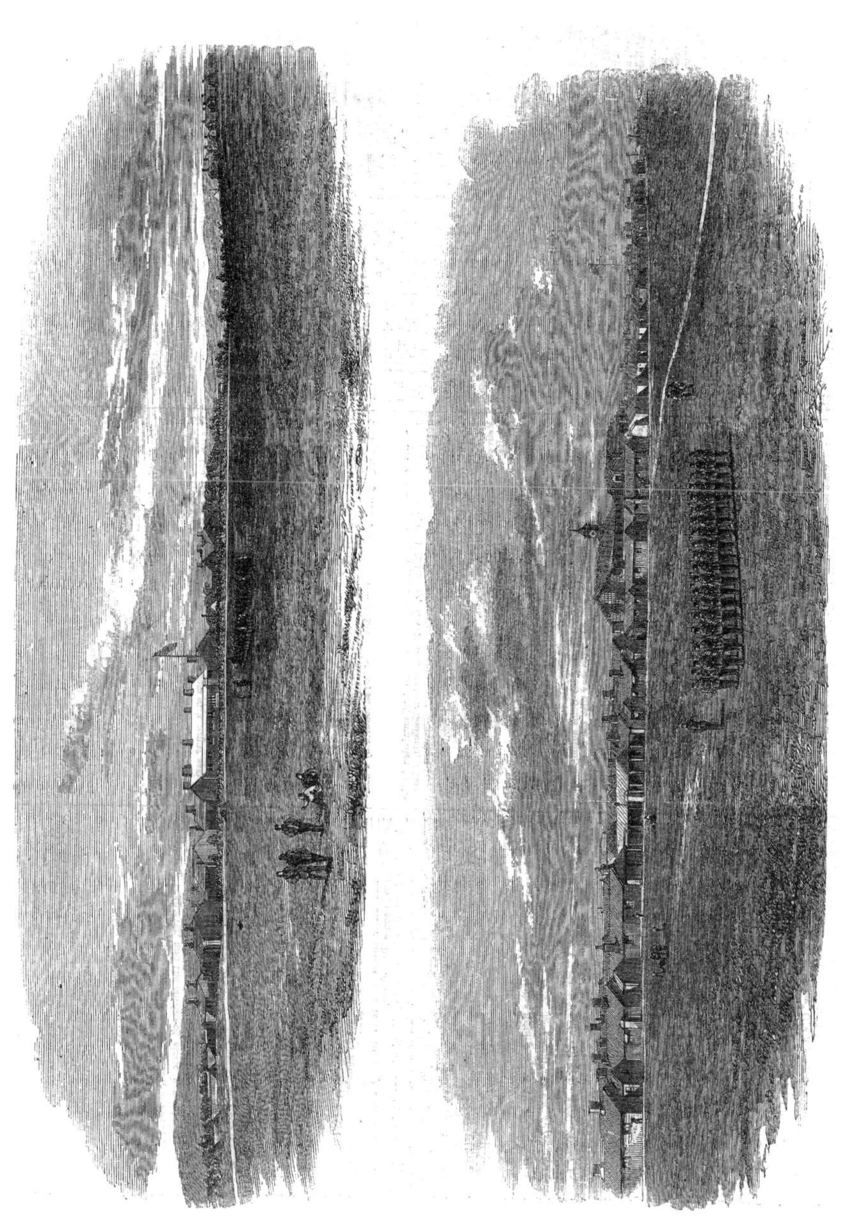

Fig. 13. *Residence of the Prince of Wales in the Curragh Camp, and a north view of the camp, from* The Illustrated London News, *13 July 1861. The colours fly over the headquarter hut, which was normally reserved for the commanding general, but made available to the royal heir.*

to keep under an umbrella, was obliged to have her carriage closed'. As the prince marched passed he was cheered by the crowd, described as 'a vast concourse of spectators from Dublin, Kilkenny, Carlow, etc.'. Prince Albert and Prince Alfred braved the elements to the end and saw 'the sham fight which was a spectacle of more than usual grandeur, for, on account of the large force of cavalry, that arm was made to play a very important part'.[9]

A contemporary account of the review described the prince as having 'headed his company bravely . . . the infantry formed in a line nearly a mile long'. Having advanced they fired a volley, then the great guns boomed causing the crowds to panic. The finale of the day was the cavalry charge, and the journalist was euphoric in his description: 'The grand charge when performed by British dragoons is probably the finest open air spectacle that can be witnessed. The horses are so fast and so high mettled, the men such horseman and so fearless, the trappings and appointments so first rate, that the eye becomes dazzled as the immense force dashes past in a perfect whirlwind of splendour and the spectators shout with excitement and delight. The charge which was finished without a single accident was a perfect success.'

Before the queen and the three princes retired for lunch to the prince's hut[10] the queen and her consort expressed their great satisfaction at the proceedings of the day to Gen. Sir George Brown.[11] During her visit to the camp the queen was entertained by a piper from the locality, and as a gesture of thanks she presented him with a set of silver-mounted Irish pipes. [55 years later, in a local paper, the question was asked: where are the pipes now? No one seemed to know].[12]

If Sir George Brown's interest in the soldier prince had pleased the queen, he was *persona non grata* to the Turf Club. Brown had shown little regard for the racing men from the time he had arrived as general commanding. Indeed, on the day of the royal review the race-course was so cut-up by the military that, to enable it to be ready for the September Royal Meeting, the Curragh ranger, at his own expense, had to employ a man to repair it. But that was not the end of the matter. On the last day of the September Meeting racing was delayed for an hour after the general 'commanded the battery of Armstrong guns to be planted on the course and the cavalry charge over it, and thereby reduced it to the condition of a ploughed field'. Subsequent requests from the Turf Club for compensation produced but a derisory sum.[13] The annoyance of the club did not abate, and in its evidence to the commission of inquiry in 1866 the conduct of the late General Brown was again condemned.[14]

However, it was not all work and no play for the young prince at the Curragh. In his free time Edward visited locally, to Poulaphouca to swim in the deep pool beneath the falls there, and to the Hill of Allen to see the building of Aylmer's tower. One of the masons working there recalled that 'the prince took out a black scut of a pipe with a shank half a length of his finger, and a pouch. Two officers were along with himself, and they smoked and sung. The prince asked me who was getting it built, and I said Sir Jerald. One of the officers gave me a shilling. The prince did not give me anything. The prince forgot his silver match box on a stone table, and I kept it and had it for a long time, but someone took it off me'. The royal visit to Allen is recorded on one of the flagstones of the tower.[15]

Other amusements were provided for the prince also, the most celebrated of which was the smuggling of the young actress Nellie Clifden into his quarters one night. Some of his brother officers, with whom she was popular, had invited her to a mess party, and after the senior officers had retired they had their fun. Eventually the story reached London, and the queen was not at all amused.[16] The prince admitted 'that after resisting he had yielded to temptation, but that now the affair was over'. But it was not forgotten, and when his father was later endeavouring to arrange a suitable marriage for his son he was told by an adviser that 'the affair had made him an unsuitable husband'.[17] Prince Albert had been very much upset by the prince's conduct on the Curragh, and when Albert died before Christmas of that year of 1861, from typhoid fever, the queen blamed Edward for his father's decline, and she was to believe so for the remainder of her long life.[18]

Another experience enjoyed by the young prince at the Curragh was participating in a race (wearing a white silk shirt) under the name 'Captain Melville'. His enjoyment of the turf was confirmed shortly before he ascended the throne as King Edward VII when he revealed that he 'was never as happy as when I can go to the races as plain Mr Jones'.[19]

During his final days in the camp the Prince of Wales presented new colours to the 36th Regiment. The chaplain, Reverend Wheeler, blessed the colours, after which Edward told the assembled soldiers, ladies and gentlemen that he should 'ever look back to my intercourse with yourselves and various corps comprising the fine divisions assembled in this camp with feelings of mingled pleasure'. Then he was entertained to lunch by the officers of the regiment, and afterwards he visited the soldiers and their families in their quarters. In the evening the sergeants of the 36th Regiment entertained the sergeants of the Grenadiers to dinner.[20]

On the day of his departure from the camp all the troops were under arms and he proceeded for the first time to manoeuvre a regiment (as he afterwards wrote to his mother 'with expert assistance I came through the ordeal'). At the end of the exercise Edward was cheered by the men of the infantry brigade, and he 'was evidently affected'. Before reaching the hutments his battalion, 1st Grenadier Guards, halted and formed a square which was photographed and later drawn for the *Illustrated London News*. It was a gloriously fine day on which to conclude the three month royal visit to Kildare. At 3 p.m. in the afternoon he left his quarters, the national anthem was played, and he drove off to Newbridge.[21]

A memory of the prince's time in the camp has been passed down in the Gillman family. Daniel Gillman recalled in 1992, when he was aged 81, that his maternal grandfather was a captain with the 19th Regiment (Green Howards) at the Curragh during the royal visit. He was 23 year old T.H. Kirby, a Roman Catholic from Limerick, and an athlete. In the month of August 1861, and in the presence of the prince, Kirby won a half-mile race in two minutes and four seconds. He was presented with a silver cup by the Prince of Wales, which was to remain a treasured possession in the Gillman family. Two years later Capt. Kirby, then of the 22nd Regiment, to which he had transferred on his marriage, at the behest of his father-in-law who did not want him transferred overseas, was the amateur one mile champion of Ireland.[22] But his memory of the prince's sojourn in the camp was that it 'caused a nuisance, there were extra guard duties, and the prince sometimes escaped from his escorts and caused problems'.[23]

The visit of the crown prince to the Curragh was visually well documented; Dr Jones Lamprey, Army Staff-Surgeon and a graduate of Trinity College Dublin, painted a topographical oil painting of the encampment. It came to notice in modern times when it was exhibited at the Gorry Gallery, Dublin, in 1989.[24] Measuring 24 × 61 cms, it is the earliest known painting of the hutments, then just seven years old. Lamprey viewed his subject from the hillock between the firing points of the rifle ranges. The features which he indicated on the skyline are identifiable: the round tower and part of the ruined tower of St Brigid's cathedral in Kildare (which was to be restored a decade after) are on the extreme left; then extending to the right, are the Red Hills – the Chair of Kildare, Dunmurry, Grange and the Hill of Allen. Aylmer's tower on Allen was not then sufficiently raised to be noticeable as it was not completed for another two years.

The camp on the Long Hill spreads around the clock tower and the flag staff behind, close to the Protestant church. The ten squares of huts in which the men were billeted occupy the highest point of the ridge, and behind them, on the right, the headquarter buildings are just discernible. In the foreground, from the right, is the bakery with its tall chimney, then the clustered huts of a barrack store, and No. 2 hospital. The central group of huts is an engineer park and barrack store, with the racket courts behind. A few bell tents, and sutlers' huts (including one which is thatched) are also identifiable.

Surgeon Lamprey's impression of the Curragh can be compared with drawings of the camp in the *Illustrated London News*,[25] which were made from the photographs taken by Capt. E.D. Fenton 86th Regiment. The photographs are now in the Royal Archive at Windsor Castle in an album called *Souvenirs of Soldiering at Camp Curragh*, and it was compiled either for Queen Victoria and Prince Albert or for the Prince of Wales. The album is bound in purple velvet, with a gilt line border and three metal emblems fixed to the front cover. They are a crown, a smaller crown with a lion on top, and the Prince of Wales' feathers.[26]

The album is a unique record of the royal visit and of training on the plain, as well as a valuable record of the encampment and surrounding places in 1861. The approach to the camp is heralded by a photo of the Camp Inn, a hut with the name of the establishment on a board above the door. A group of soldiers sit on a bench outside, and a four-wheeled horse drawn car is on the road. There is a south view of the camp, very similar to Lamprey's view, and also with a decorative group of soldiers in the foreground. There is a distant and a close view of the Head Quarter Hut, where the Prince of Wales was accommodated. A general is shown in the latter picture, with his family. Two photos are captioned 'Camp life in fine weather' and 'Camp life in wet weather'. There are lounging officers, croquet mallets and shot guns in the first picture, and a group huddled beneath umbrellas, and muddy ground, in the other. A picture of an officer lying reading on a camp bed outside a hut marked 'D' is labelled 'After a Field Day', and one of a soldier resting, with the huts and rows of tents in the distance, is called 'Soldier contemplating his Cantonment'. A photograph entitled 'Sunday afternoon, Cavalry Camp' has a party of ladies and gentlemen being entertained by a brass band with the tents behind. There is a series of pictures of the military exercises showing hundreds of men in action: 'An advance to skirmishing'; 'Squares of Wings, Squares of Companies'; 'A very sham fight the 36th Regiment prepared for a very desperate resistance'; and 'The Queen's Field Day 24th August 1861'. The latter records the coaches, cars and other vehicles, crowded with ladies and gentlemen, congregated for the occasion. The Prince is not identifiable in a picture of a 'Square of the Queen's Company 1st Bn Grenadier Guards H.R.H. the Prince of Wales in command, 11th Sept 1861'; but another picture shows the prince presenting colours to the 36th Regiment.

The main street through the camp, looking east, is clearly shown on a photo titled 'Very slow indeed!!' (see Fig. 17). Seven men sit or lounge outside a canteen while two side cars proceed towards the clock tower. That the prince visited places in the neighbourhood is shown by photographs of Kildare town, with a row of small thatched cottages, and the ruined cathedral; of the old bridge at Kilcullen; and of the bridge and falls at Poulaphouca where the prince and his friends bathed and lunched. Other photographs at Windsor, but not in the album, show the prince by his hut, and with groups of men, and of military bands, cricket games, and soldiers relaxing with pipes and tankards.

From the impressions of the thousands of massed troops and the onlookers which can be seen in the royal photographs it can be understood why at the end of the summer drill period on the plain the sward was extensively damaged by the horses, gun carriages and the vehicles of the onlookers. This naturally effected the grazing of the flocks, and caused annoyance to the commoners as well as to the Turf Club.

The racing fraternity and shepherds had further cause of displeasure in the spring of 1862 when General Sir George Brown proposed to have the furze bushes on the plain removed by his men as the gorse obstructed the training in mass formations and movements. The golden furze which sheltered the flocks was a distinctive feature of the sward, and the horse trainers believed that not alone would the beauty of the plain be destroyed, but that the sharp roots of the cut bushes might be injurious to the horses. The Master of the Kildare Hunt objected to Brown's plan as he knew that the cover was also necessary for the chase. The combined complaints caused the Office of Woods and Forests to prevent the general from carrying out his plans.[27] The Royal Engineers had also shown disregard for the plain when they removed six acres of sod. Then the vigilance and reports of the ranger to the Office of Woods and Forests caused the War Office to intervene and agree that the damage was to be undone, but after a while the stripping of the sods was repeated.[28]

In Newbridge the prince's visit was commemorated with the naming of a street and an hotel in his honour. Edward street is still there, but the Prince of Wales Hotel is not. Three years after the royal sojourn on the Curragh the hotel brought dishonour on itself. At the County Kildare Quarter Sessions in October 1864 Mr F. Page, the owner of the premises, applied for a renewal of the spirit licence. It was opposed by the crown solicitor and the police on the grounds that it was improperly conducted by being open after hours. The local landlord, Eyre Powell J.P. made a recommendation on Page's behalf, and the licence was granted.[29] It may be speculated that the prince would have approved of the justice's decision.

The visit of the royal family to county Kildare in 1861 and the military instruction there of the crown prince, was to make the Curragh camp known throughout the British Empire. The documentation of the camp and district, and of the military training, in paintings, drawings and photographs was to ensure a permanent record of the camp in its original state, and of the impressive sight of the vast army in the field. The thousands of infantry men in a variety of formations, the great arrays of cavalry, and the guns in action, as well as the ceremonial parades, created lasting images of a war machine which was one of the most powerful of its time. The pictures also make it possible to understand the attraction of the Curragh field days for the civilian population, and the glamorisation of the soldier who was part of such magnificent displays of pageantry, skill, endurance and feigned bravery. Undoubtedly, the army profited greatly from its training on the great open plain which did not only result in having a more efficient field force, but it also gained from the propaganda effect of the massive formations and their disciplined manoeuvres on the observers. The unionists found comfort in the strength of the military, while the nationalists saw only the power of the oppressor.

Public disquiet was caused in the county in the spring of 1865 when a rumour circulated that the camp at the Curragh was not going to be formed that year, even though in February the depot of the 53rd Foot had joined the 1st Battalion, and the 61st Depot had joined the 2nd Battalion of their regiments at the camp, where both had been since the previous August. Happily, the *Leinster Express* in its edition of 8 April was able to confirm that as large a force as ever would be there in the summer months, and a week later it announced that the Curragh was showing signs of summer with the military arriving daily. A detachment of the Royal Artillery was on the way, having left London to march to Liverpool. They were to spend three Sundays and two weeks on the road to embarkation. The camp on the Curragh lasted until the end of September when the troops started leaving; the 10th Hussars were reported leaving on the 29th, and the 12th Lancers and batteries of Royal Horse Artillery and other regiments on the following day. The local newspaper carried the information that there was to be a complete change of quarters throughout the whole country, with troops being posted to places where they had not been for years. The depot battalions were also to be moved. The paper thought the changes were 'a very judicious step on the part of the authorities'. No doubt, but the danger of Fenian infiltration was feared by the government.[30]

When the annual inspection of the Royal Dublin Fusiliers was held in Naas barracks in May 1865 it was a great social occasion. The colonel, the Marquis of Drogheda, and the county gentry were present in force to be entertained to lunch, and in the afternoon the public was allowed in to enjoy the unit sports. But in the autumn of that year when the barracks was vacant, except for half a dozen men for security duties, there was concern, on account of the Fenian conspiracy, for the safety of the 600 stand of arms of the Kildare Rifles which was stored there. In September the town commissioners petitioned for Naas to be made a military station, and it was proposed that the 5th Fusiliers should be billeted there, when the troops from the Curragh camp were being distributed through the country. Instead, in mid-October it was announced that 300 constabulary, with eight officers, were to occupy the barracks. However, a year later the constabulary depot was moved out of the barracks, and the commissioners again passed a resolution that then the military should be stationed there. They made the point that until the last few years there had been troops in the barracks, and now they hoped to be granted the security of a military force.

A new general commanding the forces in Ireland had been appointed on 1 July 1865. He was Hugh Henry Rose who, as a young officer, had already served in Ireland with the 19th Foot and, afterwards, with the 92nd Highlanders. From 1857 to 1859 he had commanded the Central India field forces, and from 1860 until his posting back to Ireland he was commander in chief in India. He was created baron, as Strathnairn of Strathnairn and Jansi, in 1866, and in the following year he was promoted to general. The Fenian John Devoy described Strathnairn as 'the man who blew the Sepoys from the cannon's mouth in the great Indian Mutiny of 1857', and he believed that Strathnairn had been sent to Ireland 'because his character for ruthless sternness seemed to fit him best for the work of putting down the Fenians'.[31] A later historian's view of the general was that he 'confronted the Fenian conspiracy, by a good organisation and disposition of troops under his command and acting completely in accords with the Irish Government, he succeeded in keeping the country under control, and prevented rebellion'. That he was highly thought of amongst the academics at Dublin University earned him an honorary doctor of laws when he relinquished his Irish command after five years in office.[32]

In the month of June 1865 the 3rd Royal East Kent Regiment (The Buffs) had arrived at the camp from Sheffield,[33] and within a few weeks of their coming there was a rumour that many of the men were Fenians, as were 'the greatest part of the soldiers at present in the Curragh Camp'. It was believed that Fenian agents came from Dublin to the camp with the intention of recruiting from the ranks.[34]

John Devoy, from his home near Kill, was a frequent visitor to the Curragh, and at the June 1865 races there he met three colleagues from Ballytore, Leixlip and Mountmellick. They shared a jaunting car with soldiers returning to the camp, and Devoy later recalled that as they drove along the soldiers sang *The Rising of the Moon*: 'They continued to sing till the car swung into the streets of the camp and there were approving smiles from scores of soldiers as we passed the doors of the huts'.[35] In the autumn of that year Devoy was placed in charge of Fenian organisation in the British army in Ireland, and in that capacity he made several visits to the Curragh Camp where one of his contacts was Daniel Byrne of Ballytore, who worked in one of the canteens and 'knew personally the best men in the camp and in Newbridge barracks'.[36] In October it was reported in a local newspaper that a soldier of the 3rd The Buffs who was drinking in Canavan's pub at the Standhouse announced that he was a Fenian and 'Up the Irish Republic'. He was taken to the cells in the camp.[37]

A military correspondent who had soldiered in the camp from 1865 contributed a feature to the magazine *All the Year Round* in May 1867. He described having seen in the camp some time before 'an American Celt, bearded, bronzed, and swaggering', and with 'golden dollars' attempting to seduce the troops to his cause. While it was accepted that, in drink, some of the men might have 'kissed the book', it was believed that the 'numbers of soldiers misled was greatly exaggerated'.[38] The 'American Celt' was probably William Francis Roantree from Leixlip who had served in the American navy and returned home in 1861 (Fig. 14). He was arrested in 1865.[39] Devoy believed that Fenianism was widespread in the army, and in good shape, up to the Spring of 1866. Thereafter it declined, he thought as a result of the postponement of fighting.[40] In the Curragh camp Devoy estimated that there were between 1,200 and 3,000 men sympathetic to the cause, but a recent appraisal of the Fenians and the army by A.J. Semple found those estimates to be doubtful.[41]

Semple reviewed 70 courts martial associated with Fenianism, between May 1865 and November 1867, six of which were in the Curragh.[42] The charges for which the men in the camp were tried ranged from theft, insubordination, loss of kit and drunkenness, to threatening with a gun, all of which were common offences. The trials were reported in detail in the newspapers, creating an impression 'that a considerable portion of some regiments was disaffected'. This, according to the military correspondent already quoted, was an erroneous impression as the men 'who had been induced to drink by foreign agents, would have blown out the brains of a comrade who dared to act the traitor'.[43] The fear of Fenianism in the ranks of the militia caused the cancellation of the annual training camps in 1866 and 1867[44] though, as R.V. Comeford concluded, 'it is impossible to discover how many militia were infected by Fenianism'.[45] The absence of the militia from the Curragh for two successive years would not have pleased the local traders, nor nurtured their sympathy with the Fenian cause.

As the authorities became more concerned with the danger of infiltration of the ranks commanding officers were asked to collect information on any Fenians in their charge. A police constable, who posed as a dealer in jewellery and stationery, was sent into the barracks in Newbridge and the Curragh on an intelligence mission. His report indicated that he spent a lot of his time in the soldiers' huts and the canteens, but 'did not find any disaffection'. Nevertheless, the canteens were believed to favour the propaganda of sedition, and for a time they were closed to civilians.[46]

That the threat of Fenianism disrupted the social life of the camp was regretted by a writer in *All the Year Round* in May 1867: 'But the Fenians had broken up our society rudely before the crisis came. Detachments were ordered off continually, few officers remained, and then ladies departed on visits to relatives and friends. The camp became still and silent; the pickets were strengthened; we were in a fortress which might be assailed, and the men "kept within the lines", ready to march.' The greatest danger to the camp was in the month of March when the strength there was down for three or four days to less than 300. It was not feared that the camp would be taken, but that it might be burnt as, before the rising, the 'Prophecies of St Columcille' had been re-issued and circulated; one of the prophecies was that 'the Curragh camp should be burned in the spring of 1867'. His estimation of the situation was that 'a thousand really determined men might have gained some prestige for the conspiracy had they even made the attempt and failed'. He recalled having then said to a sergeant of artillery that 'five hundred resolute Fenians who would not quail if half their number fell, might do great mischief'. 'True, sir', replied the sergeant, 'but this sort of cattle do not like the open'. The correspondent, who was obviously an officer, commented 'in a few words he described the Fenian tactics', a remark which typified the regular soldiers' abhorrence of irregular warfare.[47] Later in the year a member of the Royal Irish Fusiliers welcomed a heavy fall of snow on the plain as it would prevent the Fenians from concentrating in dangerous numbers.[48]

The Fenian rising in Munster and Dublin in February and March 1867 had necessitated vigorous activity on the part of the garrisons (Fig. 14). The security of installations was being monitored and in reply to a query from headquarters the response from the Curragh was that the 'magazine was entirely surrounded by a *chevaux de frise*, and that sentries were posted, one by day, two by night, within the compound. The building was secure, and that similar precautions had been taken at Newbridge'.[49]

During the crisis the military telegraph to the Curragh was kept busy, and the railways were found to be ideal for the movement of troops. One commentator thought that the railway lines 'seemed almost to have been planned in anticipation of the rising, so directly do they touch the very centres of sedition'. From the Curragh siding 'troops were conveyed secretly and almost silently at an hour's notice from quarters to any part of the disturbed districts. Regiments arriving from England in the early morning were paraded at the Curragh on the same day, and drafted away immediately. Troops from Dublin were incessantly passing up and down the line to Newbridge and the camp, and from both Limerick Junction, Tipperary, Mallow etc., as required'.[50] The importance of the camp as a centre for the movement of troops was obvious, and the proximity of the railway made it strategically vital. Pilot engines were used with the trains going in aid of the civil power, but when in March 1868 the pilot engine was used to precede trains travelling from the Curragh to Cork with the 92nd Foot the secretary of state for

Fig.14. *William Roantree 'an American Celt, bearded, bronzed, and swaggering', as the Fenian was described in a contemporary journal. He frequently sought recruits to his cause from amongst the soldiers in the Curragh camp.* Photo: *National Museum of Ireland.*

war wanted to know why the engine was used, and complained that it was expensive, costing £34.12s.6d.[51]

A footnote to Fenian activity in county Kildare was the observation by M. O'Connor Morris, hunting corespondent for *The Field*, in the autumn of 1876. He remarked that when the Kildare Hunt met at Naas that *en route* there he saw 'a green and gold band of musicians, the last phantom of Fenianism, which hopes *aere ciere martemque accendere cantu*'. [translation: 'in stirring men with his bugle's blare and in kindling with his clang the god of war (Mars)'. Virgil: *Aeneid* 6 lines 161–165].[52]

The threat of Fenianism had hardly receded when agrarian agitation again brought the military into the public arena in aid to the civil power. The inability of the R.I.C. to prevent outrage, or to enforce evictions, necessitated the accompanying presence of a military force in what were invariably unpleasant duties. A study of the land war in county Kildare by Thomas Nelson found that in the county 'it was not a simple struggle for survival on the part of the poor oppressed tenants against rack-renting landlords . . . it was a much more subtle exercise in local politics . . . and with the exception of one or two individuals in Kildare town, the rest of the county was very slow to respond to the land war'. There were fewer evictions than in other counties, and Nelson concluded that there was not nearly as much economic hardship as in other parts of the country. Just before Christmas 1880 Michael Davitt spoke at a Land League meeting on the Curragh. When he mentioned the large amounts of money collected in rents by the duke of Leinster there were groans from the 'especially large assembly', but no other unpleasantness was reported in *The Nation* in its edition of 25 December.[53]

Flying columns were formed throughout the country, to be located at strategic centres, and in January 1881 two columns were based in the Curragh. Each column consisted of a troop of cavalry (30 sabres), a division of artillery (two guns), 12 men of the Royal Engineers, four companies of infantry (200 bayonets), a detachment of the Army Service Corps with four service wagons, and a hospital corps unit with an ambulance.[54]

The hazards and unpleasantness of military duties in aid of the civil power were evident at Hackettstown, county Carlow, in November 1881 when up to 50 families were being evicted from the estate of Mr C.D. Guinness. For the evictions a force of 200 police was accompanied by a detachment of 21st Hussars from Carlow.[55] During the proceedings the arms of the soldiers were stolen, later to be found inside the Catholic church at Tynock. Jennie Wyse Power, who was observing the evictions on behalf of the Ladies' Land League, later revealed that she had been approached by the sergeant in charge of the Hussars who offered her 'his heart and hand' if she told him were the guns were, but she was not able to avail of his offer as she did not know where the arms were hidden.[56] The humour of the sergeant's approach masked a serious concern as the loss of arms would have had dire consequences for the men concerned. The incident also highlighted the fact that the training of the soldiers was for war, not for peace time involvement in the oppression of people largely of their own class. One military officer, expressing his rememberance of his unit's participation in evictions, wrote 'we all loathed the work, and most of us deeply sympathised with the poor ejected ones'.[57]

The odium generally reserved for the police force might then also be directed at the military, as was the case during elections in Drogheda in 1868 when a detachment of the 2nd battalion Norfolk Regiment was attacked by a mob; in the fracas four officers and 14 men were injured, and one civilian was killed when two of the soldiers opened fire.[58] An indication of the aversion of the military to their participation in policing duties is found in a letter written by a field officer who had found service in Ireland unpleasant. He wrote a letter to *The Times* in 1881 in which he proposed that regiments should not remain in the country for more than two years.[59] Clearly his service here had not been as enjoyable as was that of many of his brother officers, and his distaste for the use of the military in a policing role was shared by other gentlemen.

However, the opinion of a resident magistrate in January 1882, was that disturbed districts should be patrolled by combined patrols of military and police.[60] That year when soldiers on special service in aid to the civil power were allotted to the disturbed parts of the country only 44 were posted to the No. VI division which included Kildare, Carlow, Meath and the King's and Queen's counties, with the entire compliment going to the latter two counties.[61] If the troops from the Curragh might find themselves dispatched to assist the police in the neighbouring counties, at least their immediate station location was calm. An important opinion on the military role in Ireland in 1900 was that of the chief secretary, George Wyndham. He believed that the strength of the Irish garrison could be reduced, and it was his opinion that 'the garrison of Ireland is not there to keep the peace but, first, because the barracks and training grounds are there'. That view he shared with his predecessor A.J. Balfour, and it suggested that even then no very clear role for the garrison existed.[62]

Nevertheless, that the public perception of the military was not that of a punitive force was demonstrated at a meeting of the executive of the Gaelic Athletic Association in February 1887. Then a resolution was passed instructing the affiliated clubs that members of the R.I.C. not be enrolled in G.A.A. branches, nor were they to be permitted to enter for sports or tournaments run under the auspices of the Association 'in consequence of their action towards the people throughout the country . . . this resolution not to apply to the army or navy'.[63] John Wyse Power, a former editor of the Naas based nationalist newspaper *Leinster Leader*, and husband of Land Leaguer Jennie Wyse Power, was a member of the executive which showed sympathy with the soldiers and sailors. He would also have had some knowledge of the status of the military in county Kildare, and their integration into the life of the people. However, by 1903 soldiers, sailors and militiamen were banned from participation in gaelic games, and rule 8 of the association in 1914 excluded police, jail warders, soldiers, and sailors of the R.N., militiamen or pensioners of the constabulary, army or navy from membership of clubs.[64]

Chapter 6

The Health and Welfare of the Troops

The army, in the interest of the welfare of all ranks, displayed an almost paternalistic interest in the daily lives of the men. In addition to providing their shelter and clothing, in the form of billets and uniforms, their spiritual and physical well being was catered for by chaplains and medical officers. Protestant clergymen of the different denominations and Roman Catholic priests were appointed to the chaplaincy, and in the case of the latter, as early as 1795 the lord lieutenant had ordered that chaplains should be appointed to all the regiments of the militia encamped at Laughlinstown, in county Dublin. The attendance of the military at services in the parish churches of Naas and Kildare, where there were no garrison churches, and of the men and their families living outside the Curragh camp at their neighbouring churches, were beneficial to those parishes, and the clergymen officiating to the troops were paid a per capita allowance for the numbers of men on church parade.

Comfort for soul and body was also provided for the soldiers in the various institutes and homes established by the churches or missionaries. The moral health of the other ranks was also improved by the admission of a certain number of men into the married establishment, and their wives and families were benevolently treated by the authorities. Schooling was also available for the families of the men on the married establishment, but sectarian difficulties were inevitable as the ethos of the schools was protestant, and the parents of catholic children did not always agree with the religious instruction imparted in the schools.

The War Department provided for the spiritual welfare of the soldiers with the provision of chaplains and places of worship, and a designated time for compulsory attendance at church services. The officially recognised religious persuasions were four in number, Church of England, Roman Catholic, Presbyterian and Wesleyan.[1] In addition to the weekly services the chaplains also devoted time to the moral uplifting of their charges through after-duty meetings and lectures, and in the promotion of voluntary christian societies. In the Curragh camp, on instructions issued on 24 March 1855 by the general commanding in Ireland Lord Seaton, two churches were constructed, and barrack rooms and schools were provided for the two [protestant] chaplains of the army.[2]

Services in the two newly built churches in the Curragh camp were held for the first time on Sunday 16 December 1855. The chaplains were the Rev. McGhee, Church of England, from Dublin, and Rev. Henderson, Presbyterian, from Lisburn, who officiated separately in the west church, while Father McCarthy, said mass in the east church, referred to as a chapel.[3] Their respective stipends for the year 1856 were £292, £175 and £150, and payment was based on the parade strength at the religious service.[4]

In May 1856 the War Department approved of the appointment of Rev. Mr Henderson as Presbyterian chaplain,[5] and he was still there three years later when official correspondence shows that he was applying for payment for car hire to travel to officiate at the barracks in Newbridge. It was granted, but on condition that it did not interfere with his duties in the camp.[6]

Travel expenses claimed for that period show that a return journey from Dublin to Newbridge of 52 miles was costed at five shillings, and the three miles return trip from Newbridge to the Curragh one shilling and sixpence .[7] In that year also the Rev. R.J.S. McGhee sought to be commissioned, but he was told that it was not possible.[8] Mr McGhee demonstrated his interest in the education of the soldiers' children in the regimental schools in 1859 when he succeeded in having an unsuitable text book withdrawn from the curriculum.[9]

The Catholic chaplain, Father Daniel McCarthy a native of Cork, but educated at Carlow College, and ordained for the dioceses of Kildare and Leighlin in 1831, was the first such appointment to the forces at the Curragh in 1855. He remained there until 1857, when he was also described as parish priest of the camp.[10] He was succeeded by Fr James Hamilton, a Kerryman, but also a Carlow graduate and priest of the dioceses. He opened the first baptismal register of the garrison church in January 1855.[11] A decade later, in 1865, Fr Thomas Molony was chaplain there,[12] and from 1875–1882 Fr J.J. Doyle was chaplain, and attached to the parish of Suncroft. It became traditional for the Kildare and Leighlin Dioceses to supply the chaplains to the camp,[13] a function which continued with the advent of the Irish forces in 1922.

It was not until 1860 that the Wesleyan Methodists had a local place of worship, though not in the camp. Their iron chapel at Lumville, on the edge of the plain, was opened in September of that year,[14] and there was a Meeting House, Soldiers' Institute, parsonage and school.[15] The minister was Rev. John Frayer Matthews.[16] In 1876 the foundation stone was laid for a new church in the centre of the camp, designed to hold between 500 and 600 people, and later a Soldier's Home was attached to it.[17]

That the cloth was not unduly influential at the War Office is demonstrated in correspondence between that office and the Rev. McGhee. Early in 1866 he was seeking an assistant, due to the increased number of troops in the station.[18] The Rev. M. Crooke was appointed, and consequently Lieut. Col. Lugard, then at the War Office, discontinued the service of the Newbridge rector as an officiating clergyman to the barracks there. As the senior chaplain had a forage allowance it was expected that McGhee would minister in Newbridge where there was a strength of 265 mounted troops, 77 dismounted troops, 30 recruits, and 16 sick men.[19] But he was not happy with the War Office proposal as he felt that he himself should stay at the headquarters, and he recommended that his assistant should be given forage allowance to undertake the Newbridge duties. This proposal brought an official query as to why the Newbridge duty should not be performed by the chaplain who had the forage allowance for his horse, and subsequently an order was issued that the service there would be given by the chaplain with the allowance.[20] But McGhee was not satisfied and he persevered in his quest for a more acceptable solution, and he found it when the 59th Regiment of Foot arrived in the camp. Then it was necessary to hold a second service at 11.30 a.m., the first being at 9 a.m. The service at Newbridge was at 10 a.m., with one in the hospital afterwards. The senior chaplain insisted that neither he nor Mr Crooke felt equal to two full services on a Sunday in camp, and he asked that the Rev. Foster of Newbridge should be engaged to officiate there until another chaplain was appointed. This time McGhee's proposal was accepted.[21] The Roman Catholic priest in Newbridge who officiated to the troops was paid an annual allowance which amounted to £70 in 1865.[22]

Sometimes the families of deceased soldiers commemorated the departed with a memorial or gift to his garrison church; thus the Curragh Protestant church got a new organ in February 1885. It was presented by Miss K.G. Clarke, in memory of her brother Capt. F.M. Clarke, 5th Dragoons, who had been killed by a horse falling on him while he was on duty in the Curragh. The instrument was made by Messrs Telford of Dublin. The principal chaplain, Rev. J. M. Ritchie, in dedicating the organ expressed a wish that subscriptions might be forthcoming to 'enable the enjoyment of a regular organist capable of doing justice to the instrument'. A soldiers' choir participated in the dedication ceremony.[23]

The spiritual welfare of the camp population was not confined to a weekly church service, and the various denominations organised retreats and missions for their flocks. Nor were the religious services confined to the churches, and when annual retreats for the women and children were given they took place in the barrack recreation rooms. For example, the families of the men of the East Yorks attended their retreat in the recreation room in C Lines in February 1896. It was conducted by Rev. S. McPherson, Senior Chaplain, and a Magic Lantern Show, followed by tea, was part of the programme. As an indication of his interest in his regiment Col. R.E. Allen attended that evening .[24] The chaplain's encouragement of suitable recreational activities was demonstrated in the month of May that year when McPherson led the annual outing to Poulaphouca in west Wicklow of the Church of England Good Templer's Association. They travelled by ambulance waggons and at the waterfall the party enjoyed tea provided by Mrs Jefferson of the Officers' Club, another example of the expected interest of the officer class in the other ranks.[25]

The Roman Catholic soldiers, of whom there were usually about 1,000 in the camp (but considerably more when the militia came during the summer training months),[26] attended mass voluntarily on holy days (other than Sundays),[27] and they had an annual mission. In October 1896 Fr Butler O.P. gave the mission, which was also open to civilians from the neighbourhood.[28]

As well as having consideration for the spiritual needs of the troops the War Office had to provide for the interment of deceased men and their families from the encampment. Initially burials were made in the cemetery at the Grey Abbey in Kildare,[29] but early in January 1856 discussions were taking place between the headquarters staff at the Curragh, the chaplains, the Curragh Ranger and the rector of Ballysax concerning the opening of a graveyard for burials from the camp. Lord Seaton was of the opinion that a site near the rector's house at Ballysax would be suitable. There it could be more easily supervised than at the old fox covert near the encampment, which had been proposed by Capt. Rich R.E.[30] It was finally decided to locate it on a hillside at the Walshestown side of the plain.[31] It was consecrated by Richard Chenevix, Archbishop of Dublin, Primate and Metropolitan of Ireland and Bishop of Glendalough and Kildare on 14 October 1869.[32] Burials of all denominations took place there.

The presence of the military and their families in the county benefited the churches in the towns by increasing the congregations, and the partaking of the sacraments of baptism and matrimony therein. In the developing town of Newbridge the St Patrick's Church of Ireland was built in 1828. A reminder there of the military nature of the town are the stained glass windows presented by Lieut. George Playfair R.H.A. in memory of his 22 year old wife who died in 1877, and a memorial to the 26 year old wife of an officer of the 19th Princess of Wales' Own Hussars who died a decade later.[33] The church registers also give ample evidence of the military attendance at St Patrick's.

There was no garrison church at Naas and the soldiers were paraded to the parish churches. The Roman Catholic priest officiating to the troops was paid an allowance, which in 1865 amounted to £10.[34] By 1826 the presence of the soldiers in the parish was reflected in the entries in the church registers of the medieval Protestant church of St David; that year, of the 17 burials recorded, ten were of the children of soldiers. The marriages and baptisms of men from the barracks were likewise recorded.[35] Captain William St Leger Alcock 23rd Regiment, stationed in the barrack, was not a regular church goer, but following his first attendance there on 6 January 1833, he made an entry in his diary that 'he was not at all edified by the sermon of a Mr Burg'. [Walter de Burgh of Oldtown, vicar of Naas from 1830–1859].[36]

The increase in the congregation at St David's caused the church to be crowded, and in 1833 the church vestry received a letter from the government asking that better accommodation might be provided for the military. It was suggested that 50 new places should be provided, to bring the total to one hundred. In response the vestry undertook to have a gallery erected. In the 1840s the vestry books regularly recorded the attendance of the military at services, and of the purchase of prayer books for the army band, for the military seats, and for the children of the soldiers. In the church are several fine monuments to local military men, such as that to Maj. General Thomas, a native of the town, who died in India in 1824. He left a bequest for the poor of the parish.

The Curragh itself was in the parish of Ballysax, and during the ministery there of Rev. Dr George Wheeler (1865–1877) it is known that soldiers from the camp regularly attended the church. He was a great favourite of both officers and men, and on Sunday evenings St Paul's presented 'a truly brilliant spectacle from the crowd of red-coats who came to worship there and hear Mr Wheeler's simple and telling discourses'. On Monday the reverend gentleman used to return to Dublin where he held the position of editor of *The Irish Times,* and it was on a journey from Newbridge to Ballysax that he died, when his horse-drawn car overturned. He is buried at Ballysax. [37]

The diocesan cathedral likewise welcomed the military. As the *Kildare Observer* reported in November 1889, when the restoration of St Brigid's had commenced but was not progressing satisfactorily: there had been an objection to the effect that it was not worth while restoring the cathedral for the sake of only 120 parishioners, but the counter argument was that 'it [the cathedral] was close to the Curragh Camp, and its present accommodation was only sufficient to seat 100 persons, a fact which should be remedied when they considered the large number of troops quartered at the camp, many of whom attended the services at the cathedral'.[38] The building of the hutment barracks in Kildare town in 1900 ensured a more regular congregation for both the cathedral and the Catholic church of St Brigid as the plans for the barracks did not include a place of worship. The status of the military in the Protestant congregation was shown in the fact that the cathedral organ was especially tuned to play with the military bands.[39]

Similarly the Baptist church at Brannoxtown, near Kilcullen, which was opened under the patronage of John La Touche of Harristown in 1882,[40] was attended by soldiers from the Curragh. Mrs La Touche, who did not share her husband's enthusiasm for the Baptists, wrote in May 1892: 'I get a good deal of entertainment, living next door to the Brannoxtown Baptist Chapel. They had a service last night, and no less than 70 people attended! Two car loads of soldiers came from the Curragh. As we sat at dinner I saw all the arrivals, and heard the opening hymn "Onward, Christian Soldiers", taken so fearfully slowly that one felt the Christian soldiers did not mean to hurry themselves, and evidently wished to remain non-combatants.'[41]

The most substantial evidence of the religious affiliation of the soldiers, and their attendance at the various churches, is to be found in the appropriate registers of births, marriages and deaths. Recent compilations of entries from the registers in the Curragh area, undertaken under the sponsorship of the County Kildare Archaeological Society, found amongst its conclusions that 'the Curragh, which had a large British population is of significance for many undertaking genealogical research in Co. Kildare. Many soldiers brought their families with them and settled permanently, while others married local women, establishing roots here with the descendants still living in the area. The preponderance of English names in Kildare attests to their presence.'[42] As an old resident of Newbridge once remarked 'the British army had a major impact on the town, and there is a legacy of remarkable names as a result of their occupation. In one street alone you could always find a Legge, Arm and Foot, Love, Luck and Comfort'.[43]

The church attendance of the soldiers cannot be taken as a true reflection of their piety as the compulsory church parades left them with no option, other than disciplinary proceedings. In Naas, and from the early twentieth century in Kildare town, the weekly church parade, with its accompanying brass band, would have meant an early Sunday morning awakening for the towns' people. The congregations in the Protestant and Catholic churches in both towns would have become accustomed to sharing their places of worship with the men in uniform, and their families. But as the church-going habits of the men's wives and families is not known it may be assumed that they were comparable to that of their civilian neighbours. However, the fact that neither the soldiers nor their women were entirely either military, civil or church law-abiding folk was regularly demonstrated in army medical reports, the military and civil courts, and the pages of the county newspapers. In the parish of Suncroft, south of the Curragh, an amusing story is told concerning two soldiers from the camp who passed through Suncroft on their Sunday morning stroll. As they walked past the chapel they saw the parish priest standing at a table outside the gate. He was putting away money placed on the table by the faithful into a black bag, and the soldiers (in their ignorance of church matters) assumed that some kind of a game was being played. One of them remarked that 'the bloke in black seems to win all the time'.[44]

If the chaplains could administer to their respective flocks without hindrance there might sometimes be sectarian squabbles amongst the men, as in 1855 when the Roman Catholic soldiers prevented the band of the Clare Militia from participating in a church parade of Protestant troops.[45] The regular soldiers too could tease their comrades of a different faith to themselves, but such banter was usual on all sides. Sergeant Major Peter Rogan A.S.C., a Roman Catholic from county Down, was said, as a young soldier, to have 'occasionally ran the gauntlet of a shower of army boots from his fellow soldiers when he knelt by his dormitory bed in the barracks to recite his prayers'. When he retired from service Rogan was employed as a canteen manager on the Curragh, and he lived at Brigade Lodge at Castlemartin. With his wife from county Waterford he reared a family of five boys and a girl, who became a nun. All of the boys joined the Mill Hill Fathers, and Peter, the second son, following a chaplaincy in the British army during the Great War, was in 1939 consecrated as the first bishop of the Cameroons. He died in 1970. In Kilcullen a residential estate is named in his honour.[46]

But tolerance between denominations was not always evident, and proselytism was much abhorred by the Catholics. The children of soldiers on the married establishment had the privilege of attending the regimental schools, but this was not always perceived as such. In Naas, for example, in 1824 there was friction between the garrison and the parish priest. The cause of the problem was that Fr Gerald Doyle was concerned that the children of the garrison, who were mainly those of Roman Catholics recruited in Ireland, and who attended the regimental school in the barracks were 'obliged to do so, or remain uneducated, to hear a Protestant master, to read texts and prayers not of their own persuasion, and that they were marched on Sunday to the Protestant church, and detained there during service'. The priest's request that he might be allowed to visit the school was refused by Maj. Arbuthnot, 36th Regiment. The dispute escalated when Fr Daniel Lalor, coadjutor to the parish priest, gave a sermon which was considered offensive in Naas chapel to the officer and troops of the regiment. When this was notified to army headquarters instructions were given that the 36th Regiment would not in future attend chapel when Lalor was officiating.[47]

There was correspondence on the subject between the parish priest and the bishop of Kildare and Leighlin, Dr James Doyle. The latter suggested to the priest that he should write to the commander in chief asking him to correct the matter, and if he did not prove to be helpful that all correspondence should be sent to the bishop to be made public, or to be brought before parliament. Dr Doyle wrote to Fr Doyle recommending a cautious handling of the affair. 'A collision with a mighty power' should be avoided, and the soldiers were able to 'worship God after the manner of their fathers', even if their liberty had been 'somewhat infringed upon as regards their children'. It was advised that Mr Lalor should not 'address soldiers as distinct from other faithful, nor to comment to them on the conduct of their officers. To do so is exceeding wrong; it is to sow in the army the seeds of insubordination'. While the bishop proposed that any encroachment on the rights of conscience of the men should be reported to the commander of the forces, the soldiers should never be advised 'to assemble and deliberate as Mr Lalor is reported, I hope untruly, to have done'.[48]

But that the difficulties with the garrison remained unresolved in Bishop Doyle's time was reflected in the correspondence of his successor Bishop Francis Haly. In the latter's papers there is an undated letter from the under secretary Sir Thomas Nicholas Redington, the first catholic to hold that post (from 1846–52), to his lordship. It referred to the regulations of the army in Ireland which required that soldiers should attend their respective places of worship armed. Fr Doyle objected to the men bringing their arms into the church, and to avoid the issue he celebrated mass for the soldiers in the barracks. But this was not satisfactory to the Secretary of State for War; he held that as the regulations were that if there was a place of worship in the neighbourhood the troops should attend there, and that no remuneration would be given to Fr Doyle for his attendance in the barracks. As the under secretary believed that the parish priest would not have the men in church with their arms the consequences could be that the 'troops might be deprived of the consolation of the holy sacrifice as their religion requires'. Redington understood that in other parts of the country no difficulties were made by the clergy, and he wished that 'as in catholic countries they should render military honours at the most holy part of the sacrifice, yet in this mixed country such a proceeding is not within their orders. I hope the fact of their attending church armed will not be regarded as offensive to the feeling of the pastor, his flock or our religion, and that the Rev. Doyle may forego insisting on his objections to such a course'. The Under Secretary emphasised to Bishop Haly that the general commanding was simply obeying regulations in enforcing them, and that no disrespect to to Fr Doyle was intended. 'Perhaps, my lord,' he concluded,' your prudent advice would be able to smooth over the difficulty? I am not acquainted with Rev. Mr Doyle and I feared a direct approach to him might appear an officious interference . . .' [49] The pastor's reaction to the situation is not known, but Bishop Haly was again to be reminded of the military presence in his diocese in 1855.

That year the bishop received a letter from Archbishop Cullen of Dublin, dated 6 August 1855, enclosing a letter from Daniel O'Connell which referred to the circulation of a work entitled *Simple Truths collected from the Bible*, by the Scottish theologian George Gleig, in the newly opened camp at the Curragh. O'Connell was concerned at the propriety of allowing catholic soldiers to read the book, and the archbishop had told him that Gleig should not be employed to instruct catholic soldiers. The principle involved was as to whether Protestants were to be allowed to become religious instructors to catholics. He believed that the only fair solution was that the denominations should be given their religious instruction separately. As he felt that Bishop Haly's views on the matter might be sought he wanted them to be of one opinion, and he advised 'let the care of Catholic souls be left to the Catholic chaplain or a person appointed by him, and Protestants do the same . . .'[50]

Archbishop Cullen had already expressed his views on the education of catholic children in military schools in an address to the Catholic Association in Dublin on 29 January 1852. He believed that the protestants had united in their efforts to 'rob the Irish people of their faith', and one of their plans was that laid down by Primate Boulter a century before. It was 'to seize the children and educate them as Protestants . . . there is the unhappy state of poor Catholic children in regimental, military and nautical schools. The Catholic soldier or sailor persisting in defence of his country bequeaths his children to her care; she places them in schools maintained by funds collected from Protestant and Catholic alike and honours the fathers' service by robbing them of their faith'.[51]

That such sectarian difficulties could arise in garrison areas was not surprising. The sensitivity of the catholic church to proselytizing was then acute, and the intermingling of catholics and protestants in the ranks, as well as the presence of the latter in dominantly catholic areas, caused concern. Regimental schools and Institutions operated by religious groups in barracks were viewed with suspicion, and even as late as the 1950s the Sandes Soldiers' Home in the Curragh was suspected of missionary endeavour with the Irish army.[52]

But that the priests might also have some influence on the military was demonstrated in Newbridge in 1844. Though the ardour of the parish priest in trying to control prostitution in the town caused offence to one English officer, it also showed that the priest could enter the barracks and procure a posse to demolish the women's shelters.[53] But the Catholic Church feared not alone the danger of perversion of some of its flock by the military, but also the moral danger of fraternisation. In the autumn of 1891 a circular issued by the Bishop of Ossory was read in the churches of Kilkenny city. It prohibited the faithful from attending entertainments to be given by the Middlesex Regiments in the barracks, and emphasised the dangers to which families were subjected by going to such places. When Fr Bowe commenced to read the circular at St John's Church at the 'Soldiers' Service', Lieutenant Ross ordered his men in the congregation to fall in, and marched them out of the church. The report in the *Leinster Express* of 3 October concluded with the information that 'the matter has been reported to the Horse Guards [in London]'.

Fifteen years later in Monasterevan a similar attitude towards the soldiers was demonstrated during an inquiry into the public lighting of the town and reported in the *Kildare Observer* 28 July 1906. Reverend E. Kavanagh P.P., 'looking on the question from a moral point of view, said he hoped he was not casting a reflection on the English army, but there were some soldiers from Kildare and the Curragh camp who visited the town at night, and were a great source of danger to flighty young girls. The lamps would also keep young men from the country districts from spending weekends in Drogheda Row . . . '

Church influence was also shown in the vigilance of a Protestant clergyman on the Curragh who succeeded in having what he considered to be a sectarian textbook removed from the army schoolrooms. Reverend R. J. S. McGhee, Episcopalian chaplain in the camp, in the autumn of 1859 was responsible for having a book on physical and historic geography which was on issue to army schoolmasters withdrawn. He had complained that it contained objectionable sectarian passages which had not been detected during publication.[54]

The education of the children of the men on the married establishment was done on a regimental basis, but some of the children attended the Wesleyan school in the Curragh. Such attendance was questioned by the Assistant Adjutant Military Education in January 1878; he wanted to know why they did not go to the army school?[55] The offspring of soldiers living outside the camp attended the local schools, where at Kildare they were found to be below average. In a report on National Education they were classified, with the children from the local horse-training establishments, as 'below the town's average in both work and deportment'.[56]

If an army child required special schooling the responsibility for that was not with the military, and when an application for the admission of the child of a sergeant on the Curragh to the Institute for the Blind came before the Naas Board of Guardians in December 1886 it led to a heated discussion. The Guardians of the Union were asked to pay £11 a year for the maintenance of the child at the Institute, and the father was to give £4 a year for expenses. The colonel of the regiment gave a guarantee that the latter payment would be made monthly. When a member of the Board remarked that the Curragh paid rates voluntarily, Mr Fenlon, who was in the chair, commented that 'the government had taken up the Curragh which belonged to the people, and they were told that they paid the people a compliment by paying rates! What a beautiful institution they lived under!' The chairman said he would oppose the application for aid. Even though the child had been born on the Curragh, of an English father and an Irish mother, the Baron de Robeck said that it was possible that when the colonel would go away that the Union would be left with the expense. Again the chairman gave his thoughts, saying that 'The English government ought to care not only for its soldiers, but for the families, because they brought them up as soldiers'. The argument was resolved when another Guardian said 'This one will not, it is a girl'. That revelation raised laughter, and the motion to send the child to the Institute was passed.[57]

Augmenting the pastoral care of the clergy in every large military station were the various Soldiers' Homes and Institutes. Both the spiritual and physical welfare of the men were nurtured by such establishments in the Curragh Camp and at Newbridge. In 1882 Mrs Perry had opened a Home at Lumville, on the edge of the Curragh, as near as she could get to the camp as it was not possible to procure a site on government ground. When she died in 1899 she left her establishment to Miss Elsie Sandes from Tralee who, 30 years earlier, had opened rooms in which the men from the town garrison could meet in the evenings for prayer and bible readings. Miss Sandes saw her ideal of a place of refuge for the men come to life in 1876 when a benefactor provided a premises in Cork city. It was to be 'a Home full of light and gladness, and music, and free from the blasphemies and horrible songs which were polluting the air around me; a Home where these men could find warm human hearts, always ready to welcome, to help, and befriend them; a Home where they would hear of the only One who could free them from sin, and make their lives glad, and useful and victorious'.

By 1898 her Homes had spread to eleven other army stations in Ireland and four in India. The Sandes Homes provided home comforts to the men, there were tea and reading rooms, libraries and billiard rooms, and time for prayers and hymns. Voluntary lady helpers staffed the institutions. Many years before she inherited the Curragh Home Miss Sandes had been encouraged in her work by her friend Lord Wolseley, who had been appointed commander in chief in Ireland in 1890, and who had expressed a wish to see a Home opened on the Curragh.[58] Major General Morton offered Miss Sandes a site in the camp about 1903, but it was to be another eight years before the Home was built and occupied. It became the headquarters of the Sandes Homes (Fig. 15).[59]

During the summer when regiments gathered in the Glen of Imaal for training Miss Sandes opened a home for her men. She had successfully managed canvas homes in South Africa during the Boer War, and she continued them at Coolmoney in the glen. In two canvas marquees each evening 'the men could be seen patiently

Fig. 15. *Recreation Room at the Sandes Soldiers' Home in the Curragh Camp* c. *1914. As the gentleman soldier Horace Wyndham observed, the leisure time of the men is occupied with letter writing, draughts and reading, but the billiard table is not in use!* Photo: *National Library of Ireland, Eason Collection.*

waiting their turn at the writing-tables, and every evening happy groups gathered round the games, and found real pleasure in looking at illustrated papers, and in reading interesting stories'. Then they clustered around the harmonium to sing, after which Elsie Sandes told them stories.

The booming of the guns and the shriek of the shells reminded her of the war, but when the Last Post was sounded and the lights went out in the tents and the mist settled over the mountains, and the only sounds were of the sheep grazing nearby, or the neigh of a horse, she would muse 'who would think there are hundreds of men here quite close to us?' She hoped that 'some word spoken, some book read in our canvas home, might help the bright lads who nightly filled it to be ready if their country called them to lay down their lives'.[60]

In the first decade of the twentieth century, though her Home was being still referred to in an illustrated book on the Curragh as The Perry Soldiers' Home, it was noted that Miss Sandes had '31 similar Soldiers' Institutes in Ireland and India'. The Curragh Home was described as 'a rallying place for men from the Curragh Camp and Newbridge. They use it as a club, and as all soldiers are honorary members, and pay no entrance fee or subscription, Miss Sandes looks to the Public to help maintain it'. The coffee room was open from seven in the morning to late at night, and in the lecture room voluntary services and meetings were held every night 'for the motive which constrains Miss Sandes and her Lady Helpers (of private means) to give their lives freely to this work is the great desire that many men in the army should lay hold of the One Almighty Friend who wants to save and help them'. There was a recreation room with games, reading and music, and there were 'no rules or restrictions. The men come in and out as they like, and bring their dogs and their pipes'. This latter invitation was in response to notices posted on places of amusement in Belfast during the last decade of the nineteenth century which warned that 'no dogs or soldiers admitted'.[61]

As her work prospered Miss Sandes oversaw the opening of her new Home in the Curragh at the end of October 1911. She wrote to *The Irish Times* asking that illustrated papers and magazines might be sent there, or to the new Newbridge Home.[62] That year too an indication of the public support for her work was a legacy of £500, left to Miss Sandes by a widow from Bray, county Wicklow.[63] The illustrated papers received at the Home augmented the library, surviving volumes from which, though bearing the book-plate of the garrison library, suggest that the reading material there was staid in character. Sets of bound volumes of *Macmilian's Magazine* and of *The Argosy*, dating from 1880 to 1893, and one volume of *London Society* 1889, inscribed 12th Lancers Curragh, show signs of having been well thumbed over in their day.[64]

The good intentions of Elsie Sandes and other missionary ladies, or the contents of their libraries, were not always fully appreciated by the rank and file. One man who expressed his views on a home in the Curragh was the literate

Private Horace Wyndham. He served in the camp in the early 1890s, and he was not entirely happy with the overtly religious tone of the institution which he patronised. Admitting that he rather abused the hospitality of the home by surreptitiously bringing in a novel and reading it for an hour or two in a comfortable armchair, with his pipe, he found the strains of the wheezy harmonium and the hymns from the adjoining room distracting. Though the reading room in the home was comfortable the reading material was poor with 'the most innocent description of popular literature [being] rigorously tabooed therefrom. Sporting papers were here naturally considered *anathema maranatha,* and even *Punch* was felt to be less helpful in its tendency than the [governing] committee would have wished. Cards were not mentioned within the precincts of the institution, and a billiard-table was held to be a sure and certain road to perdition for those who habitually made use thereof; bagatelle [a minor billiard game], however, might be indulged in without unduly imperilling one's future'.[65]

The rebuilding of the Curragh camp in the last decades of the nineteenth century also enabled the various promoters of soldiers' welfare to expand or improve their endeavours with the building and opening of new premises. Field Marshal Lord Roberts performed the opening ceremony of the Church of England Soldiers' and Sailors' Institute in the Curragh on 19 June 1896.[66] The home, which was 'open to all who wear the King's uniform', provided sleeping accommodation, which was always in demand, in addition to the other services which included a bar (where intoxicating liquors were not served), a library, reading, writing and billiards rooms, and baths. The garrison branches of the Royal Army Temperance Association and of the Independent Order of Good Templars met in the Institute.[67]

Lord Roberts was also the patron of the Rainbow Bazaar, concert and gymnastic display held in aid of the Wesleyan Soldiers' Chapel and Home later that year. Lady Roberts was to open the Bazaar.[68] A decade later that Home was described as 'commodious . . . the result of several enlargements' due to its popularity. There the Methodist church was providing 'the best equipment possible in the interests of the troops' since it had opened in 1893.[69]

Another 'long felt want' was also being remedied in April 1896 when Lady Anne Kerr, wife of the General Officer Commanding Major General Lord Kerr,[70] had laid the foundation stone for a Catholic Soldiers' Institute. She used a silver trowel, decorated with shamrocks and with a bog-oak handle, and which was afterwards presented to her. Already half of the £1,500 required for the building had been subscribed with £10 coming from the bishop of Kildare and Leighlin, and £107.7.4 from a matinée which Dan Lowry had given in the Star Theatre in Dublin. The Institute was to have a café, chaplain's rooms, bedrooms for visitors and a large meeting hall. The building, which was designed by the district military surveyor, was to be built of Athy brick, and the builder was Patrick Sheridan from

Birr, who had also been contracted to construct other buildings in the camp.[71] It was opened by Field Marshal Lord Roberts on 29 April 1897. Fr Joseph Delany, R.C. chaplain to the forces in the camp, was commended for his part in the opening of the Institute. He had visited the parishes in the diocese of Kildare and Leighlin collecting funds sufficient to defray almost one-third of the expenses of the building.[72]

A home for Presbyterian soldiers was opened in 1904, the foundation stone of which had been laid on 21 January of that year by Lady Morton, the wife of Major General Sir G. de Courcy Morton, General Officer Commanding the 7th Division and 14th Brigade at the Curragh. General Morton performed the opening ceremony, on behalf of H.R.H. the duke of Connaught. An extension was added to the Institute in 1906.[73] That year, shortly before he was due to retire, Major General Morton died suddenly at his headquarters in the camp. His funeral to the military cemetery on the plain was an impressive ceremonial parade, with his charger (which had his empy riding boots hung from the saddle) following the gun carriage bearing the coffin. Lord Grenfell, general commanding in Ireland, led a cortege of almost 400 officers, and 13 minute-guns were fired as a farewell salute. General Morton had been born in India, and had a distinguished career there, including a term as adjutant-general of the army in India for which he was much decorated. A monument to his service in India was erected in Christ Church Cathedral in Simla, also recording the fact that he died at the Curragh on 20 April 1906. The obsequies were fully reported in the *Kildare Observer* of 28 April 1906.

The army also provided comforts for the troops in the form of the Garrison Libraries in the camp and at Newbridge.[74] The books were labelled and numbered,[75] and an annual report was made on each library.[76] In May 1896 it was arranged that the last editions of the Dublin evening papers would be on sale at the Curragh library at 9.30 p.m. each evening.[77]

By the end of the nineteenth century social conditions for the soldier stationed in the Curragh had greatly improved from those of the earlier years. With a choice of an evening in one of the five Institutes or the regimental canteen, and a variety of associations and clubs to join, it was no longer necessary for the men to travel to the neighbouring towns for entertainment. The dreariness of the camp, as had been described by Lieutenant Tulloch, Royal Regiment, in 1856, was no longer complete, even if the atmosphere of the Institutes did not appeal to every individual, as Private Wyndham found later in the century. The rank and file were also welcomed in the churches of the various denominations in the camp and in the district. Men living outside the camp, or those billeted there, could in their free time, if they tired of the sermons and services of the army chaplains, instead find comfort in the neighbouring parish churches. However, the majority of the men being unmarried and isolated from normal home life, and hardened by the training, restrictions and habits of army life, would not have been overtly religious, even if they were so inclined. The very fact that attendance at church on Sundays was compulsory would have antagonised many of the men, and made them as captive worshippers less receptive to the admonitions of the preacher and the Word of God. But with the provision of the facilities for worship the War Department was fulfilling its role as legislators for the army, and the services of the chaplains and other charitable people were also available to the soldiers.

That men of the calibre of Sergeant Major Rogan while in army service and employment could successfully raise his entire family to have vocations for the religious life may have been remarkable, but it demonstrates another aspect of the moral tone of some military families. The proselytizing atmosphere of the early part of the nineteenth century, as reflected in the educational problems at the regimental school in Naas and elsewhere, waned about 1860, from when, to quote Joseph Robins in his book *The Lost Children* 'the bible society era was coming to an end, and henceforth the battle for souls was fought on a reduced and reducing scale'.[78]

No doubt but George Bernard Shaw was being facetious when in one of his plays he remarked that 'when the military man approaches, the world locks up its spoons and packs off its womankind'.[79] Whatever about the silver, womankind has always been believed to have been attracted by the uniform, and in an endeavour to regularise the status of the women of the regiments the British army created the married establishment.

But the soldiers were not much encouraged by the authorities to get married, and the circumstances in which those men permitted to marry might occupy quarters were laid down in the Warrant for the Regulation of Barracks, issued in 1824. It specified that 'the Barrack Master General and Board of Ordnance, may if they think fit, and when it in no shape interferes with, or straitens the accommodation of the men, permit (as an occasional indulgence, and as tending to promote cleanliness, and the convenience of the soldiers) four married women per troop or company of 60 men, and six per company of 100 men to be resident in barracks, but no article on this account shall be furnished by the Barrack Master except beds and bedding for the regulated number of women, and if the Barrack Master perceive that any mischief or damage arises from such indulgence, the commanding officer shall, on their representation, displace such women, nor shall any dogs be suffered to be kept in the rooms of the barrack or hospital'.[80]

Early in 1855 an official committee, which had been appointed by the London government to report on army accommodation in general, had issued its findings. These included the observation that in general the barracks were inadequate for the comfort and convenience of the troops, and would not encourage a higher tone of social habits in the men. The fact that the soldier ate and slept in the same room was deplored, and the provision of dining rooms was recommended. More disturbing to the committee was the general practice by which soldiers' wives admitted into barracks had to share the billets of the unmarried men 'with no means of separation except a curtain around the bed.'[81] In some regiments married couples were placed in barrack rooms apart from the unmarried men, but both circumstances were condemned by the official committee. It suggested that either married women should be excluded altogether from barracks, or accommodation should be provided beyond the barracks.[82]

The other ranks married quarters in the Curragh, as designed and built in 1855, were of a high standard compared with those in the older barracks. There was accommodation for 160 married couples in huts, each of which held eight couples. There were two huts for married soldiers in each of the ten barrack squares, each soldier having a room with a separate entrance and a fireplace. There was a common cook house, and there were privies for the women, and a wash-house with boilers and wooden troughs.[83]

The families of the troops stationed in the Curragh camp, if they were on the married establishment, lived in the limited quarters there, or if not they lodged in the locality. But the men married with leave did not necessarily always have their families provided for. Early in April 1857 half rations and fuel were refused to the wives of soldiers who were absent on temporary election duties.[84] However, in January of the following year compassion was shown to the women when it was decided that 'clothing left over from the hospitals in the east during the Crimean war was to be given to the wives of soldiers sent overseas now, who are badly off and not accompanying their husbands'.[85] But families at home could also be distressed, such as in 1859 when the wife and children of Private Boyle 2nd Bn 18 Foot were found to be destitute. Boyle had enlisted six weeks before and was then with his regiment at the Curragh camp.[86]

By August 1860 matters were being better organised and when the 63rd Foot moved into the Curragh camp it was decided that their women and children left behind were entitled to an allowance in lieu of the half or quarter rations which they got, and it was ordered that such an allowance was to be made in future.[87]

It was not unusual for a girl to follow her soldier to a station with the intention of being wed. The marriage could be performed by licence or after bans, but in either case the prospective bride had to reside within the boundaries of the parish in which the ceremony was to take place for eight clear days. At the Curragh she might find lodgings in one of the small thatched mud cabins on the edge of the plain where an airless bedroom must be shared with several other married or unmarried women. The soldier, if he married without leave, did not have much choice of wedding day; but if he was in good standing with the provost marshal, time would be made available; otherwise he had to wait for a day when there was no parade or route march to hurry to the church.[88]

One clergyman told of his endeavours to discourage girls from marrying without leave, usually to no avail. Their hope was that soon they would be placed on the married strength, and all would be well. The precarious situation of such wives was illustrated at the Newbridge Petty Sessions in June 1865 when Louisa Mills, the wife of a man in the 53rd Regiment, was before the bench for trespass in the Curragh camp. She had been found in the lines soliciting washing from the soldiers. Her plea was that she could not live on the four pence a day allowance which her husband gave her, and she was trying to make up the fare to return home to Plymouth. The magistrate gave her permission to go into the camp on a certain day, but if she wanted to go in again she was to have permission from the general.

In 1867 the number of recognised wives in the Curragh was stated in a monthly magazine to be 'nearly 8,000', an exceptionally high figure which may have been based on a compilation over a period of years. A circular issued from the Horse Guards in that year had made marriage a prize for good conduct. Seven men out of every 100, rank and file, could have permission to wed, providing they had served for seven years and obtained at least one good conduct badge. It was intended that the husband and wife would have a separate room in the barracks (as they had in the Curragh since 1855), and that they would get light, fuel and rations at state cost, but in the meantime they shared a billet with five or six other men and their families. Soldiers' pay and rations had then also improved, which combined to make the life of the couple more comfortable. There were schools for the children, and the women generally found employment by taking in washing, or as servants to the officers' families. The status of the women within the little world of their husbands' regiments was not a bad one. The officers' ladies took an interest in the welfare of the women and children, and to be recognised by the captain's wife was a social bonus to the men's wives.[89]

Another employment available to the women was in the Curragh Needlework Association where shirts were made for the troops. For example, on 1 November 1876, Capt. Hanning Lee 2nd Life Guards, secretary of the association, notified headquarters that between that date and the end of March of the following year that 'in addition to those already allotted, the Curragh Association could make 3,000 grey flannel shirts'. Industrious needlewomen could thus supplement their household budget if their family situation permitted.[90] Captain D. Matheson, Inniskilling Dragoons, and A.D.C. to the Maj. Gen. Thesiger, was secretary to the Needlework Association in 1887, and ladies interested in the work met weekly at No. 39 Hut, F Square. The Association was then described as being entirely self-supporting.[91]

The women and children of the men living in the camp were entitled to medical care at the female hospital. There they were given the regulation medical comforts, which included port wine, brandy, arrowroot and essence of beef. During the drill season, when the camp was fully occupied, a second female ward was opened in the Hare Park Hospital, and an extra nurse was engaged.[92] She was paid 1/– a day, with free rations.[93] If there was an outbreak of fever in the camp, as there was of scarlatina in May 1860, the hospital at Hare Park was also used.[94] Sometimes permission was given for the employment of a special nurse for an individual; in February 1861 'a woman of the Military Train under treatment for mania' was so nursed, but in June when a nurse was requested for a child with measles in the isolation ward which had been opened in Hare Park the deputy adjutant general wanted to know why the mother did not nurse the child, and so save expense? The nurse was appointed, on the same conditions as usual.[95]

At the Curragh lying-in hospital in 1867 the wages of the head nurse were 1/6 per day, with fuel and light. The second nurse got 1/–, and rations, but only when the hospital was occupied by the wives and children of soldiers married with leave.[96] A further improvement in the care of the families in the camp came in 1897 when the Soldiers' and Sailors' Families Association appointed the Association's first nurse, Mrs Norah Diamond, to work with the women and children in the married quarters in the Curragh. Subsequently a plaque was erected in the camp to mark that innovation.[97] By comparison with the facilities available to civilian families of the same class the army dependants on the establishment were well looked after.

The wives and children of men married without leave had to seek whatever medical care they needed locally, in the infirmary at Kildare, or the Union Workhouse at Naas.[98] They could not be admitted to the military hospitals, but it was not unknown for the regimental physician to attend such cases in their lodgings, and to administer whatever treatment and medicine he could. Statistics for the year 1867 showed that 401 in every 1,000 of the recognised wives in the Curragh had suffered from a sickness, many were minor illnesses. The death rate was 7.36 per 1,000. It had fallen since 1860 when it was 9.33 per 1,000, a ratio nearly equal to that amongst the men in 1867. It was believed that improved ventilation, drainage and better married quarters had brought about the change. A further decrease in deaths amongst the women was expected when scarlet fever, which was one of the most prevalent disorders among the soldiers' families, was reduced by the sanitary improvement. That anaemia, or poverty of the blood, was a common affliction was blamed on defective nutrition, and the attention of the military authorities was drawn to that fact by the writer in a monthly magazine who asked if there was not someone in the Curragh who might open a small house for sick soldiers and wives married without leave. Or, he pleaded, 'will the time ever come when marriage shall not be a military crime?'.[99]

In the garrison towns the women married without leave found accommodation amongst the civilian population, as best they could. Similarly in the cities, and especially in Dublin where there were several barracks, they could have difficulty in finding lodgings. A man in Dublin who was aware 'that soldiers whose wives were married without leave, or who were dishonest, turbulent, quarrelsome, slovenly or habitually intemperate, were not allowed to bring such objectionable characters into the regimental quarters'. He decided that he might speculate on the situation, and he had a large building erected near Richmond barracks, in which he rented accommodation to the women. He was not worried about their habits, once they paid the rent. He was not disappointed. The apartments were 'no sooner vacated by the incorrigible termagants of one regiment, than a succession of vixens was supplied by another to fill the unedifying edifice'. The proprietor had not appropriated any particular name to the building, but it quickly became known in the district as 'the she barracks'.[100] In Newbridge a row of cottages in Francis street, across from the barracks, was also known as 'the she barracks', but local tradition is that it was a haunt of the prostitutes.[101] Some clusters of houses in Naas and Kildare, and on the verge of the Curragh, might well also have qualified for such a distinction.

Single officers occupied huts in the camp, and married officers, if their families accompanied them, and the limited married quarters were full, either took houses or lodgings in the district. The committee on barrack accommodation of 1855 had recommended that quarters for married officers should be available, and that there should be a reading room in each officers' mess.[102] In the headquarters area there was living accommodation for the general commanding and his *aide de camp*, and a drawing dating from the 1860s shows the interior of the hut occupied by the latter officer. It depicts Lady Fremantle, the wife of Captain Arthur James Fremantle, Coldstream Guards, *aide de camp*[103] to Major General Ridley General Officers Commanding Dublin and North District (including Kildare), sitting at a cloth covered table beneath a small window. A large window at the gable end of the room looks out over the plain, and the sloped ceiling of the hut is cracked in places. There is a fireplace, surrounded by a fire screen, and with a hearth rug. On the mantle piece are a clock and a pair of candles, with a mirror above. The furniture consists of a sideboard with shelves over, a chaise longue and a pair of ladies' and gentlemens' chairs (Fig.16).[104]

But even generals could sometimes have difficulty in finding appropriate accommodation. Major General Chatterton, who occupied 'A' house at Newbridge barracks, was about to take over the cavalry brigade in 1857. He had been granted the usual lodging allowance of £200 per year by the Secretary of State for War in preference to the appropriation of the barrack quarters, but the general said he had furnished the quarters at great expense, as he could not find a suitable house in the neighbourhood, except at a distance from the brigade and the Curragh, and that would be good for neither himself nor the service. The general commanding recommended that Chatterton should be allowed to stay in the barracks, and 'to be given as much accommodation as the crowded state of the barracks admits'.[105]

Nevertheless, by 1866, the recommendations of the committee of a decade before had not been implemented everywhere, and when a young soldier, recently married, brought his bride to Chatham he was 'quartered in a room with three

Fig. 16(a). *Lady Freemantle, in her hut in the Curragh Camp in 1886. Her husband, Capt. Arthur Freemantle, was* aide de camp *to the general officer commanding in the district.*

(b). *The shoemaker's shop of the Highland Regiment, Curragh Camp. May 1888. The soldier cobbler, surrounded by boots and accoutrements for man and beast, is busy at his last. Drawing courtesy of Lieut. Hubert Hamilton.*

other married couples. There is no partition between their beds of any kind. His wife shrinks from dressing and undressing in the presence of three strange men, and they are now living on dry bread in order to save money to buy a screen by which some sort of privacy may be obtained at the expense of ventilation'. Swearing that he would never rejoin the army, the soldier said that his complaints were on behalf of the women and children of his regiment, 'who are treated by the military authorities with less consideration than a farmer would treat his swine'.[106] It is likely that such circumstances also continued in some of the Irish barracks, and families posted to the Curragh were the lucky ones.

The re-building of the Curragh camp at the end of the century included the construction of a large variety of married quarters to cater for the different military ranks, and the size of family. At Newbridge, Naas and Kildare quarters were also built, and the occupants became accepted members of the local communities. The consequent increase in trade in the towns was welcomed, as was the participation of the families in church and social activities. However, as the character of a minority of individuals led to lawlessness and court appearances, and reports in the local press, the public perception of the soldiers and their dependants was not always a positive one.

When the county Wexford traveller, Thomas Lacy, had visited the new encampment on the Curragh in the autumn of 1855 he had been impressed by its healthy situation, and the fact that the natural slope of the ground ensured that it would not become waterlogged.[107]

The success of the attention given to the welfare of the men is evident from a memorandum by the director general of hospitals in 1857. Dr Maclean considered the elevated camp site to be exposed and unsheltered, but having good drainage and no dampness. Even though the water supply was abundant, he thought it was excessively loaded with lime, but it was not injurious to health. During the two years of the camp's existence it had been occupied in the first year by militia, with a small number of line regiments, a total of 4,590. From June 1856 it consisted entirely of regiments of the line, with a small detachment of the Military Train, a mean strength of 5,266. In the first year amongst the regular troops there had been 28 deaths, in the second 16, and total admissions to hospital respectively were 5,635 and 5,126, or 1,228 and 974 per 1,000 of mean strength. Of those figures 2,142 in the first year and 1,978 in the second were with venereal diseases, the next highest illnesses were disease of the lungs, phlegmon and abscess.[108]

Men suffered from diseases other than the those mentioned above. Official returns of the numbers of men flogged in the army record that on the Curragh in 1856 seven soldiers, one each from the 16th, 57th, 59th, 94th, 96th Foot, the 3 Bn. of the 60 Rifles, and the Military Train, received up to 50 lashes for crimes of theft, disgraceful conduct or offering violence to a superior.[109]

The health and welfare of the troops in the Curragh camp, as elsewhere, was a constant concern to the officers, as the fitness of the men was a vital factor in their professional efficiency. The medical care of the soldiers, and of the families of those on the married establishment, was vested in a surgeon of the army medical staff[110] and the personnel of the regimental and female hospitals. In the men's wards there were soldier medical orderlies, and nursing sisters who could be posted to other military hospitals at home or overseas.[111]

If there were infectious diseases such as small pox the patient was confined in a tent, which was afterwards destroyed.[112] In cases of enteric fever or diphtheria a board of officers was assembled to investigate the circumstances of the infection.[113] Preventative medicine in the form of vaccinations was practised, and regular inspections 'of the whole of the warrant officers, non-commissioned officers and men, and families' were ordered at the station hospital.[114] Naturally there were frequent deaths in the camp, such as that of Sapper Tracey 11th Company Royal Engineers, a young soldier, who died of scarlet fever at the infectious hospital in Hare Park in March 1896.[115]

With such a large number of personnel moving through or resident in the Curragh area there were accidents either on training or in off-duty hours. A recurring accident was that first reported in August 1855 when a somnambulist from the 11th Hussars fell to his death from a third floor window in Newbridge barracks. In August 1876 an inquest found that 21 year old Private Alfred Smyth 7th Dragoons, died the same way.[116] Twenty years after that, but at the new Cavalry barracks in the Curragh, 22 year old Trooper Metcalfe 10th Hussars, and who was described as an intelligent and promising soldier, died from a similar fall. He had got out of bed to go to the toilet and climbed out the window to reach the verandah, but he was on the wrong side of the building and fell to the ground. It was the fifth accident of that sort in the camp, and it was recommended that bars should be put across the windows in such billets.[117]

There were also regular cases of suicide or attempted suicide, as in August 1885 when 26 year old Sgt. Major G. Woodward 5th Dragoons, Newbridge barracks, shot himself.[118] A few months afterwards Cpl. William Stewart, of the same regiment and barracks, shot himself in the billet at reveille; the inquest found him to have been temporarily insane.[119] Sgt. W. Taylor, a clerk in the Curragh Brigade office, was found bleeding in his bed, with an open razor, but he recovered from the wounds.[120] The cause of the death of Sgt. Hamilton of the Highland Light Infantry on Christmas Eve 1887 was believed to have been a combination of depression, drink and a censure from a superior officer. Hamilton cut his throat with his razor.[121] In another rash of suicides by the razor in 1900 a gunner at Newbridge, and a man of the Dragoon Guards died, while another member of the latter regiment failed in his attempt. A sergeant-major of the engineers used his gun to die.[122] The boredom of barrack life and routine, the absence of family, or loneliness, were other factors which might urge men to desperate actions.

Amongst other deaths recorded at the Curragh in 1896 were those of Pte. Howell 1st Munsters who burst a blood-vessel while washing himself and died instantly,[123] of Sapper Goods 11th Company Royal Engineers, who contracted illness while on manoeuvres,[124] and of Driver B. Holden A.S.C. who died of pleurisy.[125] There were also accidents, sometimes due to jesting, like that which led to the death of Private Earnest Frederick Windgrove 11th Hussars in the barrack room at F square in the camp in July 1886; a popular jest was to take the powder out of a cartridge and put the bullet back, and then to fire at somebody; but sometimes enough powder remained to project the bullet. The week that Windgrove died there was another soldier in the hospital as the result of a similar prank.[126]

Military honours were given to deceased soldiers, including retired officers. When Lord Clonmell from Bishop's Court, Kill, died in June 1896 he was given a military funeral to Maudlins, Naas. The officer commanding the 102nd Regimental District, Naas, and many other officers attended, and there were wreaths from Clonmell's former unit, the Rifle Brigade, of which he was an honorary lieutenant colonel, and from the Grenadier Guards. The Union Jack-draped coffin was topped with the deceased's sabre and busby.[127]

The documentation of medical matters was done with precision, and a detailed Army Medical Department Report was made annually. In addition to reports on the sanitary condition of barracks it carried statistics of diseases, illnesses and deaths. From the 1860 report on venereal diseases, which was based on the admissions per 1,000 mean strength of a station, it was shown that the figure for the Curragh was 373 cases.[128] The report for the year 1869 gave statistics for the same disease over a three year period, based on the admissions per 1,000 of mean average strength; at the Curragh for the years 1867, 1868 and 1869 the totals were 104, 85 and 85 [in Dublin the figures were 129, 139, 180, and at Cork 72, 61, 73].[129] In October 1873 regulations were introduced which made it an offence for a soldier to get a venereal disease, and he forfeited his pay while under treatment. The order was cancelled six years later when it was found that while the number of admissions to hospital was reduced, the men continued to suffer in secret.[130] The hospital admissions in the camp for secondary syphilis for the years 1881 to 1888 showed a reduction as they averaged 22 per 1,000 strength of troops.[131]

If the incidence of venereal diseases amongst the troops was of concern to the regimental and medical officers, the health of the women with whom they associated was also of concern to them. But the military authorities did not see themselves responsible for the plight of the prostitutes, and medical facilities for the unfortunate women was not to be officially provided until 1868.[132]

During the year 1869, Surgeon Major Evans reported, the amount of sickness and mortality on the Curragh was small, and a decrease on previous years.

However, the medical officer of the 66th Regiment believed that the high proportion of chest affections in the regiment was due to the cold of the huts, and the exposed position of the camp. Surgeon Major Evans disagreed. His view was that a large proportion of the afflicted were recruits and young soldiers who were in training for India; in the other regiments in the camp during the winter months he found that there was no unusual amount of chest affections. The sanitary condition of both the Curragh and Newbridge was described in 1869 as good.[133]

In 1887 the sanitary condition of both stations was classified as 'very good', and the principal medical officer said that 'everything is done that is possible for the welfare of the soldiers in the district'. He had no recommendations to make beyond proposing that the drainage system in the Curragh might be improved. The report also noted that the building of No. 2 Station Hospital was continuing, with all of the ward pavilions being completed and equipped, and the nursing sister's quarters finished and occupied. The cook house and patient's dining hall, with a connecting covered corridor between the pavilions, were also under construction by the end of that year.[134]

While the Army Medical Report for 1896 showed a decrease in the invaliding and mortality rate, and the general health was very good, seven cases of enteric fever were treated, one of whom was an officer. The locations where the fever was contracted were mentioned; at a village near the Curragh, on military works at Athgarvan, on manoeuvres in Kilkenny, and in the case of the officer, at Island Bridge barracks in Dublin. Indeed, just at that time a gentleman in the Curragh hospital was pioneering a new clinical application of X-Rays. Surgeon Major J. C. Battersby, a graduate of Trinity College, Dublin, was later in his service in Egypt and the Sudan to play a key role in advancing the use of Röntogen rays in military surgery. After the battle of Omduran in 1898 he suceeeded, with the use of X-Rays, in locating embedded bullets in 20 out of 22 wounded soldiers.[135] However, when men retired or were discharged from the service and they were no longer subject to the attentions of the army medics they frequently drifted into ill health, often through poverty. But the army did give support to ex-soldiers through the insertion of regular advertisements in the local papers inviting discharged soldiers or army reserve men who were living in the county, and who needed employment, to apply to a named officer. There were also several charitable organisations to assist the needy, but there were always some casualties amongst the old soldiers.

The same newspapers sometimes carried reports on the deaths of ex-servicemen who had failed to settle in civilian life. Martin Dunne, who died in Naas workhouse in the winter of 1876, was such a case. After his discharge from the army he managed to loiter in the Curragh Camp amongst his comrades, and escape the vigilance of the authorities, until he fell ill. Then the officers discovered his presence and had him sent to Kildare Infirmary, but he was not admitted there as it was not a fever hospital, and he was directed to Naas. The carman who had brought Dunne to Kildare refused to transport him to Naas, and the patient was left waiting until it was arranged to send him by train to Straffan, and then back to Naas workhouse. Following Dunne's death questions were asked in the press; why was a man in fever allowed to wait on the street in Kildare, and in the waiting room at the station, where he might infect others? The indifference of the military authorities and their ignorance of the Poor Laws were commented on, especially as there were 'a number of camp followers and hangers on [such as Dunne] who frequently find their way inside the lines' at the Curragh. It was thought that both the police and the relieving officer should have been notified of the case.[136]

The high standard of medical care, sanitation and hygiene which was part of the army system had beneficial effects for those sections of the population which were influenced by the military. Barrack routine required that all living quarters, toilets, bath and wash houses and cook houses were regularly cleaned and inspected. Swill and garbage was efficiently disposed of. The men themselves were constantly inspected for health and cleanliness, and their clothing and bedding fumigated when necessary. Civilians employed in barracks observed the habits of the military, while the soldiers themselves carried with them into civilian life hygiene and sanitation practices inculcated by their service. Army wives and families who lived in married quarters were obliged to maintain their homes to a high standard, and under the influence of their husbands, or of the officers' ladies if they were in contact with them, acquired some semblance of military behaviour. In many instances the army families resident outside barracks brought some of the better qualities of the military system to their neighbours, demonstrably in the cleanliness of their homes and persons.

A negative health effect of the presence of the army in a district was the possible infection of the local population with contagious diseases, such as in 1906 when troops exercising near Athy were said to have contracted scarlatina there, but civilian medical opinion was that they had introduced it to the district from Kildare barracks. More virulent was the contagion of venereal diseases between the soldiers and the local prostitutes, and to native girls. If the character of the diseased men did not cause dismay to their Kildare neighbours, it and the health of the prostitutes did. This was a problem which, though reduced, was to remain in the locality of the Curragh even after the British evacuation of the camp in 1922. The health and care of old soldiers and their families living on their pensions in the county was also of concern to the general public as their welfare and maintenance had to be subsidised by local government and charity. The saying that 'old soldiers never die, they only fade away' was proven daily in the back streets of Naas, Newbridge and Kildare and along the edge of the Curragh where the veterans of the various wars retold their epics in the local public houses.

Chapter 7

CRIME AND PROSTITUTION

One of the inevitable consequences of the establishment of a major military station on virgin ground in a rural area such as the Curragh was an increase in civil disorder in the locality. This might be caused by off-duty soldiers, their families, civilian employees or sutlers in the camp, or by male and female camp followers (Fig.17).

In the very first months of the occupation of the camp rumours of internal feuding between militia units stationed there circulated in the locality. During August there had been a report of a riot at a canteen between the County Dublin Militia and the militias from Longford and Westmeath; bayonets were drawn, brick-bats were thrown, several men were injured and the ringleaders taken into custody. A few weeks later there was an *emeute* when the band of the Clare Militia was prevented by Roman Catholic soldiers from accompanying Protestant soldiers to church. Four officers were knocked down in the affray, and murder was threatened. It was considered prudent to yield to the threats, and the band did not march.[1]

Within a week there was more trouble and two troops of cavalry were called in from Newbridge. Six regiments had been having a clodding match when stones were thrown; the Dublin, Longford and Westmeath militias had sided against the Limerick, Clare and North Cork units. In the row huts were damaged, windows broken, and the officers failed to stop the fighting, and one officer was knocked down. Towards evening the men got tired, and when more military and police reinforcements were called it petered out. Exaggerated rumours of the affray quickly spread through the country as they were relayed to people travelling on the train through Newbridge. To prevent further incidents the militia units were dispersed to distant stations, the Longfords to Limerick and Clare Castle, and the Dubs to Waterford.[2] The military reacted promptly to quell the rumours, and in the *Standard* the affray was described as 'so silly, with a small number involved, and no life lost'. To demonstrate the normality of camp life a public dinner for invited gentlemen was given in the mess room of one of the encampments, and

Fig. 17. Very Slow Indeed! *Officers relaxing outside their huts in the camp in 1861. Two side-cars with their top-hatted passengers proceed towards the centre of the camp at the Clock Tower.* Copyright: *Royal Archives, Windsor Castle.*

on the following day 10,000 men took part in a field-day 'amid great hilarity and acclamation'. The newspaper even went so far as to suggest that Irish and English regiments should swop places, to add to the education of both![3]

Friction between the militia units seems to have been a common occurrence, no doubt based on regional rivalries. At Naas, five weeks after the Curragh affray, the Leitrims and the Kildare Rifles caused trouble with the locals in the streets of the town. Bayonets were out, stones thrown and people in shops and houses were terrorised. That the trouble was taken seriously by the army authorities was shown by the sending of Lieut. Col. Pinder 15th Regiment to Naas 'to investigate the outrageous behaviour of the military there'.[4]

To counteract lawlessness, both within and outside the encampment, there were military and civil law enforcement officers. Within the Curragh camp the Provost-Sergeant, aided by an assistant in each brigade, and a non-commissioned officer in each regiment, was responsible that the sutlers complied with such regulations as were issued from time to time.

To facilitate the maintenance of order during the building of the encampment Robert Browne, who was both the Curragh ranger and a magistrate, was given the use of a hut in the camp, and from which he also administered his usual order on the plain, that was the protection of the grazing rights and game, the prevention of encroachments, and duties with the Turf Club. Following the publication of the

Curragh Act in 1868 the position of the ranger was firmly established, so that with his deputy and bailiffs, he sought to ensure that the newly promulgated bye-laws for the Curragh were observed. From 1873 the bye-laws were amended to include the prohibition of night-walkers or common prostitutes from trespassing on the sward.

Police barracks for the Irish Constabulary (given the title Royal from 1867) were located on the plain, and the Resident Magistrates and Justices of the Peace, who were local notables, sat at Lumville Court House, on the north-eastern edge of the plain, to deal with committals, minor civil and criminal cases, applications for licences and such matters. Offenders apprehended on the Curragh might also appear before the courts at Newbridge, Athy or Naas, depending on the circumstances. A report of the Kildare Petty Sessions in July 1855 suggests that the constabulary were kept busy. Amongst those offenders arraigned were Cathleen Cushion and Anne Brian, charged with theft, and William Edgar, a tradesman who was employed at the encampment, with assault on a fellow worker. Charles Blakney and Thomas Hughes were arrested for stabbing Private James Woods, 8th Hussars, at the camp.[5] With the increasing population in the locality, and the regular rotation of troops, it was not surprising that the tranquil atmosphere of the district should be disturbed by nuisance and lawlessness.

A decade before the building of the camp on the Curragh a police barracks, to be known as the Stone barracks, had been constructed in the middle of the plain, beside the roadway from Newbridge to Kildare. However, the plans for the Curragh camp in 1855 included the building of police barracks at Brownstown and Lumville. The huts for the head constable, with a cook house, and four huts to accommodate 27 constables, cost £407.10s. The wooden court house erected at Lumville, cost £339.5s., and the making of a lock-up and cells there amounted to another £211.16s.2d.[6]

An indication of the types of crime which might be committed by workers in the camp, or by soldiers, can be found in a report on the Athy Quarter Sessions of October 1855. One Phillip Tighe was sentenced to three months in jail for stealing doors from one of the camp contractors, and Michael Reilly, a clerk, was accused of embezzling £5, the property of a canteen keeper in the camp. Private Donal McInnes 3rd Bn 60th Rifles was apprehended on board the Glasgow steamer for the theft of £9, a gold chain, a gold eye glass, a blue cloth, a clock, a gold brooch, three suits of clothes and three pairs of boots, the property of Patrick Kelly, shoemaker to the 60th Rifles. He was given three months hard labour. Michael Melia got 6 months hard labour for being concealed under the bed of Erasmus Green, the Regimental Sergeant Major of the 5th Dragoons at Newbridge. At an Inquest held in the same month no cause could be found for the death of James Vickers 96th Foot who cut his throat at Hare Park in the camp.[7]

Trespass was a constant annoyance to the Curragh ranger, and amongst those charged with the offence were jarvies who plied for hire in proximity to the camp.

Lieutenant Colonel Shaw R.M. dismissed two charges against carmen in July 1860, and made a recommendation that a proper stand for all cars should be appointed at or near the camp. But all of the offenders were not adults and on 1 April 1861 Shaw also dismissed a charge against a ten year old orphan, James Kelly, for trespass.[8] Two years later Kelly was termed an 'old offender' when given a month for the unlawful possession of a soldier's trousers. In 1862 the issuing of summons for trespass in the queen's name was questioned, and Lieut. Col. Shaw R.M. sought direction; he was told by the military secretary that the Under Secretary to the Irish Government said it was correct to do so.[9]

To assist in the maintenance of order, and on the suggestion of Maj. Gen. Ridley, Commanding the Division at Dublin and the Curragh in June 1862, a number of special constables for the baronies of east Offaly, Connell and Kilcullen [which encompassed the Curragh, Newbridge, Kildare and Kilcullen] were chosen from reliable non-commissioned officers in the camp. They were appointed regularly during the following years on the parchment authority of the duke of Leinster, lord lieutenant of the county.[10]

Of grievous concern to the military authorities in 1865 was the murder of an officer in the Dublin and Curragh Division, who had also been robbed by a servantman near Parsonstown in Kings' county. The victim was Lieut. Clutterbuck, 1st Bn 5th Regiment, and the military secretary, on behalf of General Sir Hugh Rose, Commander of the Forces in Ireland, sent a letter to Col. Masters, the lieutenant's commanding officer, requesting that Mr Hackett, Moore Park, Newbridge, Mr Curran R.M. and Mr Moriarty, chief of police, should be thanked for their assistance in the matter, and for the excellent work of Constable Sheehan. It was, he wrote, 'gratifying to think that the country people of the neighbourhood had no sympathy with the perpetrator of this horrible crime, but that on the contrary their best endeavours were freely given to the discovery of the body, whilst the shocking fate of this poor officer excites in them the warmest feelings of pity and terror.'[11] It was apparent from the remark of the military secretary concerning the attitude of the country people that he realised that the presence of the military might not always be appreciated by the people.

There were also lesser crimes committed in the camp that year, including a number of robberies from officers' quarters and from the post office. Eventually Private Harry Williams of the 3rd Buffs [Royal East Kent Regiment] was charged with the thefts which included medical books and surgical instruments from the medical officer's quarters, and a thermometer from Capt. Simms R.E. From the post office he had stolen a box containing £47 in gold. The post master's excuse for the breach of security in leaving the box in the office was that while he normally took it to his bedroom, on that evening his wife was about to go into confinement and he was distracted.[12]

In April 1865 when a rumour was circulated that the office of Curragh ranger was to be abolished, there was concern in the locality as it was feared that such a move would be injurious to both the plain, the horse owners and the people.[13] Indeed, such a proposal was made in the House of Commons in the following month when it was suggested that Robert Browne should be put on pension, and the position left vacant 'as it conflicted with the rights of the War Department'. This aroused great alarm in Kildare as it was believed that if Ranger Browne, who was also a Resident Magistrate, was retired that a military tribunal might be appointed to try civil offences on the Curragh. 'It will certainly be a strange thing if civil offenders are to be tried by drum head court martial within 22 miles of the city of Dublin', wrote an outraged journalist in the local newspaper.[14] Such speculation was dampened when Browne remained in office.

A continuing annoyance to the ranger was trespass, and especially the driving by the Doolans of their ass and cart across the Curragh to the camp market. The ranger was being pressurised in May 1868 by the solicitors of the Commissioners of Woods and Forests to try and have the Doolans convicted. The fact that Henry Hart had been fined three pence and costs for the trespass of his pigs suggested that a conviction could be got against the Doolans, but such was not the case. That month too both Catherine Troy, Patrick Doyle and B. Smith were fined a total of two shillings for allowing their swine to damage the Curragh.[15] Women of the 'unfortunate' class, or the disputing wives of soldiers, were regular litigants. In the case of the former they were frequently before the Petty Sessions on charges of trespass on the Curragh, or of vagrancy. A respectable resident was fined 7/6 for the trespass of her pig on the sward, while on one day alone five women got sentences ranging from two weeks to a month for vagrancy.[16]

However, on 24 August 1868 there had been a change in office of the Woods and Forests, and Mr Howard, the officer there who had been responsible for the management and preservation of the Curragh, no longer had that responsibility, and the ranger was directed to take no further proceedings in regard to trespass on the part of that department.[17]

The promulgation of bye-laws for the Regulation of the Curragh in December 1868 was intended to facilitate the enforcement of order on the plain. The new laws forbade unauthorised grazing, the removal of gravel, sand and sods, or the disturbance of the surface of the plain. Furze was not to be cut. Manure or rubbish was not to be put on the plain. However, sheep dung was not to be removed, but was to be scattered each morning by the shepherd in the area in which the flock had sheltered overnight. Vehicular traffic across the sward and trespassing or creating a nuisance was illegal. Fines of up to £5 might be given for breaches of the laws.[18]

Amongst the charges before the bench in 1876 was that against Frank Anderson, who was caught with 22 pounds of oats near the artillery stables in the camp, and two men were charged with taking 'certain gold, silver and copper coins' which they had found while clearing the ruins of a canteen which had been destroyed by fire. A pair of youths from Newbridge were fined for being found to be in possession of blank ammunition, and at Naas two men were charged with the theft of two hair-cutting machines from the sergeant-major of the R.D.F.[19]

Sometimes there could be a conflict of responsibility between the military police and the Royal Irish Constabulary in the administration of the law locally, as in February 1876 when it was found necessary to hold an inquiry in the camp into the scope of their relative duties; it was afterwards reported that as a result of the inquiry 'great changes are being made amongst the provosts, all tending to the advantage of the civilians'.[20]

That it was not uncommon for the soldiers from the Newbridge barracks to be in breach of the law in the town was borne out by their regular appearance before the Petty Sessions at the Curragh, or at the Quarter Sessions or County Assizes at Athy or Naas. Sometimes the offences were comparatively trivial, such as that for which two men from the 50th Regiment, stationed at Newbridge, appeared before Col. Hon. W. Forbes R.M. at the Curragh Court in May 1876. They were charged with the theft of an umbrella and a jacket at Newbridge Railway Station, and were remanded to the Quarter Sessions at Athy.[21] But there were frequently more serious charges. Three years hard labour was the punishment given at the Quarter Sessions in October 1876 to a private of the 4th Regiment of Foot who burgled the dwelling of a lieutenant of the same unit at the Curragh. He stole a watch, which was later offered for sale in the cook house by Rebecca Dawson, a 'woman of ill fame'. She was given a five year sentence.[22] Another woman was charged with stealing clothing from the quarters of Farrier Sergeant Charles Miller.[23] The ease of access of both of the women to barracks was an indication of the failure of the authorities to maintain security, and of the constant communication between the soldiers and the women. But men living-out of barracks might suffer from the lack of the companionship of the barrack room; an Inquest in November 1876 found that 37 year old Sgt. Major Alexander Simpson 7th Dragoon Guards, living-out in Newbridge, had died from suffocation when drunk.[24]

In the summer of that same year John Norton, the servant to Capt. French 2nd Queen's Bays [2nd Regiment Dragoon Guards], was fined 10/– by Col. Forbes R.M. for assaulting an army pensioner at the Curragh races,[25] and there were also no less than five other cases of assault, indecent assault or rape in the same period. Three of the women involved were described as 'unfortunate', the euphemism for prostitute, and the others were a septuagenarian from Morristown Biller, who was raped by a man from the 53rd Regiment (and who was given a 20 years penal servitude sentence by the Crown Court),[26] and the wife of a sergeant of the 99th Regiment who was ravished by the bandmaster of the same regiment in his hut in the Curragh Camp.[27]

The military were themselves often the victims of civilian criminals, perhaps from minors like the 14 year old who stole the gold lace from three sergeants' caps and was given 12 strokes of the birch,[28] or 12 year old Thomas Healy who was given 14 days in prison and five years in a reformatory for uttering base coin at a canteen and two draper's shops in the Curragh.[29]

The constabulary at Naas were regularly employed in duties related to the military presence. Not alone was the county town a favoured place of recreation for men on leave, but a number of soldiers also lived there. That many of the women folk associated with the soldiers were troublesome was very evident from newspaper reports of the Petty Sessions; their mothers, wives or daughters frequently fought among themselves, with neighbours, or male acquaintances. In September 1885 the wife of a soldier who was away in India for almost two years summoned a man who had used abusive language to her.[30] Six months later Mrs Mary Anne Traynor was accused of using abusive and obscene language to Anne Brien; she alleged that Mrs Traynor, her daughter and her son-in-law were beating a soldier on the street.[31] While the trade of Naas and other towns frequented by the military benefited from their presence, the townspeople in return had to tolerate the rough habits of some of the men and their associates.

Soldiers were also known to attack their own wives, and in 1885 when Private W. Major assaulted his spouse for not having the dinner ready, the case was dismissed. Less fortunate at the same Sessions was Bessie Hynes, another unfortunate, who was given a month in Naas jail for stealing a bath rug from 'G' Lines in the camp, the property of the secretary of state for war and valued at 4/1d.[32]

Desertion from the ranks was another frequent offence committed by the rank and file. The consequence of such an offence was the cause of a civilian being charged before Naas Petty Sessions in February 1886; he was found in the possession of a uniform, and his plea was that the soldier had left it with his wife, and she did not know it was an offence. He was fined 5/–.[33] Before the same court a few months later a man of the Inniskilling Fusiliers was given 14 days hard labour for trying to enter the ranks illegally; at Naas barracks he had been found to be medically unfit, but two days later he presented himself at the Curragh and made a false attestation.[34] If the intention of men to leave the ranks was revealed to the authorities immediate action was taken, as in the case of four men of the South Wales Borderers who were chased by the mounted Military Police from Kilcullen to Grangebeg in September 1886. One of the men was apprehended when he hurt his ankle.[35]

When the militia reported for their annual training they too sometimes became involved with the law, but not always as spectacularly as in the Spring of 1887; at Kildare Railway Station one morning at 9.30 a.m. 70 members of the Kildare Militia rioted and caused damage. The mounted provost guard had to be called from the Curragh, and 16 men were arrested, before order was restored.[36]

It was the militia which gave Sub-Lieutenant Ian Hamilton 12th Suffolk Regiment his first experience with fixed bayonets when he was involved in clearing Militia Square of squabbling Militiamen from the North and South of Ireland in the camp in August 1873.[37]

That little had changed in the camp by 1886 was evident from the reports of the Petty Session there that year. Local residents, such as James Dunne, who lived at the Standhouse, and his wife and servants, were also regarded as regular trespassers. Despite the bye-laws, they believed that their ancient rights entitled them to drive across the sward to the Ballymanny road on their way to Newbridge or Kilcullen. The Deputy Ranger Colonel Foster charged Dunne for what he described as 'old trouble' in December 1886, and a fine of five shillings was imposed.[38] A decade later, in 1896, a soldier of the Royal Engineers was given 18 months hard labour for the indecent assault of the five year old daughter of a sergeant .[39] James Frawley's sentence for the theft of Miss Clancy's shoes, valued at 1/6d, and which she had left on a seat in the Roman Catholic chapel in the Curragh, was three months in jail.[40] Patrick Howley, an ex-soldier, and who was described as a constant loafer in the camp, was charged with trespass, the theft of 'an unfortunate's' shoes, and with the indecent assault of the wife of a sergeant from the 1st Yorks.[41]

At Naas railway station in February 1896 Constable Porter noticed a man wearing the socks of the 10th Hussars, and though he at first gave a false name he was identified and returned to Newbridge barracks. An award of £1 was given to the constable.[42] On another day the same policeman apprehended two men, one in uniform, but who gave false names. They were the servants of Major Cluff on the Curragh, and they had taken his horse and trap for a trip to Naas. They were all handed over to the military. So was Walter Woodward, 2nd Battalion East Yorks Regiment, stationed on the Curragh. Disguised in the clothes of a coachman, his military appearance betrayed him, and he was arrested at Naas Railway station on suspicion of desertion.[43] Occasionally even the officers came in for adverse mention, as in 1896 when the appearance of Lieut. R. B. Webb, Royal Dragoons, Curragh Camp, before the bankruptcy court in London with debts of £11,623, and assets of £725, was reported in the local press.[44] However, the standard of living expected from the military gentlemen was such that financial difficulties were not uncommon, and not a matter of utter disgrace.

The rumours of war in the Transvaal, and the possibility of her soldier son being sent there, reached Kate Quinn in Birr in July 1899. Although she was almost blind she set out to walk to the Curragh where the Leicestershire Regiment was based; it took her a month, and when she finally found her son and discovered that he was but going on the manoeuvres and not to war she took a celebratory drink. She swore that she had not touched a drop until she reached Igoe's corner on the plain, but the police sergeant said in evidence that 'she was not only blind, but blind drunk'. However the magistrate, himself a former military man, took a kindly view of her offence and suggested that the next time she came to the short grass she was to be more careful.[45]

Darker crimes were sometimes committed in barracks; the people of Newbridge were shocked in March 1899 when several soldiers of the 14th King's Hussars were arrested in the barracks following the discovery of three regimental horses with their throats cut. That trial was not reported in the local paper.[46] Of lesser moment were the antics of 'Tommy on the loose', as a newspaper headlined the antics of two drunken soldiers at Kildare railway station. They had taken a car following a scuffle, and they were chased by a superior officer in another outside car, causing much amusement to the onlookers, and being described by the journalist as Atkinese.[47]

It was generally recognised that the men who voluntarily joined the colours did not come from the most respectable sections of society. The poor pay and general conditions of service did not attract the better class of man; as General Wolseley once remarked 'unless we can give a very high rate of pay, we shall always be obliged to take the "waifs and strays"'. The majority of the recruits came from the unskilled labouring class, and as late as 1888 it was estimated that 60 per cent of the soldiers were illiterate, or barely literate. Taking into account the background of the average man, and the unpredictability of the soldier's life, it might but be expected that the women who were associated with them would have come from similar circumstances to themselves. One individual who wrote an account of his life in the ranks recalled that his father was infuriated when he enlisted in 1876. That his son should be a soldier was a disgrace to the family, and he would have preferred that the young man should be out of work for the rest of his life rather than to be a soldier: 'More than that, he would rather see me in my grave, and he would certainly never have me in his house again in any circumstances'.[48] Another soldier, writing in 1899, observed that 'the class of men from which the average soldier is drawn is not at all the sort to be discontented with an adequate amount of food, clothing, lodging, and pay . . . It is certainly true that the moral standard of the rank and file of the army is not of the highest quality, but this is not altogether necessary for its efficiency'.[49]

However, the military efficiency of a man of the rank and file would be reduced if he was ill, and if the illness was contracted through his own fault, perhaps as a consequence of his low moral habits, it was a serious matter in the eyes of his superiors. Under military law he could be charged with malingering, and from 1873 if he became incapacitated through venereal disease he would forfeit his pay while under treatment.[50] While the army medical department provided treatment for men affected with venereal diseases, and kept and published records of such cases, no serious effort was made on the Curragh to give medical care to the afflicted women with whom the soldiers had intercourse until 1868 when the Lock Hospital was constructed in Kildare.

If the military authorities turned a blind eye on the presence of prostitutes, who were found in the environs of every military station, they were forced to initiate action against them if they became a public nuisance. As the general commanding in Ireland had said of them during his evidence at the Curragh Commission in 1866 'those unfortunate women who traverse the Curragh more prominently than is necessary, particularly on Sundays. I think these persons should be controlled. Of course these women cannot be removed altogether, but they need not be allowed to appear prominently'.[51] The military understanding of what was regarded as the normal habits of single men of the soldier class was based on their wide experience of commanding great numbers of men in stations throughout the empire. They would have known from experience of the moral difficulties associated with the concentration of large numbers of men without female company, and of the reports, which were officially communicated to London, from the penal colonies in Australia of the homosexual and sadistic practices of some of the convicts there.[52]

The moral and social habits of the unfortunate women who lived on the Curragh caused annoyance to the military and civil police, and scandal to the local people. The women entered the camp illegally, they frequently behaved in a disorderly fashion, and they were accused of petty crimes such as theft and drunkenness both in and outside the camp. That the lives of the women were wretched was apparent, but the social conscience of the nineteenth century was not as liberal as it is today, and generally the women were seen by the local people as debased and diseased, and a scandal to everyone.

If the general increase in lawlessness distressed the local civilian population, the additional cost to the county of accommodating those persons convicted of crimes in Naas jail, and of ill prostitutes in Naas workhouse, was deplored by the Grand Jury and the Board of Guardians until such time as financial and other arrangements were made by the War Office.[53] As to the unfortunate women themselves, their contribution to disorder in the Curragh area over a period of almost 70 years will next be examined; unlike the soldiers, whose tenure on the plain is permanently marked with substantial military buildings, their lives have fallen into oblivion.

Within a few months of the occupation of the encampment on the Curragh in 1855 there were complaints about the activities of some elements of the camp followers. A 'Kildare cesspayer' voiced his views on the advent of thousands of men on St Brigid's pasture, as the plain was locally called, in a letter to the *Leinster Express* of 27 October: 'One of the great evils accompanying a great body of men is the numbers of loose and impure characters who follow in train. This evil has arisen on the Curragh, and the government found it necessary to put police there, and to fix the camp as the residence of resident magistrate Capt. Hill, late 25th Regiment. This gentleman, it appears, has all the women found loitering around the camp brought before him as vagrants and trespassers, and sent to the county jail for two months. Naas jail is crowded with the abandoned, diseased prostitutes of Ireland, and the cesspayer is burthened'. The writer went on to say that in the last week nine women were committed to jail, and as there were already ten incarcerated it brought the total of unfortunate women to 19, almost half of the total female inmates of the jail. It was emphasised that it was not only the pecuniary effects to the cesspayers, but also the moral effects of the prison, as the wretched women were mixed with the others, including girls committed for petty larceny. The cells, it was known, were crowded and unhealthy. 'It was never the intention of the legislature', the 'Kildare Cesspayer' concluded, 'that county prisons should be Lock hospitals, and in justice to the cesspayers, and to the other female prisoners, and to the officers of the jail, something should be done.'

But the problem was not a new one to the county, and it may be assumed that the irate cesspayer was a resident of the Curragh, but that he did not regularly frequent the garrison town of Newbridge, or indeed any other garrison town throughout the country.[54] Camp followers had existed since ancient times, but when they arrived on one's doorstep it was another matter.

An officer who had served in Newbridge barracks in 1844 wrote a graphic account 20 years afterwards of his observation of the prostitutes then there, and it was published in the magazine *All the Year Round*, which was edited by Charles Dickens, on 26 November 1864. The officer recalled a priest coming into the barracks one winter's day to ask the commanding officer to give him a fatigue party of soldiers to go outside the barracks and to pull down a few booths which the women had made against the barrack wall. The soldiers were paraded, and led by the priest they proceeded to burn down the 'the shelter these unfortunates had built. At this time it was quite common for the priest, when he met one of the women, to seize her and cut her hair off close. But this was not all. In the summer of 'forty-five, a priest, meeting one of the women in the main street of Newbridge, there threw her down, tearing off her back the thin shawl and gown that covered it, and with his heavy riding-whip so flogged her over the bare shoulders that the blood actually spurted over his boots. She all the time never resisted, but was only crying piteously for mercy. Of the crowd which was formed around the scene, not a man nor a woman interfered by word or action. When it was over, not one said of the miserable soul "God help her". Five days afterwards I saw this girl, and her back then was still so raw that she could not bear to wear a frock over it. Yet when she told me how it was done, and who did it, she never uttered a hard word against the ruffian who treated her so brutally.'

Then the writer castigated the Irish for their unchristian attitude towards the unfortunate women. In England, he said, 'if even an animal had been attacked in the way the woman had been, the law would have intervened, but in Newbridge there was no one man enough to take the whip from the priest's hand and of giving him a lesson. Shopkeepers would not sell to the women, nor did farmers permit them to shelter in their outhouses, for fear of being denounced from the pulpit, in a country where inhumanity of that sort was encouraged'. The refusal of help and charity to the fallen resulted in 'noon-day immorality and drunkenness, and nightly licentious revellings. When all the vice is out of doors wandering shameless and defiant through the streets of Newbridge, the by-lanes of Cahir, and the purlieus of Limerick, Buttevant, Athlone and Templemore, it becomes more mischievous than it can be in the cellars and courts of the back streets of Dublin'. Strong words indeed, and 20 years later when he was again back in the area, the officer was to have equally critical views of what he saw on the Curragh.

The priest in Newbridge was not alone in his determination to cleanse his flock of the pernicious females. At Templemore in the November of 1855 Reverend. J. Fennell C.C. regularly went around the town at nightfall to seek out and chastise any of the females 'who so much infest garrison towns'. On one such excursion he mistook the wife of Private Logan for one of that class and he struck her. The soldier brought the curate before the Petty Sessions for assault, and he was fined £1 and costs. The case, it was said, caused a great sensation locally.[55]

The Curragh camp medical returns for (a) July 1855 to June 1856, and (b) July 1856 to June 1857, show that the principal disease affecting the troops was venereal. Of the 5,635 and 5,126 admissions, 38.0% and 38.6% were from venereal disease in each year (respectively). That the annual mean strength was 4,590 and 5,266 in each year suggests multiple admissions of some men.[56]

The incidence of disease, and the concern expressed by responsible residents of the county, encouraged the military to try and alleviate the problems of illness amongst the troops, and of the women. In the autumn of 1859 the War Office was in communication with army headquarters in Dublin on the matter of a Lock hospital 'to prevent to some extent the prevalence of V.D. among soldiers'. It was not considered feasible that in places where soldiers were quartered that all women should be examined and sent to hospital, but commanding officers were recommended to compel the men to use the baths for private ablutions which, at 1%, had been installed in most barracks. Later that year, when the annual Return of Sick and Wounded was being considered the formation of a Lock hospital was still being discussed, but nothing definite had been decided on.[57]

The vulnerability of the women to attacks other than of disease was evident in the case of Hannah Dixon, 'an unfortunate of the Curragh Camp' who was admitted to Kildare Infirmary in August 1860. She had been wounded by Pte. Joseph Hannon, 2nd Battalion 14th Foot, and he was being held liable for the payment of the 2/– car hire from the front of the camp to the infirmary.[58] At that time the medical treatment of the diseased prostitutes was at the Naas workhouse, and in July 1860 a confidential letter from the military secretary at the Horse Guards to the military secretary at Dublin sanctioned the payment of an allowance of £100 per annum to the Guardians of Naas Poor Law Union for the maintenance and treatment of abandoned women from the Curragh Camp.[59] Records of similar payments for subsequent years are also to be found in the letter books of the army headquarters at Kilmainham and the minute books of the Naas Union Workhouse.[60]

At the meeting of the board of the Naas workhouse on 28 July 1860 it was noted that H.M. Treasury had sanctioned the allowance of £100 a year for the treatment of prostitutes from the Curragh Camp suffering from venereal disease, and that the Poor Law Commissioners would 'sanction the building of a hospital near Newbridge for this purpose'. However, while the board unanimously agreed to accept the money, the suggestion that the guardians should build a hospital was unacceptable to them and it was agreed that, 'if that was the condition connected with the £100, then it should be declined'.[61] When the board met for the first time in the following month they decided that while they were not prepared to build a hospital they would, if the government provided the £100 a year, set aside a section of the workhouse for the care of the prostitutes.[62] The army medical report for 1860 revealed that 373 men in the Curragh Camp were treated for V.D.,[63] while the census taken in April 1861 gave a figure of 70 prostitutes in the county Kildare, 26 of whom were not born in the county. The military strength then in the camp was 5,103.

That year the total number of prostitutes in Ireland was 1,057, and in Leinster 590. Of those in Leinster 518 were Roman Catholics, 68 Established Church, and four Methodist. There were 147 brothel keepers in the country, seven of whom were men. By religious profession 130 were Roman Catholic, 14 Established church, and three Jews.[64]

The ever present annoyance caused by the women to the Curragh ranger was reflected in correspondence between himself and the Office of Woods on 4 January 1861 when the proposal of a doctor from Cork to open Turkish Baths in the old Stand House at the Curragh was being discussed. Ranger Browne supported the proposal as otherwise he feared that the 'place would become the haunt of prostitutes'.[65] On a more positive note, that the proposal to provide shelter for the women in Naas workhouse materialised is evident from an entry in the minute book of 12 January 1861. It recorded that the lord lieutenant had been informed that the alterations to the auxiliary workhouse had been made, and that the building was then ready for inspection by the military authorities.[66]

The predicament of the women on the Curragh was brought to national notice in the autumn of 1863 when a girl named Rosanna Doyle died soon after admission to the workhouse of Naas on 19 October. An inquiry was held into her death, and the circumstances it revealed caused public outrage. On 16 October one of the caretakers of the Curragh named James Greany was in the shop of a woman named Dolan who told him there was a girl in the kitchen who was sick. He met Rosanna Doyle, and promised to report her, but he did not do so until the next day when he went to the police, who in turn were to notify Patrick Cosgrave, the relieving officer who lived at Kildare. But he was out, and it was late when he got the note saying that the girl had been found in a furze bush opposite Greany's house (a witness described a 'bush' as a space among the furze, protected at the side by sods and bushes, but without a roof) and that she was sick and destitute. Cosgrove waited until the following morning to arrange transport for the girl to Naas, but Bergin, the usual carrier, said he had been out late the night before, and would go on the following day. The relieving officer agreed to that arrangement, and did not go to visit the girl. About 8 o'clock on the morning of the nineteenth Bergin went to collect Rosanna Doyle. He found her lying in a bush 'with very bad clothing on, and a sack over her; and that she was as wet as if she had come out of a river; and that she was not able to stand for weakness'. Another woman, Bridget Lyon, told the inquiry that Doyle had been in the bush, exposed to the very bad weather, for two nights. After she had eaten a little the sick girl was placed in the donkey cart on some straw, and covered with a sack and she was taken on the three hour journey to Naas. Within ten minutes of her arrival at the workhouse she died.

The subsequent inquiry found that Cosgrove had been negligent in his duties, and that he lived too far away from the Curragh. He was dismissed from office. It was proposed that the relieving officer should be obliged to live as near as possible to the camp, or at Newbridge, and that he should have an office at the Curragh, and visit there three times a week. The purchase of a covered spring-car to be based at Newbridge was approved. It was to be horsed by a contractor who was bound to send it out at any hour of the day or night on the order of the relieving officer or the master of the workhouse. From Mr Cosgrave's Application and Report Book it was seen that during the year ending on 31 October 1863 there had been 422 applicants for relief, of whom 163 were from the Curragh.[67]

For the same financial year the expenses for the relieving of prostitutes from the Curragh camp in the Naas workhouse were £354.4s.8d. The statement of expenses was recorded in the minute book on 7 November of that year:

Maintenance of an average of 20 prostitutes @ 3/4d each weekly		£178.15.6
Clothing for same		£19.19.9
Medicines		£12.0.0
Conveyance		£7.5.5
Rent of auxiliary buildings		£10.0.0
Assistant matron, salary, and rations		£34.12.0
Nurses' rations		£3.12.0
Extra payment for chaplains		£40.0.0
Relieving officer's salary		£30.0.0
Covered vehicle		£18.0.0
	TOTAL	**£354.4.8**

That the Board of Guardians was not subsequently to be satisfied with the government grant of £100 per annum was reflected in the minutes of a meeting on 4 August 1866 when a case was made that the prostitutes should be entitled to at least one half of the cost of their maintenance.

On Tuesday 24 November *The Irish Times* printed a long report on the Doyle case, and of the appearance of Cosgrove before the petty sessions at Naas on 23rd on a charge of manslaughter. Rosanna Doyle was described as belonging 'to a class, unfortunately too numerous in the neighbourhood of the camps'. The newspaper concluded its leading article on the death with the comment: 'but surely it is the bounden duty of the workhouse authorities to arrange some system by which the wretched human beings who frequent the Curragh should be relieved as soon as possible. We trust that early next session a bill will be introduced to give the crown complete authority over the Curragh, and to prevent, if possible the vice and misery which now infects the place.'[68]

The gentleman who had described conditions in Newbridge in 1844 made a return visit to the area in 1863, but this time he concentrated on the situation in the vicinity of the Curragh Camp. Admitting that a similar problem existed around the large camps in England, he claimed that it was not at all as extreme as what could be found daily in Ireland. Upon the Curragh there were hundreds of women, 'many of them brought up respectably, a few perhaps luxuriously, living in all weathers in the open and in conditions worse than those of 'the Simaulees and Hottentots of Africa'. They received no charity, and if they asked for a crust of bread, christians would refuse it'.

In Newbridge the writer had seen Mr Tallon refusing to sell one of the women a half ounce of tea with the words: 'no, I'll not serve you', and when he questioned the shopkeeper he replied 'were she dying, I would not give it to her, or any like her'. The gentleman himself bought the half ounce of tea and brought it out to the woman. Nevertheless it was found that Newbridge was not the worst part of the country as there a charitable farmer who owned some small fields near the barracks had allowed the women to live in a five foot deep dry ditch by the roadside. This they had covered with hay and branches to form a shelter.

On the Curragh it was much worse. There, in rain and snow, the prostitutes lived in the furze bushes, and if they attempted to make a shelter it was pulled down by the police or the deputy ranger. Often, as he went on an early morning exercise, he had seen the women lying by threes and fours in a ditch or by the lee-side of a bush. One morning as he was riding from the Stand House towards the Camp Inn he observed four women lying in a hole they had scooped out of the ground. One of them begged for a shilling, then she thanked him – 'long life to you, this will get us a drop of whiskey'. He noticed that one of her companions was ill and he reported her to the police station. She was, in fact, Rosanna Doyle, the girl who died on admission to the Workhouse a few days later.[69]

Despite the prosecution of the relieving officer following the Doyle case, and the good intentions expounded by the Poor Law Commissioners, there was no immediate improvement in the situation. Mr Browne, the Curragh ranger, protested to the guardians of the Naas Union, but when soon afterwards he himself saw a girl who was ill, and he could locate neither the relieving officer nor a poor-law guardian in Newbridge he again made a formal complaint to the commissioners. Despite their orders, the relieving officer did not live in the town but at Milltown, several miles away. Finally the ranger wrote a letter to *The Irish Times* exposing the condition of the unfortunate women whom he likened to 'the esquimaux in that, when it snowed, they lay with their backs upwards to form a temporary support for the snow to rest on, and which, when accumulated, kept them partially warm'.[70]

'I am well aware,' continued the observer from England, 'that these women are the dregs of society, also that some mistaken christians will say that "any pity shown to them is at best an encouragement of vice", while others, like Scrooge, will inquire "whether the workhouse and prisons are still in operation?" To such it is useless to make any appeal. But to those who can feel for the poor and homeless, who, to the best of their ability, attend to the divine commands to feed the hungry, clothe the naked, visit the sick, and raise the fallen, I appeal for at least a thought of christian mercy towards the wretched outcasts, who exist on the Curragh, and around barracks in Ireland.'

That compassionate army man sought to explain to his readers how the pomp and glamour of the military could seduce impressionable and innocent girls. 'The dress of the soldiers, the gilding on the uniforms, the regular step, and the martial bearing of the men, are as if specially contrived to carrying the feelings and good wishes of spectators away captive.' The camping grounds with their tents, flags and bugles, the galloping to and fro of mounted orderlies, the plumes of the staff officers, and the neatness and regularity of the guards were all very impressive. But around every barracks there was 'an outlying circle of misery and sin . . . it has never been otherwise'. 'When the troops come to a town they attract the girls, and when they march out some of the girls will follow them. But soon they will be in trouble, their little money gone. With her clothes in tatters, she cannot return home. The man she followed cannot support her, and she deteriorates into a ditch by the barrack wall or in the furze on the Curragh.' Frequently the soldiers tried to raise money to send the girls back home, if they believed their families would receive them. In India, he explained, the camp followers were placed in the care of a female overseer who was paid from the canteen funds, and the system was satisfactory. He recommended that the system, modified to suit moralities here, might be introduced in an attempt to stem the spread of disease 'which fills our army hospitals, and ruins the health of our soldiers'.[71]

The ruination of the women was displayed at almost every sitting of the petty sessions and assizes in the county Kildare. For example, in the spring of 1864 at

Naas, Susan Donnelly 'one of the unfortunate women from the Curragh camp', was charged with offering for sale a quantity of butter which was believed to have been stolen, but when a man swore that he had given it to her the case was dismissed.[72] Ellen Hannon 'who appeared to be one of those unfortunate females who frequent the neighbourhood of the Curragh' was found not guilty of stealing money from John Nolan.[73] There was a mention in a local paper at about that time that Naas jail was crowded, 'mostly with female prostitutes from the Curragh camp'.[74]

Official action on the prevalence of venereal disease in the army came in 1864 with the Contagious Diseases Act, which was enacted for three years. It was intended to give the army some control of the sources of infection in the garrison towns, and a senior medical officer was to oversee the administration of the act. In England the places covered by the act included Aldershot, Woolwich and Colchester, and in Ireland, Cork and the Curragh. The Curragh area included the parishes of Ballysax, Kildare, Kilcullen, and Great Connell and Morristown Biller, which included Newbridge. The legislation allowed for a fine of £10, or imprisonment, for the owner of the house in which a woman was apprehended, and a constable could take the suspected woman before a justice of the peace who had the power to have her admitted to hospital for medical examination. If found diseased she would be kept in hospital for three months, or until cured. It was also possible for a woman to seek to have herself admitted to hospital. For the treatment of the disease certain hospitals were certified, and in county Kildare an auxiliary ward in the Naas Union Workhouse was so designated.[75] The cost of the preparation of the auxiliary building was £488.[76]

Concern was also being expressed for the spiritual welfare of the Curragh women, according to a report in the *Leinster Express* of 2 February 1865. It said that it was 'gratifying to learn that a female mission is in contemplation at present among ladies of this county on behalf of that wretched class of our fellow creatures, "fallen women", 500 of whom, it is calculated, are every summer in the vicinity of the Curragh and Newbridge.' The inspiration of the county ladies came from John La Touche, the banker and landowner of Harristown, near Kilcullen. He had undertaken evangelical work amongst 'the lost' in London, and now he proposed that a mission to the homeless and friendless in the neighbourhood of the military barracks should be started. To awaken public interest in the mission he gave a series of discourses in Naas Town Hall.

Though the attendance was small at his discourse in early April 1865, it was influential and interested. He explained that he was already on a committee which managed a house for fallen women in Marlborough street, Dublin. There it was hoped to reform them, 'to help them to abandon their evil course, and to submit to religious and moral training with a view to being restored to their friends and provided for'. It was known that many influential people living in the neighbourhood of the Curragh who had been impressed by the magnitude of the social evils in the neighbourhood, had recommended that a hospital for the bodily relief of the women should be established, but Mr La Touche believed that their spiritual well being should also be looked after. His audience wished him every success with his work.[77] A week later he was able to report that a lady had been appointed to commence missionary work, and was then in Newbridge. She was going to labour around the town and in the environs of the Curragh Camp. Contributions were invited for the mission, and they could be sent to Rev. C. Moore at the garrison chapel in the camp.[78]

Despite the efforts of the missionary lady the women carried on with their normal activities. In June, Margaret Noble, Johanna Ryan, Anne Kelly and Cecilia Loftus appeared before Newbridge Petty Sessions charged with stealing straw during the night from the stables of the Military Train in the camp. In the morning the police followed the hay trail to the women's hut at French Furze. The sentries on duty at the stables were punished for not apprehending the women.[79] A couple of months later Private John Thomas 8th Regiment, who was in the company of a number of unfortunate girls at the Curragh, accused one of them of stealing his purse.[80]

In the meantime the Naas workhouse remained as the only refuge for the unfortunates. But as winter approached the plight of the women was again highlighted when Mary McCarthy, who lived in a ditch, got ill. She was coughing up blood and had a bad heart. While she got medicine from the dispensary, and an admission ticket to the infirmary at Kildare, there was no transport to take her there. Her companions Anne Dillon, Kate O'Meara and Margaret McGuire, did their utmost to help her. They went repeatedly to the home of the relieving officer and to the police to try and have her taken to Naas workhouse, but the former was away on duty in another part of his large area, and by the time he returned she had died in the ditch. Then the head constable took charge of the body, soliciting the comment in the local paper that 'alive no official would come to her, but when dead they did'. At the subsequent inquiry before the poor law inspector Capt. Robinson at Naas workhouse the public heard from Mary McCarthy's friends that she had no place of residence, and had never slept in a house for the past four months but in a ditch, without shelter, in a lane at the back of Newbridge barracks.[81]

Two weeks before she died Mary McCarthy had been sentenced to seven days in Naas jail for drunkenness, and while there she was very ill. William Noble, a medical student and assistant to Dr O'Beirne, who had examined her said she had a bad heart and spat up blood. The relieving officer, Andrew Fitzpatrick, explained that he was away at Allen on the day she died. He described her as never having been any trouble, and more orderly and better conducted than many of her class. It was established at the inquiry that the van for transporting patients to Naas workhouse was not available for taking them to Kildare infirmary. A letter to Naas Union from the chief clerk at the Poor Law Commissioners office in

Dublin recommended to the Board of Guardians that the union should be divided into three districts. Andrew Fitzpatrick, the relieving officer, was not found guilty of negligence, and the appointment of a third relieving officer to the union was suggested. However, the guardians decided that two relieving officers would be sufficient, and so it was. The whole tragic incident inspired 'Madeleine' to write a letter to the *Leinster Express*; she thought it was 'shocking to read of another inquest in Kildare from exposure and misery. Is this a civilised country, or are we lapsing into barbarism? Treating women, Irish females, as wild beasts and allowing them to die like rotten sheep, lying about in mid-winter, forbidding anyone to shelter them, is infinitely worse, it is simply slow murder'.[82]

At last, officialdom had begun to do something. The founding of a Lock hospital[83] to cater for the afflicted women was being actively considered, but at the beginning of February 1865 a private meeting was held in Kildare to protest against the proposal that the hospital might be located in that town. A memorial was sent from the meeting to the duke of Leinster and other governors of Kildare infirmary (which had been established in 1767).[84] The memorial was adamant that a 'hospital for discarded women . . . which would bring the vices of Newbridge and the Curragh to our doors' should not be located in the town. The memorialists continued that 'they had no connections with either Newbridge or the Curragh, or responsibility for their sins, but that either place would be more suitable for the establishment.' They opined that the number of prostitutes in both places, winter and summer, averaged 300, while in Kildare there were only eight. A rumour was then circulating that the Turf Club Hotel was going to be taken over as the hospital. If so, the memorialists held 'our wives and children could not leave our doors without coming in contact with [the women]'. They implored his grace to support them in their objections.[85]

At the February 1865 meeting of the governors of Kildare infirmary it was agreed that there would be no objection to their assisting in the management of a Lock hospital, but that it could not be located in the infirmary. Newbridge was suggested as a location, as it was close to both military installations.[86] The governors agreed that the infirmary surgeon might superintend the Lock hospital for a couple of hours in the afternoon.[87] Further pressure came on the authorities in the same month when Dr Leonard, inspector under the Contagious Diseases Prevention Act 1864, reported on the accommodation and treatment of the abandoned women from the Curragh at Naas Union. He found that there was insufficient accommodation, and that the treatment in some particulars was wrong.[88] However, in the minute book of the union it was noted that the inspection found the auxiliary ward was 'clean and sufficient as a reception room for the prostitutes', and a week later the payment of the usual £100 was acknowledged.[89] But there must have been further reaction to the condition of the ward because early in August the chairman, under-chairman and deputy-under-chairman of the union had a meeting with the chief secretary, Lord Naas. They were seeking additional assistance for the auxiliary hospital and they also claimed that 'the prostitutes in the auxiliary were entitled to at least one half of the cost of their maintenance'.[90]

A correspondent who signed himself *Pro Bono* expressed his views on the Lock hospital in the local paper on 4 March. Welcoming the views of the infirmary governors, and the attention being given to 'the poor fallen creatures', he went on to regret that the disease was on the increase, spreading through rural districts miles around the camp. He said that he had to send many cases from his area for treatment, including one female, an ex-servant, who had been instantly infected by a kiss. The poor girl was ruined, and her prospects for employment in the county gone. *Pro Bono* thanked the governors for undertaking the management of the proposed hospital and for making the medical officer available. The duke was praised for 'his usual liberality, in making the site available rent free forever'.[91]

That letter shows that the site for the building of the hospital had already been decided on, and another letter in the same paper a couple of weeks later confirmed the fact. The writer, who signed himself 'K' agreed that the site was suitable as the officials could easily attend there after infirmary meetings, rather than have to drive two to three miles across the bleak Curragh in bad weather. It was also convenient for the medical officer and the constabulary. 'K' could not see what objection there could be to having the hospital a mile in one direction or the other, so far as regards morality.[92]

But the debate continued with the *Leinster Express* in its edition of 25 March inviting discussion on the location. Its leading article found that very few circumstances are purely beneficial, and that the advantage of a large military encampment had great drawbacks for many people with its social evil. As the Kildare infirmary had no accommodation for syphilitic patients, the proposal to have a hospital to 'cure the diseases consequent upon the infamous careers of the camp followers' was a good one. The same issue of the paper published a letter from *Veritas*, who described himself as one of the inhabitants of Kildare who had signed the memorial against the hospital. Now he regretted having signed it as he had an erroneous impression of the plan. He knew that it would be well conducted by the governors of the infirmary, and that there would be nothing offensive which could wound the sensitive. The outlay of money on the building, and the employment given, as well as the annual expenditure of from £600 to £800 a year in contracts was to be welcomed in the town.

In his evidence to the 1866 commission on the Curragh General Lord Strathnairn, commander in chief of the forces in Ireland, expressed his view that 'these persons [the women] require to be controlled. Of course these women cannot be removed altogether, but they need not be allowed to appear so prominently'.[93] James Savage, provost sergeant in the camp, said in his evidence that 556 prosecutions for trespass or drunkenness had been taken against the prostitutes during a period of six months, and he was of the opinion that there were generally about 100 of them on the plain in the summer, and in winter between 30 and 40.[94] Lord Strathnairn was again concerned with the unfortunate women and the proposals for a Lock hospital before the Christmas of that year. In correspondence with the secretary of state for war he referred to a War Office letter which had given four options for the hospital:

1. Establish a Lock ward in connection with the Kildare infirmary.
2. Convert part of Hare Park hospital in the Curragh camp to a ward for diseased females. (This was not recommended as the ward would be required for troops within a few months.)
3. Erect a building on the north/west corner of the Curragh. (Plans and specifications for a building to hold 32 patients were enclosed, and the cost of the hospital would be £1,000 if in wood, or £1,400 in brick.)
4. Use Athy barracks as a hospital. (However, as that town had little or no connection with the camp it was expected that the townspeople would object.)

General Lord Strathnairn recommended that a ward in Kildare infirmary would be the most suitable solution.[95] However, on 19 February 1867 the military secretary wanted to know if the people of Athy had any objection to the use of the empty barrack there, as it was still being considered for use as the hospital.[96] By April it had been established that there were objections from Athy, but that the Stone [police] barrack in the middle of the Curragh plain could be available, if the military found it suitable.

The Royal Engineers reported that it was unsuitable from the point of view of construction and position. In the meantime the offer of the duke of Leinster of a one and a half acre site near Kildare at the rent of £5 per annum, and a bonus of £30 to the tenant, was recommended for acceptance by Lord Strathnairn to the secretary of state for war, and in September 1867 a draft lease for the site at Kildare was being prepared for approval of the war department; it was to cover the transfer of land from the duke of Leinster to the military authority.[97]

In the meantime the afflicted women continued to be admitted to Naas Workhouse, and Dr Leonard was reporting on their situation there.[98] After Mary Conroy and Anne Nevin, who were ill and destitute, were admitted to the workhouse in August 1869, the relieving officer was asked to state under what grounds he had given tickets to the prostitutes, and he cited their circumstances.[99]

Attention was again focused on the scandalous plight of the women when a journalist from the *Pall Mall Gazette* in London visited the Curragh in September 1867. He thought that to ask for accurate knowledge from official persons 'would be answered by a gift of a stone, as it always is'. His sensational findings were published not only in the gazette, but as a pamphlet entitled *The Wren of the Curragh* that same year.[100] Following the establishment of the military camp on the Long Hill in the Curragh he had heard 'mysterious little stories which were wafted to England, hints and glimpses of a certain colony of poor wretches there who lived as nobody else in the three kingdoms lived, and died like most people who do come within the bills of mortality, tramps and others, when they happened to perish of cold, want, and whiskey upon that vast common'.[101] The journalist crossed over to Kingstown, and went to Kildare by train. There he established himself in the Imperial Hotel before engaging a carman to take him on a tour of the camp and its environs. He spent a few days, and one night, discovering a number of the women who were locally known as 'the wrens of the Curragh, because they live in holes in the banks'. The plain was found to be 'not an unfrequented nook in some distant land, but near a capital city, an encampment wherein thousands of Englishmen as well as thousands of Irishmen constantly live . . . and where scores of strangers, visitors who go there for no other purpose but to see what is to be seen, peer about every week of every summer season'. Driving across the Curragh the visitor encountered several cars of 'well-buttoned military men', and other soldiers appeared as 'patches of red amongst the dark green masses of the furze'. His guide could understand his interest in the military, in the new camp, and in Donnelly's Hollow, but he would not accept that 'them wrens' might be worthy of notice.[102] It was almost dusk when, as they drove close to a patch of furze, the first 'wren' was seen, and the driver exclaimed 'and there's a nest'. The journalist found that there were ten 'nests' in all, accommodating between 17 and 25 women, some of whom had been there for up to nine years. Located in a clump of furze, and known by a number given to it by the inmates, who numbered from three to eight, each nest consisted of a shelter measuring some 9 feet × 7 feet and 4½ feet high, made of sods and gorse. With a low door, and no window or chimney, and with an earthen floor, the 'nest' had for furniture a shelf to hold a teapot, crockery, a candle, and a box in which the women kept their few possessions. Upturned saucepans were used as stools, and the straw for bedding was pushed to one side during the day. At night the fire within the shelter was covered with a perforated pot, and the women undressed to sleep in the straw. In summertime the 'nests' gave some shelter, but in winter the wind whistled through them.[103]

The women, who were all Irish, came from different parts of the country. [A recent study of prostitutes in Ireland suggests that some of the women may have been 'involved in long term standing common-law relationships with some of the soldiers', due to the army's restrictions on admissions to the married establishment.[104] However, it is most likely that those women, who would have some support, would be lodged locally rather than resident in the furze]. Some of them had followed a soldier from another station, others came to seek a former lover, like the girl from Cork who sought out the gunner who had seduced her; when she found him and said that she was pregnant he told her 'go and do like the other women did'. Her baby was born on the plain, and she remained there. Another girl's story was that, at the age of 13, she had been seduced by 'a rale gentleman', an officer in a rifle regiment. Migrant harvesters, who had grown accustomed to the outdoor life, settled in the 'nests', but one woman was honest enough to admit that 'it wasn't one man brought me here, but many'. Nevertheless, it was obvious to the journalist that the majority of the women were simply trying to make enough money to keep themselves alive.[105]

They lived, received their families (if they came to visit), gave birth and died in the 'nests'. Their clothing consisted of a frieze skirt with nothing on top except another frieze around the shoulders. In the evenings when the younger women went to meet the soldiers in the uninhabited gorse patches, they dressed up in crinolines, petticoats and shoes and stockings. The older women remained behind to mind the children, of whom the visitor counted four, and to prepare food. All the takings of a 'nest' were pooled, and the diet of potatoes, bread and milk was purchased on the few days a week when the women were allowed into the camp market. Otherwise it was out-of-bounds, but an army water-waggon brought them a regular supply. The gentleman from the *Pall Mall Gazette* decided that, contrary to popular opinion, the women did not live in the furze because they loved vice. They were there because it was known that those who sought refuge in the workhouse at Naas lived in even worse conditions.[106] The 'camp sweethearts', as the writer termed the women, had only enough money to buy by the day, and as well as going to the camp market they also frequented a small shop where 'the good-natured woman behind the counter . . . was very considerate to the degraded creatures who flocked to her shop, and they seemed to be thoroughly appreciative of its spirit'.[107]

The reporter was thorough in his investigation of the lives of the women, and he journeyed to Naas to inspect the Union workhouse. This is what he found: 'The workhouse itself was half empty; but these [women from the Curragh] were not allowed to enter it and share such comforts as it might easily have given them. The whole fifty, with four children, were turned into a range of low hovels separated from the main building by a high wall, and so ruinous as to be totally unfit for human habitation; and this was winter. The beds were bags of foul straw, and two or three women slept on each of them, huddled, sick and sound together, without any attempt at separation; and more than one-fourth of them were sound.[108] The measurement of one of these hovels was as follows; length 28 feet, breadth 14 feet, height 9 feet. Imagine a room, a broken hovel, of these dimensions; imagine twelve such beds in it as we have described; imagine those twelve beds occupied by twenty-three women and two children; and ask whether you also would not rather have lain out on the common. That is a fearful picture, nor need anything be added to it, except that those despised and certainly very wicked women were not even allowed to worship with the other paupers; they had to thank God by themselves, and listen to the exhortations of his minister in their hovels apart.'[109]

Back again on the plain the Englishman asked one of the women how they survived in the open in the wintertime: 'We'd get the biggest little bush we could find, and lie under it, shifting around the bush as the wind shifted.' If they woke soaked from rain or snow they 'had to wait for a fine day to dry out'. At night the visitor saw the girls returning from their trysts with the soldiers, and many of them were drunk. One violent woman 'her hair streaming down her back, she had scarcely a rag of clothing on . . . made at me with a large jug, intending to be smashed against my skull'. Knowing the ferocity of her drunken companion, another woman protected the visitor, and advised him to run.[110]

Terrible and all as the observer saw the situation, he was aghast to hear that five or so years before the situation had been even worse, then, on account 'of the extravagant behaviour of one of the women in the presence of a lady (related to the general officer) who was riding on the Curragh, they were all driven from the common, and their hovels destroyed. A ditch in Furlane, leading to Athy, was for some time their only home. When some seven of them established themselves in a field belonging to a Scotsman on the south side of the plain he allowed them to remain there providing they prevented trespassers; they were said to be 'better than twenty policemen'.[111] While the London journalist thought the 'wrens' to be 'as foul as any Hottentots', he believed they were deserving of care and understanding. 'I suppose,' he mused, 'it is not possible to allow such things to continue in a christian country?'[112]

Meanwhile the Naas Workhouse continued to shelter the women and at Christmas 1867 Thomas Fraser M.D., surgeon major 10th Hussars, acting provost marshal officer, Curragh Camp, visited the Naas Union. He found the 'five wards that were set aside for the unfortunate women, all of which were clean. There were 34 women occupying the wards and twelve were sick and put in two separate wards apart from the others'.[113] But that they were not all confined to their beds had led to a motion at a meeting of the Naas Union on 13 April 1867. It referred to the fact that some of the 'women of immoral character' were going in and out of the workhouse as they pleased, and that this should not be allowed.[114]

Early in the January of 1868 the board met to consider the suggestions of the union committee of 21 March 1867 that a portion of the workhouse block should be allocated to the women of bad character.[115] In the summer of that year the women were moved to the new department, and it was ordered that 'the building they were in was to be knocked down and the materials used to enclose the burial ground attached to the workhouse.'[116]

A report from the committee on the pathology and treatment of V.D. decided in 1868 that the evidence which they had considered was conclusively in favour of the absolute necessity of subjecting prostitutes to compulsory examination, and of their immediate separation from the community, and their seclusion in hospitals until they were cured. This decision foresaw the opening of special hospitals to deal with the disease.[117]

That the journalist visiting the Curragh plain in 1867 was not exaggerating in his horrific descriptions of circumstances on the plain is obvious from some of the evidence given before the Curragh commission at Newbridge courthouse in 1866. The nuisance of the women to the local residents was then expressed clearly, and the inadequacy of the law in relation to the whole question of trespass and abuse of the plain apparent.[118] The necessity for proper legislation was widely accepted, but another two years were to pass before it was enacted.

If the sexual immorality of the soldiers was an accepted part of army life, the morals of the officers could also be questioned. The well documented incident of the Prince of Wales and the actress in the Curragh camp quarters of the general commanding in 1861 exposed the recreational activities of some of the gentlemen. It is still traditionally believed in the Curragh area that a brothel for the officers operated for a time at a named country house near the camp, and that immoral activity was also known to be practiced in other country houses in the vicinity. The attitude of the county gentry to such behaviour is not known, but the ordinary people were said to be scandalised by it.[119]

Ten years after it was first proposed, work on the building of the Lock hospital on a one and a half acre site leased forever from the duke of Leinster on the lands of Broadhook farm, close to Kildare town on the road to the Curragh, began in the summer of 1868.[120] It was estimated that the project would cost £6,048, and the accepted tender was for £5,200. In July of the following year when the work was completed the total cost was found to be £6,105.1s. Very soon the main road, from which the avenue to the hospital opened, was to become known as Hospital Road.[121]

The hospital consisted of a group of one-storey slated buildings, linked with corrugated-iron roofed corridors, and with a separate wash-house and laundry (Fig.18). There were coal sheds, a water tank, and a dead house. A sewerage pit was made some distance away. The two wards, to accommodate 20 and 16 beds, the matron's quarters, medical stores, examining depot, segregation building, porter's quarters and policeman's waiting room were built of Athy brick. Part of the segregation building was adapted for use as a Roman Catholic chapel. The lighting of the buildings was by oil. Two outdoor recreation areas were laid out, and there was an appropriate number of baths and toilets attached to the wards.[122]

The plans for the hospital had been made by the School of Military Engineering at Chatham for the War Department. To administer the institution a staff consisting of a matron, three nurses, a steward and a porter was appointed. After only 20 years of use the hospital was closed in 1887, and the buildings remained in the ownership of the War Department.[123]

Just before the Christmas of 1869 the usual certificate necessary for the payment of the annual grant to Naas workhouse was processed. Staff Surgeon Evans from

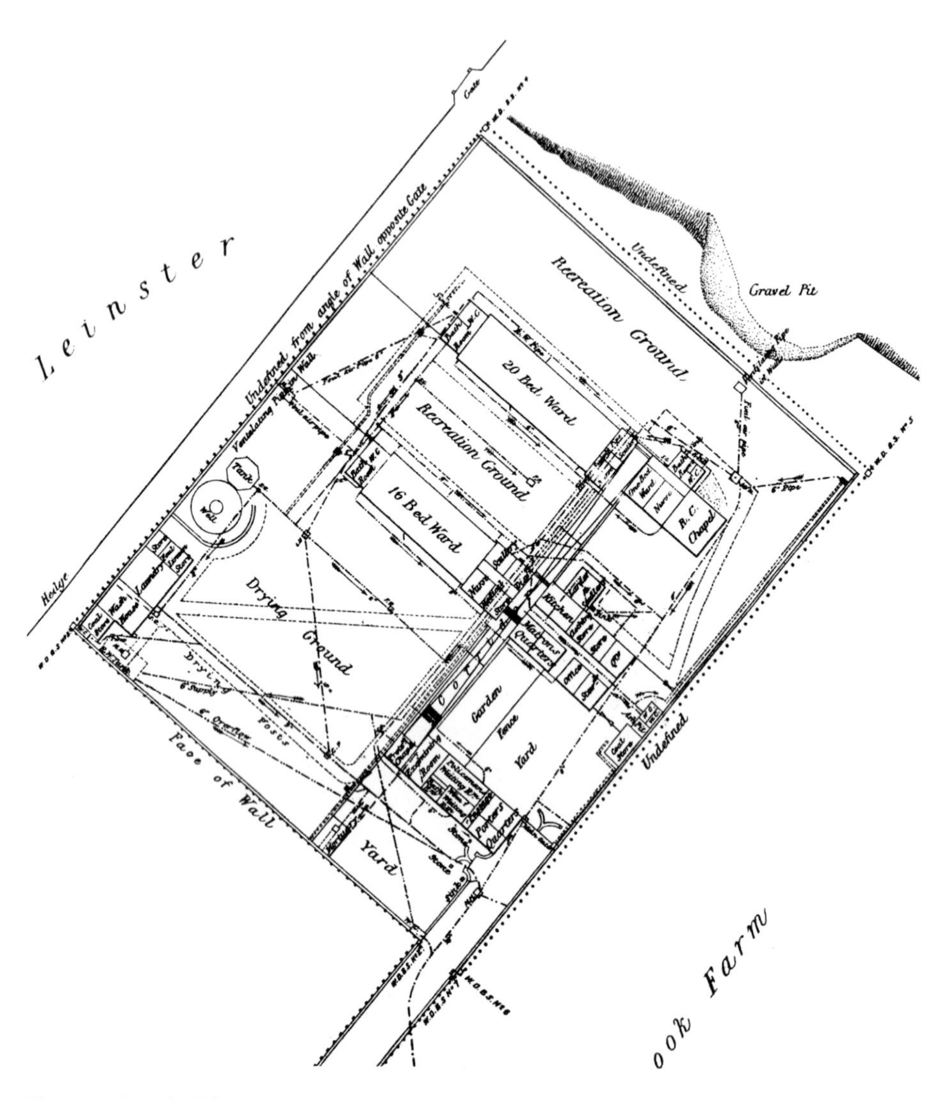

Fig. 18. *Detail of the plan of the Lock Hospital, Kildare 1882. The two wards face on to the recreation ground with the R.C. chapel in an extension from the larger ward, and the matron's quarters from the other ward. Courtesy Military Archives.*

the Curragh had inspected the wards occupied by the unfortunate women, and he found them satisfactory.[124] Whether the accommodation in the workhouse would be still required when the Lock hospital was functioning was being questioned by the deputy quartermaster general. On Christmas eve the military secretary replied that there was no information of the opening of the Kildare hospital,[125] and on 1 January 1870 it was confirmed that the new hospital did not do away with the need for the wards at Naas. They were wanted for the destitute, immoral women, not those who were diseased, and whose presence in the county was attributable to the military. In severe weather there was great need for the Naas wards, and only the diseased women were admitted to the Lock hospital which was finally opened on 6 December 1869.[126] The fact that London did not know of the opening of the institution by Christmas was probably due to the delays in the postal system.

When the army medical report for the year 1869 was published it tabulated the hospital admissions per 1,000 mean strength at the Curragh for venereal disease patients from 1867. In that year there were 104 admitted, in 1868 the figure was 85, and for 1869 it was 88. [For Dublin the figures for the three years were 129, 139, 180, and for Cork 72, 61, 73.][127]

Three years after the hospital opened a pamphlet titled *The Moral Utility of a Lock Hospital* was published by William McGee. It set out to examine the benefits of the Kildare Lock hospital and other reformatories in Ireland for fallen women. The moral and physical condition and the behaviour of the patients admitted to Kildare were considered. These were found to be discouraging. The women, with few exceptions, were found to be 'Curragh Wrens', and these were described as 'a type of woman it would be difficult to form any fair idea of, without personal observation of their appearance and habits. For the most part the very dregs of their degraded class, from all quarters of the Kingdom, congregated together in the vicinity of a great military camp, without house accommodation of any kind, with no recognised place of even casual admittance, except the lowest wayside public-house, or the workhouse.'[128]

Then their huts and shelters in the furze bushes were described as 'such as would disgrace the aborigines of any country'. The women's habitual debauchery and drunkenness was deplored, and the efforts of a few clergymen and other interested people to reform them had proved to be futile. Over the years the inhabitants of the neighbourhood of the Curragh had become apathetic to the condition of the 'wrens', even though they were naturally distressed when it was reported that a woman was found dead in the snow, or from violence or neglect. But despite their sympathy for the unfortunates, the local people 'seemed to agree to speak of these abandoned women as a kind of dreadful and scandalous necessity, as being beyond the pale of human sympathy or help'. Indeed, it was even feared that the 'youth of the neighbourhood of the humble class', especially the girls, 'were so far influenced by the habitual aspect of vice thrust on them, that they had lost at least that modest simplicity of conduct and appearance that distinguished them a dozen years before, and in but too many instances had outstepped the line that divided them from their wretched neighbours.'[129]

The writer of the pamphlet then considered the opening of the government hospital at Kildare 'in the midst of a great deal of ignorant prejudice, as well as honest and intelligent opinions, for and against the movement'. The women were compelled to enter the hospital, and there to obey the rule, cleanliness, order, respectful behaviour, and employment of the institution. But the inmates fought against those conditions and the rule was found to be inadequate for the 'exceptional class of woman' there. The only real mode of punishment available was committal to gaol on a magistrate's order, and this was rarely resorted to. It had never been intended that the hospital should be a prison, and the women at every opportunity climbed the walls and roof to hail the passers-by on the public roads. Most of the women had not been in a place of worship for years, except while jailed, and they had a fear of medical treatment, resenting it as 'a series of experiments made for the purpose of tormenting them'.[130]

Unaccustomed to living inside a house, or to normal daily routine, the women resented being restrained or being asked to work; they would not eat at table, or keep their persons or clothes tidy, and they showed little respect to others. They found it difficult to relate 'to the respectable of their own sex and were in turn abashed or insolent, sullen or abusive.'[131] Gradually, however, 'the forbearance of the officials' brought about an improvement in the women; the patience and fairness of the staff and the kindness to the sick encouraged them to behave better. One characteristic of the women which was remarkable was their kindness to each other in sickness; even virtuous women, the pamphleteer believed, could take a lesson in tenderness from the poor despised wren. In time the women learnt to sew and knit, arts which they had forgotten; they used less bad language and they began to take a pride in keeping their quarters clean. In fact many, even some of the more hardened ones, seemed sorry to leave the hospital, and other women from the Curragh sought entry to the institution.[132]

The anonymous writer of that pamphlet had an intimate knowledge of the Lock hospital, and may have been one of the governors or the staff. The christian understanding shown of the predicament and character of the women was exceptional, and the use of the terms 'patients', 'unfortunates' or 'wrens', rather than that of prostitute, indicated the writer's strength of feeling for the women. The final paragraph of the treatise appealed 'to the charity and common sense of every thinking person' to support the establishment of places such as the Lock hospital 'to raise a lost and fallen and too numerous class'.[133]

An official report on the operation of the Lock hospitals issued a year after the publication of the last mentioned pamphlet echoed many of the sentiments of that document. In fact it might have been that the writer was one of the gentlemen involved in preparing the report. The report accepted that following admission to the Kildare hospital 'even the wrens have been humanised and softened, and many who, until their admission to the hospital had scarcely ever known the shelter of a roof, and who three years hence were fierce as wild animals, liable to give way to fits of uncontrollable passion or sullen obstinacy on being requested to assist in the work of the wards' had been taught to become useful domestic servants. Some of the women discharged from the hospital had emigrated to New York and from there they wrote back to their friends and companions offering advice, and assistance for them to emigrate to America.[134]

The report noted that the majority of the inmates of the Kildare and Cork Lock hospitals were Roman Catholics, and that priests had been appointed to those institutions. Mass was celebrated on Sundays and holy days in the chapels of the hospitals which had been fitted up at the cost of the War Department. Protestant women in the hospitals were visited by a clergyman of the parish.[135]

In the hospitals the women were employed in making-up clothing or knitting, and 'an abundant supply of long woollen stockings had been knitted by the women, who had never used a needle in their lives before being taken into the [Curragh] hospital'. When they were discharged the women were sent to refuges, or into the care of relatives or friends.[136]

An important step was taken in the removal of the women from the Curragh in January 1873 when the lord lieutenant issued an order in council making a bye-law which made it unlawful 'for any night walker or common prostitute to locate herself or trespass on or resort to the Curragh of Kildare for the purpose of prostitution'. A fine not exceeding £5 could be given for each offence. The application of the law was to bring about a transformation of the plain, as the women moved into nearby areas. Combined with the removal of women to the Lock hospital, the new bye-law was the beginning of a reduction in the number of prostitutes through official control.[137]

That the women who continued to live around the Curragh had not changed their habits was obvious from the court reports in the local press. In the summer of 1876 a soldier charged with the rape of Anne Cullen was discharged, but a new summons was served for assault. There were several other cases concerning the women, including a serious assault on Maryanne Williams,[138] and one against a woman found in the officers' kitchen in the camp. She had been a patient in the Lock hospital, as an official from that institution testified, but clearly not one of those who was reformed.[139]

In 1878 the workhouse at Naas was still receiving the women from the Curragh, and the minute book of January of that year records that General Seymour, commanding the Curragh District, noted that the accommodation there for the unfortunate women was good and sufficient, and recommended that the guardians should get the usual payment.[140]

By the end of 1879 General Seymour had found it necessary to report to higher authority that the prostitutes had returned to French Furze, 'a few hundred yards from the open camp, from which they were successfully evicted some years ago . . .' Their presence there he regarded as bad for the discipline of the camp as the men were breaking bounds, and the women 'invaded the camp nightly'. Prosecuting them for trespass did little or no good. An order had been given that the soldiers should not speak to the prostitutes on the public roads, and the mounted police and the picquets endeavoured to enforce that dictum. Seymour was of the opinion that the civil authorities were responsible for the matter as they were not hunting the women off the plain. He suggested that the Irish government should be asked what they were doing about the situation as 'the proximity to the camp of these women was most objectionable'.[141] The reaction of the government to the general's complaint was the issuing of an order from the lord lieutenant that the bye-laws of 1873 should be enforced.[142]

In a report on the nuisance caused by the women from the Marquis of Drogheda, Curragh Ranger, dated 11 February 1880, he held that steps were being taken to enforce the bye-laws. But he believed that 'the Curragh was the best place on which those unfortunate women could be located'. Drogheda was convinced that it was impossible to banish them from the neighbourhood of the camp as ever since the bye-laws were passed in 1873 the women lived in ditches in the lanes and by-ways in the vicinity of the Curragh, and chiefly in the direction of Kilcullen where they had been a grievous nuisance to the persons living near. In the autumn of 1879 the magistrates at the Kildare petty sessions gave orders that the women should be prosecuted if they remained where they were and the consequence was that they returned to the Curragh.

Lord Drogheda believed that the prostitutes did not create annoyance to the parishioners of the Reverend John Nolan [parish priest of Kildare who died in December 1880.][143], and he was certain that they caused less inconvenience to the general public if left where they were than if they lived in the neighbourhood of the camp. It was his understanding that in every country in the world, except England, the presence of prostitutes in the neighbourhood of camps was sanctioned under proper supervision. 'This,' he said,'is also the case among our own troops in India, and there was reason to believe that even at Aldershot their presence is connived at. At the Curragh, however, Reverend John Nolan refuses to allow any of his parishioners to give shelter to any of those unfortunate women'. His lordship emphasised that in making that statement he was in no way questioning the conduct of the reverend gentleman. Requesting the chief secretary to lay his letter before the lord lieutenant, he asked that he might also be informed that his instructions would be obeyed, but with regret, as he 'was sure that the interests of public morality and decency are better served by these women being allowed to remain on the Curragh, rather than in its vicinity'.[144] His views were supported by the commander of the forces who said that he had 'no objection to these unfortunate women remaining where they are at present located as long as they do not become a public scandal'.[145]

In a confidential memo the deputy adjutant general wrote that 'as it is impossible to banish [the women] altogether, if any remote and secluded spot on the Curragh could be found to which these unfortunate women could be restricted. The commanding General Sir John Michel felt that such an arrangement would be better than that they should occupy the approaches to the Curragh'. The general officer commanding the Curragh should be asked for his views.[146]

Major General Fraser, who commanded the Curragh brigade, communicated to say that during his command he received no complaints on the subject. He now understood that the women were quietly behaved, and healthy. His suggestion was that a secluded place beyond the railway line could be appointed for them, and that the present state of affairs should be allowed to rest. The deputy adjutant general agreed with Fraser's views, and he knew that privately Fraser had asked that the authorities should know that the women had been asked to be quiet, and they would be hard pressed if not. Further, the poor inhabitants of the Curragh borders were harbouring the women, 'which is the best thing that can happen, and will give the priest the chance of reclaiming them!'[147]

Early in April 1880 the chief secretary's office was asking if anything had occurred on the Curragh to alter the opinion of the commander of the forces with respect to the women? He was told that a notice had been posted at French Furze on 21 February prohibiting their presence there, and the police were enforcing it. The Curragh bailiffs were held to be the proper persons to initiate proceedings for trespass against the women.[148] On 13 April the commander of the forces let it be known that he had no occasion to alter his opinion as expressed in his minute of 1 March, and that as long as the women did not become a public scandal they were best left as they were.[149]

If the military authorities were satisfied with the plight of the women and the inevitability of their presence, the guardians of the Naas workhouse were not happy with their charges. Some of the prostitutes in the workhouse were discharging themselves every second or third day 'for improper purposes, and who remain out for the day and part of the night'. Then they sought re-admission, and soon repeated the discharge process. At the board meeting of 7 April 1880 it was asked if the relieving officer could limit the issue of tickets of admission to those women who were abusing the system.[150] The outcome of that proposal is not known, but in January 1881 the accommodation at Naas was again found suitable, and certified for payment of the government grant.[151] A comment made in public by Patrick Driver, a member of the Naas Poor Law Union Board, has become part of the folklore of the county. Driver described the many sexually diseased British soldiers in the Curragh as having been 'crippled under Venus, rather than Mars'.[152]

The conduct of the women ensured that they were constantly in conflict with the law, such as the two 'of the unfortunate class' who were before the county Kildare assizes in July 1885. They pleaded guilty to stealing a purse containing 15/–, from Pte. Arthur James Alsopp of Naas barracks, who had the purse attached to the inside of his belt. One woman was acquitted, but Mary Ryan got two months imprisonment. It was stated that she had been 40 times in prison and had been convicted on about 50 various different offences.[153]

In October 1886 the Naas Board of Guardians held a special meeting to consider the proposal from the war office that the Lock hospital at Kildare should be closed.[154] The official decision was that it would be closed on 1 January 1887, when the average number of women then in the hospital was 24. The guardians were not pleased with the decision, and the medical officer of the Naas workhouse gave as his opinion that 'there was nothing as objectionable as to bring in these persons to the [work] house. It was not right for anyone to force those people into the house, no matter who they were'. He thought it would be most objectionable. If they came to the workhouse they would have to be treated separately and would require trained and skilled nurses. New wards would have to be built. He believed that the women's presence in the house would deter many poor people from coming to the establishment. One member of the board thought that the suggestion that the women might be placed in the workhouse was an insult to the board, and the Baron de Robeck agreed that the women were the responsibility of the government. An unanimous resolution was agreed 'that the Naas Board of Guardians express our entire disapproval of in any way taking care of women patients in the Lock hospital'.[155]

By the repeal of the *Contagious Diseases Act* in 1886 the government Lock hospitals at Kildare and Cork were closed, and the Westmoreland Lock hospital in Dublin was then the only one left in the country. Commenting on the possibility that the Dublin hospital might also be closed *The Lancet* said 'it would be a great disaster, since the Lock wards of the general hospitals are so few, so far between, and contain so few beds as to be a mere "drop in the bucket"'.[156]

The persistence of the problem of prostitution was again apparent in the evidence given before Newbridge Petty Sessions in January 1887 when Mary Costelloe had two privates of the Royal Artillery tried for indecent assault. One of the men, Robert Condon, had brought the woman into the barracks on Christmas night. Costelloe said that 'in the morning he dragged her out of bed and put her out of the barracks', and in the process of which she had lost her money, which amounted to 6/8. Evidence was given that she was drunk, and the case was dismissed. The chairman of the sessions said that 'statements from women like the prosecutrix required corroboration, and he thought that she should be charged with trespass in the barracks'.[157] The attitude of the chairman towards the woman did not seem to take into account that access to the barracks was controlled and that when she entered there it must have been through a gate guarded by a sentry or policeman. The responsibility of the military in the matter was not questioned.

Almost a decade later three men of the 1st Yorkshire Regiment in the Curragh camp were charged before Capt. Forbes R.M. at Lumville on a somewhat similar offence. The evidence was that on 5 November 1896 while Bandsman Bush, Derbyshire Regiment, was on duty in 'K' lines he heard noise from the lines occupied by the Yorks. When he investigated he saw about 30 of the soldiers with a female who was screaming and being dragged along towards the stables at Donnelly's Hollow. She was almost naked, and there was blood on her clothes which were nearby. About ten of the men were arrested, and three of them were remanded by the magistrate. The girl was then said to be in an acute state in Carlow asylum.[158] Unlike the barracks at Newbridge, there was no wall around the camp, and ease of access by civilians was a constant problem. But in the instance mentioned, the soldier on duty was alert and took action on the trouble which he encountered.

An encounter with some of the 'wrens' was recalled by Tom Garrett in recent times. He was born near Brownstown in 1901 and he remembered that one morning when he was aged about eight or nine and on his way to school in Kildare, but not by his usual route, as he walked by Frenchfurze one of the women stopped him and asked that 'he bring a bottle back with him', but he went home by his normal way and so did not meet her again. His recollection was that about five or six of the women lived in an old house near Corrigan's Cut or in the open at Frenchfurze, and that sometimes they were admitted to the poor house. He believed that they departed when the British army left.[159]

That such may not have been the case was remarked on by a retired Irish army officer who had served in Newbridge in 1922. He referred to patrols which the army was required to make around the town, and which he disliked on account of the encounters with the prostitutes who loitered there.[160] Some of the houses in the back streets of the town had been placed out-of-bounds to the troops, while an area beside the town at Morristown Biller where the women congregated was known colloquially by the equestrian slang term 'the bucket'.[161]

The presence of the prostitutes on the Curragh, in its environs and in Newbridge, was decried from at least 1844. In county Kildare evidence of the condemnation of the proclivities of the women by the parish clergy exists from early in the nineteenth century, and again from the later decades. The reform of the women was attempted by well-intentioned individuals in the county in 1865, with little success. But when the Kildare Lock hospital was opened in 1868 not alone was it believed that the venereal diseases were being controlled, but that many of the women were being reformed.

By 1871 branches of two of the English associations formed to force the repeal of the Contagious Diseases Acts had been established in Ireland, and in the following decade the social purity movement, which was aimed at reforming men's attitudes, was active in Dublin. It was promoted by the Church of England, while in the Catholic church the confraternities supported the encouragement of virtue.

The moral effect of the existence of the prostitutes was believed to lower the status of the neighbourhood, to scandalise right-living people, and to be an offence to the sensibilities of the officers' ladies if they came within their vision. The physical effects of association with the prostitutes were apparent from the high incidence of venereal disease among the troops, and 'the apparent weakness of the British army in the aftermath of the Crimean War'. The contagion of women and men of the locality was also feared.

The nuisance of the women was expressed in evidence before the Curragh Commission of 1866; landholders beside the Curragh plain complained of trespass, theft, intimidation, depreciation in the value of their property, injury to their stock, and moral damage to their families. The military themselves wearied of the invasion of the camp by the women, and the impossibility of excluding them from the open streets and billets, and their bad behaviour, drunkenness and crime within the camp itself.

While the commanding officers of the district deplored the behaviour of the women, they accepted that their presence was inevitable. It was felt that if they were less brazen in their appearances, especially on important occasions such as field days and reviews, and if they confined themselves to an obscure area of the plain, that it would be tolerable.

Proposals that the women should be under military control, as in certain foreign stations, were not adopted. The constant appearance of the women, especially those that were regularly apprehended, before the local courts was a bother to the magistrates, and an unending task for the police.

The contagious nature of the venereal diseases was of concern to both medical and other residents of the area. It was believed that local women who associated with soldiers were becoming infected and thus the civilian population was becoming contaminated. When the opening of a special hospital for the treatment of the prostitutes was proposed, there was widespread opposition to its location, but the possible financial benefits to the town of Kildare were also appreciated.

Trespass on the Curragh was likewise resented by the Crown, and in 1862 a parliamentary paper was issued on the subject. It showed that the number of convictions for trespass in the two years from 1 March 1860 to 28 February 1862 was 277; of those 113 had been heard at the Curragh court and 164 at Newbridge Petty Sessions. Three women convicted of trespass were Theresa Barnes, Mary Anne Byrne and Anne Fitzpatrick. Both of the former were described as 'notorious prostitutes and old offenders'. Anne Fitzpatrick was arrested at 10 p.m. on 13 April 1862 near the officers' quarters by the assistant Provost Sergeant Thomas Gee; next day she was brought before the magistrate at the

Curragh and convicted of trespass, and remanded. She was described as 'an old offender and a discharged convict', and was sentenced to six years penal servitude at Naas Assizes for robbing an officer's quarters.

The local government officials of the county were concerned at the additional expense caused by the existence of the women. If they were sent for medical treatment at the workhouse at Naas, or sentenced to a term in the town jail there was additional cost to the taxpayer. The relieving officer from the Curragh area was constantly involved in the plight of the women, and when some of them died in disgraceful circumstances the officer was subjected to severe censure. Transporting the women to Naas was another extra cost, and though the War Office contributed annually to the Naas Board of Guardians there were still regular complaints associated with the women.

Undoubtedly the problem of such camp followers was part of the military presence throughout the empire. Garrisons in the cities and towns elsewhere in Ireland would also have had the resident or transient females in proportion to the strength of the garrison. The difference in county Kildare was the permanent presence of thousands of men between the Curragh, Newbridge and Naas, and later Kildare town itself. In the months of the summer training periods there could be up to 10,000 men on the plain and in times of war the totals could be higher. The number of prostitutes also varied from as low as 26 in 1861 to estimates as high as 500 in 1865. One hundred was given as the normal summer figure.

The rural nature of the Curragh camp made the activities of the women more obvious to the public, and the destitution of their appearances and the squalor of their habitations made them especially objectionable in the eyes of the local residents. If the women were variously referred to as prostitutes, fallen women or unfortunate women, the local appellation of 'wrens' was especially appropriate as the habitations of the women amongst the furze bushes on the plain were said to resemble the nests of those almost invisible little birds. When the unfortunate women retired to their shelters in the ditches or the gorse they too became invisible, but the general public was forcibly made conscious of their condition when it became known that a girl gave birth or died in such circumstances. The critical notices that such deaths received in the British and Irish press focused attention on the plight of the women on the Curragh, and inevitably influenced the opening of the Lock hospital at Kildare which was the first positive move towards relieving the lot of the Curragh 'wrens'.

As to the identity of the women themselves, they were described in 1865 as being 'Irish females' and in 1867 as 'all Irish' but in 1872 as coming 'from all quarters of the kingdom'. Of 21 women named between 1862 and 1887 18 had Irish surnames. Reports on the Lock hospitals at Kildare and Cork in 1873 suggested that the majority of the inmates were Roman Catholics, and that in the former place a room had been adapted as a Roman Catholic chapel in 1882. That the majority of the women were Irish Roman Catholics was to be expected as the greater part of the population of the country was of the same persuasion. The national census of 1861 had shown that the Roman Catholics had an absolute majority over all other religions except in four of the counties in Ulster.

Chapter 8

The Social Scene

If the breaking of the law by a small section of the soldiery and their followers was a negative social result of the occupation of the Curragh camp and the other barracks in county Kildare, there were also positive social aspects to the presence of the military, their families and the many civilians who were employed by them. The financial benefits generated by the military to the population of Kildare as a whole were considerable, but the pleasant and varied social life created by the transient army men and their families was also a major social boost to the county.

The recreational activities of the military brought excitement to all classes of Kildare folk. The county gentry welcomed the officers, many of whom they knew from their own service in the army or militia, or through family connections. The hunting field greatly benefited from the newcomers, and the race course, polo ground, cricket pitch, and shooting parties likewise prospered. The seasonal balls and other entertainments in the camp were attended by the socially acceptable families, and the reciprocal pleasure of the young ladies and the subalterns added to the excitement. Sometimes weddings ensued, or at least country house entertainments were enlivened by the young bloods.

In county Kildare in 1887 the 'Gentry, & c., of the Neighbourhood' of the Curragh Camp and Newbridge, as listed in the military monthly official directory, consisted of about 90 persons, some from the same families.[1] The list included gentlemen from as far afield as Maynooth, Carbury, Celbridge, Athy and Portarlington. All of the then resident county families, headed by the ducal house of Leinster, and the earls of Mayo, Drogheda and Clonmell, and a sprinkling of lords and ladies, led the field of military and civilian gentlemen who, out of a population of 75,804, were considered to be the social equals of the commissioned officers.

Leading the list was the duke of Leinster, the premier nobleman in Ireland, whose family, the Fitzgeralds, had been settled in the county for 700 years. The seat of the Leinsters was at Carton, Maynooth, and they owned 71,977 acres in the county. Men of the family, and of all the other county Kildare families here mentioned, had served or were serving in the line and militia regiments of the British army, and their sisters frequently married soldiers. There was also intermarriage between the families of the county nobility and gentry.[2]

Landed families settled in the county since the seventeenth century included the MacWilliam Bourkes, who had the title of earls of Mayo since 1785; their seat at Palmerstown stood on almost 5,000 acres.[3] Another family from the same root, the de Burghs of Oldtown, Naas, whose ancestor Thomas Burgh, the engineer and surveyor-general for Ireland, had settled there late in the century on an estate of less than 3,000 acres, and his brother William, of Bert, Athy, was the great grandfather of General Sir Ulysses de Burgh, 2nd Baron Downes. There were several other high-ranking military men in the family, the most recent being General Sir Eric de Burgh who served in the South African war and the First World war. The Weldons, of Kilmorony, Athy and the Borrowes of Gilltown and Barretstown, Ballymore Eustace, were respected landed and military families, also tracing their origins in the county to the unsettled times of the seventeenth century.

Though only established in the county since the eighteenth century the Clements of Killadoon, Celbridge, the Mansfields of Morristown Lattin, and the de Robecks of Gowran Grange, Naas, held respectively in the county 487, 4,542, 1,838 acres, apart from estates elsewhere in the country. The founder of the latter family was a Swedish nobleman who served with the French army in America in 1781; after marrying an Irish heiress he settled here and his descendants included many distinguished soldiers and sailors. They intermarried with other county families, and the 4th baron was a captain in the 8th Foot, a major in the Kildare militia, and Curragh ranger from 1892 to 1904. If not great military families, the La Touches of Harristown and the lords Cloncurry of Lyons, had provided brides for other officers, or had married soldiers' daughters. Examples were Emily Maria La Touche who, in 1865, married Lieut. Gen. Hon. Bernard Matthew Ward, and Valentine Browne, the 2nd baron Cloncurry, whose wife was a daughter of Major General George Morgan.[4]

At Moore Abbey, Monasterevan, which they inherited in 1725, the earls of Drogheda had 16,609 acres. They lived in great style, and were enthusiastic supporters of the turf; Ponsonby Moore of Moorfield, Newbridge, who owned much of the land developed by the military there, was a son of the 5th earl. The third marquess of Drogheda was lieutenant colonel commandant Kildare rifles, and ranger of the Curragh from 1868 to 1892.[5]

A comparative newcomer to the county was John Henry Scott, the 3rd Earl of Clonmell, who married Anne de Burgh, the daughter of General Sir Ulysses de Burgh 2nd Lord Downes of Bert, in 1838. He then purchased Bishopscourt, Straffan and added the 1,958 there to his 25,688 acres in six other counties. He had served in the 1st Life Guards. His daughter married Capt. George Fitzclarence R.N., and their four sons joined the Services. One of them, Brig. Gen. Charles Fitzclarence, was awarded a V.C. for service in Africa.[6]

Not mentioned in the 1887 directory was the important Conolly family of Castletown, Celbridge, as Thomas Conolly had died a decade earlier, and his heir was then but 17 years of age. He was killed in South Africa in 1900 when serving as a major in the Scot's Greys. Celbridge had earlier military associations when Col. and Lady Sarah Napier came to live at Oakley Park, near the home of Lady Sarah's sister, Lady Louisa Conolly. The Napier's three eldest boys all rose to the rank of general; Sir Charles was the conquerer of Scinde, and Sir William was historian of the Peninsular War.[7]

Lesser, though long established gentry families such as the Medlicotts of Dunmurry, the Wolfes of Forenaghts, the Kennedys of Johnstown Kennedy and the Sweetmans of Longtown[8] were included in the military directory, and all of those families would have socialised as equals with the officers of their own status.

Other socially acceptable persons included horse trainers such as Henry Eyre Linde and the Beaseley brothers, retired army officers, a few medical doctors and clergymen, but no farmers or trades people. The latter categories were to find their social level within the other ranks, if they wished to associate at all with the military.

But for the other ranks life was not so pleasant as it was for the gentlemen, and few of them left written impressions of their lives. Of particular interest for that reason is Fusilier Horace Wyndham's views of the Curragh where he served in the early 1890s. That he found the camp dull was understandable, as on his own admission he did not participate in the earthy pleasures of his fellow recruits. For him the excitement of an evening in Newbridge or the furze bushes was indeed *anathema maranatha*.[9] The rigours of recruit training and discipline, or the exertions of the drill season, would have made any training camp an unpopular place, but for the men, and especially for the officers, stationed in the Curragh for longer periods the situation was different. They integrated in the local society, and enjoyed the sporting opportunities of the district.

As Fusilier Wyndham aptly remarked 'reading-rooms and libraries, cricket and football matches, with shooting clubs and all other attractions enumerated therein, look very well on the Government circulars, but they are not exactly free gifts'. Just as the entertainments in the officers' and non-commissioned officers' messes were financed from members subscriptions and mess funds, the recreational needs of the men were supplemented from the monies deducted from their pay.[10] The varied sporting and social functions for all ranks, and which so greatly entertained the civilian neighbours who were invited to them, were at the expense of the military. In the officer class not alone had the soldier to be a gentleman, but he also had to have adequate private means to enable him to meet his social requirements.

The non-commissioned officers and men participated in the life of the county at a different level to the officers; they patronised the hostelries of the neighbouring towns, and which could be placed 'out-of-bounds' if they proved unruly.[11] An old resident of Kilcullen recalled that the non-commissioned officers from the camp drank in a special public house there, the men in another, and the officers in Bardon's Hotel.[12] The other ranks attended the race and coursing meetings, the cricket and football matches. Concerts, smoking parties and dances were given regularly by the various regiments in the barracks, and the invited civilians always enjoyed an evening in the mess or canteen. Regimental sports were always attended by local people, and on the cricket field the officers and men experienced a rare social mixing when they also met up with the teams from the civilian clubs. In the summer months the soldiers from the Curragh were permitted to bathe in the river Liffey at Kinneagh, but only after a notice had been published in the Duty Orders as only then would life-saving measures be in operation.[13]

That the regulations did not always prevent accidents was later proven by the fate of Robb, a fellow signaller with Frank Richards in the 2nd Battalion Royal Welsh Fusiliers in India. Robb had soldiered in China during the Boxer rising, and he afterwards boasted of having shot a terrified old Chinaman. Richards in his book *Old Soldier Sahib* tells us that after Robb was transferred to the home establishment he joined the 1st Battalion on the Curragh. There, one day 'he went with a lot of other men to bathe in a large pool. He dived into the water, where no doubt the ghost of the murdered Chinaman was lurking in wait for him. He did not come up from his dive and when his chums came to miss him they found him lying at the bottom near a big stone on which apparently he struck his head. They pulled him out half-drowned and unconscious, with a big bruise on his head, and carried him to hospital. The injury left half of his body permanently paralysed. I never heard what became of him in the end, but the doctors did not hold much hope for his recovery.'

It is not known if a swimming pool was included in the scheme proposed by the enterprising Dr Richard Barter from Blarney in the autumn of 1860 when he considered the conversion of the old Stand House on the Curragh to Turkish or Roman Baths. He emphasised 'the utility of such baths to the public as well as to the troops stationed at the Curragh and the barracks at Newbridge'. He even suggested that he might erect baths for horses which he had found 'to be very beneficial',[14] but his plans came to nought, and the military continued to do their bathing at Kinneagh.

A major source of entertainment for the civilians of all classes were the military revues, field days, manoeuvres or ceremonial celebrations for royal birthdays or other commemorative occasions held on the Curragh, and which attracted thousands of onlookers from the county and beyond.

For the officers one of the great attractions of service in county Kildare was the prospect of hunting with the county and neighbouring packs of hounds, and of participating in or attending the race meetings on the Curragh and at Punchestown, as well as the regimental and other point-to point fixtures. Kildare was a sportsman's paradise, and several gentlemen recalled in later life the idyllic nature of their stay there (Fig. 19).

The Kildare Hunt Club, which had been established in 1793,[15] was the principal sporting outlet for the officers from the Curragh, Newbridge and Naas barracks. Membership of the hunt included regular officers living in rented houses locally who might have individual membership or gentlemen who hunted on a mess subscription from their regiment. Amongst the county gentry were many men with militia ranks, or who retained their ranks from army service. The general officer commanding in Ireland, and the general officers from the Curragh or Newbridge were usually members, and listed in the honorary members at the end of the nineteenth century were the duke of Connaught and Field Marshal the Earl Roberts.[16]

During the mastership of Sir John Kennedy (1814–1841) officers from the barracks in Newbridge hunted regularly with the Kildares. One gentleman kept a hunting diary, and in it he remarked that 'it was then the custom of the members of the hunt to test the quality of the officers of every newly arrived regiment, and as a matter of course one of the [regiment] had to go through this ordeal. For a fortnight after our arrival [at Newbridge] the Kildares met at the covert nearest the barracks, and the officers, myself among the number, all well mounted, rode out with a firm determination of proving their metal . . .' The

Fig. 19. *Major Richard St. Leger Moore, Master of the Kildare Hunt 1883-1897, with the huntsmen on the steps of Killashee, Naas. He soldiered in the 5th, 9th and 12th Lancers, and commanded the 17th Battalion Imperial Yeomanry in the Boer War. His wife was Alice Tynte from Tynte Park, Dunlavin. Their son also served in the 12th Lancers.* Photo: *Courtesy of La Sainte Union des Sacres Coeures, Killashee.*

diarist wrote of the excitement 'of the peasants' as they followed the hunt, especially 'a fellow named Sugu', who, the morning after the regiments first outing, called into the barracks to compliment the officers on their conduct in the field, and who assured them that 'we might depend on being protected from harm and hospitably entertained, an assurance which was afterwards fully realized during our stay in the country'. It is not clear from the last observation whether Sugu was going to protect the military from the hazards of the hunting field, or from local hostility.[17]

An indication of the support given to the hunt by the military is evident from the club's annual accounts. For the year ending 1 May 1866 there were 12 officer members, including General the Right Hon. Lord Strathnairn, commanding Dublin and North District, and Lord Seaton a former holder of the same office, and Lieutenant Colonel Forster *aide de camp*. Donations totalling £153 were given to the funds by the Scots Greys, 5th Dragoon Guards, 10th Hussars, Col. Hon. L. Curzon Smith, Major General A. Gordon and Major General Key. The total of members' subscriptions was £1,447.5s.[18] The 9th Lancers donated 50 guineas to the hunt in 1868–9.[19]

Over 30 years later, for the year ending on 1 May 1899, there were 68 officers listed amongst the 182 members. Regimental donations amounted to £420.15s and individual officers gave £48.8s., totalling £469.3s of the £632.8s donated.[20] During the Boer War, in 1901, there were six officers on the hunt committee of 14, and 71 amongst the 199 members. Of the total of £547.17s donated that year £125.15s was from the military.[21] Regimental subscriptions in the year ending in May 1913 totalled £518.7s and there were 77 individual subscriptions from officers in the list of 165 members the total of which subscriptions was over £2,300. The fortunes of the Kildares depended, to a certain extent, on the strengths of the stations during the hunt season.[22]

The participation of H.R.H. the duke of Connaught in the Irish hunting field for the season of 1876–77 made it an especially memorable one for the ladies and gentlemen who hunted, and for the many other socialites who attended the meets and followed the chase in their carriages. The officers from the Dublin and provincial garrisons turned out in strength for the local meets, and in Kildare this was especially so. In his regular letters to *The Field* the hunt correspondent M. O'Connor Morris reported on the meets of the main hunts, including the Kildare Hunt and the Kildare and the Newbridge Harriers; in 1878 he published his reports in a book dedicated to the duke of Connaught.

O'Connor Morris realised the importance of the soldier members to the hunts, remarking that the Kildare and Newbridge hare hounds 'had proved a most valuable adjunct to the large camp at the Curragh and the cavalry regiment at Newbridge, training half the regimental horses and giving their owners a few capital gallops. As a matter of title, I believe I am correct in stating the Kildare pack claim the greater part of the Curragh as their prescriptive arena; both packs, however, drive hares over the vast plain from the surrounding border lands; and game is so scarce on the Curragh that the two packs meditating an odorous assault on the single hare of the grassy common might remind one of the two kings of Brentford distilling the sweetness of a single rose.'[23]

The opening day of the 1876 Kildare Hunt Club season at the traditional meeting place of Johnstown Inn attracted a huge field, with nearly 300 half-crowns being paid in cap money. General Seymour and Captain Lee from the Curragh Brigade led the large military contingent which included officers from the 7th Dragoon Guards, the Inniskillings, the 3rd Dragoon Guards, and the Horse Artillery from Newbridge. Captain St Leger Moore had returned from India for the season, 'forsaking his wonted pastimes of tent-pegging, lemon slicing, and all those Indian feats of horsemanship for which his regiment is so celebrated'.[24] When the Kildares met on 4 November at Maynooth a special train with horseboxes came from Dublin bringing 'half of the Dublin garrison, including the Inniskillings'.[25] Later that month, at Celbridge, the duke of Connaught was present, with his equerry Captain Lord Maurice Fitzgerald from Carton. There was again a large military presence in the 200 strong field.[26] As O'Connor Morris commented 'the Curragh and the Dublin Garrison swelled the assembly'.[27]

The soldiers were to be seen at every meet of the hounds, and when on 6 January 1877 it was at Courtown Gate, near Kilcock 'nearly all of Kildare, including the soldier population of Newbridge and the Curragh' were there.[28] 'Soldiers' drags [vehicles], flashing sunlight, war horses innumerable, and all the pomp and pride and circumstance of war's image' enlivened the meet at Rathcoole on 26 January, in what was seen as 'a field day for the garrison and staff, cavalry, artillery and rifles; but no corps muster so strong as the Inniskillings . . .'[29] Such military glamour was evident at every meeting, and especially at the hunt ball which closed the season. As always, that in April 1877 was held in the Town Hall of Naas and it was especially grand as the Duke of Connaught was there.[30]

Officers from Kildare also hunted with the fox hounds in the neighbouring counties of Dublin, Meath and Carlow, and with the harrier packs. In 1876 there were harriers at Kildare and Newbridge, and sometimes drags were laid; such was the case by the Newbridge Harriers on the morning after the Kildare Hunt ball in 1877, a meeting seen as 'a happy thought for the belated dancers of the previous night'.[31] That pack had been formed in the Newbridge garrison in 1877 under the mastership of Colonel the Hon. William Forbes, and its country was around the town and the Curragh. Two packs hunted near Monasterevan, that of Captain Brown of Riverstown, and that of Thomas Waters of Kilpatrick which were known as the Kildare Harriers. The North Kildare Harriers were formed in 1906 from Sir George Brooke's Harriers, a private pack which dated from 1869.

Thomas Conolly M.P. of Castletown, Celbridge, kept the Salt Harriers,[32] and the Naas Harriers of 1918 were believed to have originated with the Newbridge garrison pack founded 40 years before when a regiment transferred to Naas.[33] Other private and regimental packs operated from time to time and, in all, the military gentlemen could hunt either the hare or the fox four or five days a week.[34] If a meet was over twelve miles distant, and too far for hacking, or if there was not a suitable railway station close to the meet, gentlemen who would have preferred the fox hunt satisfied themselves with the harriers, if convenient.

As well as being regular attenders at meetings of the Kildare Hunt, the officers, when possible, also followed the regimental hunts or the hunts of neighbouring counties. Sometimes they entertained members of the Kildare Hunt in barracks, as in January 1865 when the 4th Royal Irish Dragoon Guards gave a breakfast to several gentlemen at the mess in Newbridge barracks. Included amongst the guests were such county notables as the Marquis of Drogheda, the Baron de Robeck, H.G. O'Kelly, and E.A. Mansfield.[35]

When Captain Thomas Gonne 17th Lancers was posted as brigade major of the Dublin Division and of the Curragh in 1868 he participated in the sporting life of the county. His daughter Maud in later life recalled living in 'a little wooden hut in Kildare with a verandah running round it and a sun porch where Tommy grew flowers . . . soldiers in red coats constantly came and went. A grey donkey lived in a paddock. There was a kennel of fox-hounds reputed to be dangerous to any but their keepers near my uncle's house. I succeeded one day in getting into the yard. No one was there and the key was in the door of the kennel. When the kennel man returned he was horrified to find me happily playing in the kennel with the hounds.' When she was older, and her father held the post of assistant adjutant general in Dublin, Maud returned to Kildare to follow the hounds.[36] In 1903 she married Major John MacBride, who was executed in 1916 after having participated in the Easter Rising.

By 1881 the political situation resulting from the land war was disrupting hunting activity. When the Kildares met at Dunmurry in the last week of November 1881 a large number of tenant-farmers were assembled there and they declared that they would allow no hunting in their country (Rathangan, Boston, Drennanstown and Allen) 'while the suspects (Parnell and others) were imprisoned'. Despite the remonstration of the huntsmen the farmers would not disperse, and the meet was abandoned. The master gave the word to move on to the safer ground of the Curragh for the days sport.[37] Such disruption of the sport continued throughout the country, and in 1882 Punchestown races had to be abandoned. That autumn at a meeting at Kildangan the Kildare hunt was again disrupted,[38] but in the following April the Punchestown meeting was again held. By then the Land Act of 1881 had come into effect, and the formation of Parnell's National League brought hope to the people.

Both the Kildare Hunt and the Newbridge Harriers had military rank-holding masters in 1885: Major St Leger Moore was the master of the Kildares, and Capt. Gunston of the harriers.[39] The attendance at meets of both hunts always included a large number of officers, and many of their wives. In 1885, when the Kildares met at Oldtown, Naas, it was noticed that 'several officers from the Curragh and Newbridge' were there,[40] and at the opening meet of the Kildares in 1886, as always held at Johnstown Inn, the participation of the officers of the 11th Hussars was mentioned. The next meeting of the hunt that season was to be at Flagstaff in the Curragh.[41] The 5th Dragoon Guards had their own pack at Newbridge in 1885, and they received hounds from the local harrier club.[42]

There was also hunting with private packs such as that of Col. Crichton of Mullaboden, near Ballymore Eustace. As well as being secretary of the Kildare Hunt he was a member of the County Kildare Archaeological Society. In the former capacity he organised a testimonial for the pack's huntsman who was retiring; the list of subscribers was headed by the officers of the Royal Horse Artillery from Newbridge, and amongst others were one major general, two colonels, one lieutenant colonel, five majors and two captains.[43] The participation of the officers in the hunt greatly increased its membership, and there was an annual Meet in the camp, that in February 1899 being at the 8th Hussars Mess in Ponsonby Barracks.[44]

The importance of the military to the membership of the Kildare Hunt was evident when they were absent in times of war. At the opening of the hunting season in November 1899 it was accepted that the hunt, like many others, would suffer as so many of the gentlemen had gone to fight the Boers in South Africa.[45] The Great War of 1914–1918 caused even greater problems for the Kildare Hunt Club as a large proportion of the costs of the hunt had been met by the military. The club found it necessary to make an appeal for funds, and the Kildare County Council supported the cause.[46] Again the hunt survived the emergency, as it did the disruptions caused by the War of Independence, and the eventual withdrawal of the army from the county in 1922.

The popular association of the officers with the hunting field was expressed in the verse of a ballad eulogising the Kildare hunt which mentioned many of the regular followers in an affectionate way. This was the verse:

> It's Captain McCalmont, I knew he'd be there,
> An officer bould in the Royal Hussars
> A suitable place for a great son of Mars.[47]

Captain Hugh McCalmont, (1845–1924), 7th Hussars, from county Antrim, was later to become Major General Sir Hugh McCalmont, and a veteran of the Ashanti War, Russo-Turkish war, South African War, Afghan war and the Egyptian war.[48]

The Kildare Hunt Club, which traditionally had held annual Steeplechase Meetings for its members, met for the first time at Punchestown, near Naas, on 1 April 1850. Four years later the Meeting was extended to two days. It was to become the premier such Meeting in the country, and over the years it attracted ever larger fields and attendance, including royalty from England, and military officers from stations throughout Ireland. It became the venue for the Irish Grand Military Steeplechase, and other military races.

The April races at Punchestown in 1868 were especially important as the Prince of Wales made a return visit to the county for the inaugural steeplechase there for the plate in his name. With memories of his last excursion to the short grass county, his mother was not pleased. 'I much regret that the occasion should be the races,' she wrote, 'as it naturally strengthens the belief, already too prevalent, that your chief object is amusement.' To which he replied, 'I am very anxious, dear mama, that you should fully understand that I do not go there for my amusement, but as a duty . . .'[49]

Three hundred police were drafted into Naas to cope with the enormous crowds that came by train to Sallins, and by road; it was a glorious day, and the royal party came by special train to Sallins from Kingsbridge. At the rear of the royal stand at Punchestown a refreshment room and other apartments had been built for the visitors. The racing was good, and the only unpleasant part of the day was 'the dreadful music of a German band', according to a report in *The Irish Times* of 17 April. The prince and princess saw four steeplechases, including the Irish Grand Military Steeplechase, in which there were ten runners. It was won by 'Juryman', ridden by Capt. Hutton, and owned by Col. Ainslie of the Royal Dragoons, who were then stationed in Longford, but came to the Curragh in August. In the main event of the day, the Prince of Wales's Plate for £500, there were 21 runners. Captain Pigott's 'Excelsior', ridden by Capt. Harford, was the winner, and another army man, Capt. Tempest on 'Curlew' was also placed. It must have been another great week of parties, not only for the ladies and gentlemen of the county and their visitors, but also for the officers from Naas, Newbridge and the Curragh who were an accepted and valued part of the social scene.[50]

Indeed the presence of the military from those stations in the social life of the county was especially welcomed in hunting and racing circles. The officers participated in all of the race meetings, but the highlight of the year was always the races at Punchestown (Fig. 20). In 1873 Lord Drogheda, from Moore Abbey, Monasterevan, was the dominant figure in the Irish National Hunt Steeplechase Committee, and it was he who was 'presiding at Punchestown when there was an extraordinary objection in the Irish Grand Military Gold Cup'. The dispute was between Captain Middleton, A.D.C. to Earl Spencer, the lord lieutenant, and Capt. Hugh McCalmont and it concerned the misdescription of the former's hunter. The military stewards finally adjudicated in favour of McCalmont.[51]

Fig 20. *'Notabilities at the Derby' on the Curragh, pictured in* Irish Life *on 4 July 1919. The portly gentleman in the centre is Lieut. General Sir Frederick Shaw, Commander of the British forces in Ireland. The lady on his left is in conversation with an Irish Guards officer.* Photo: *National Library of Ireland.*

The attractions of the Curragh as a military station were succinctly recalled by Lieut. Col. Charles à Court Repington in his biography. As a subaltern in the 3rd Battalion Rifle Brigade there in 1880 he enjoyed 'hunting, racing, polo, racquets, cricket, and plenty of society. General Charles Fraser was in command then, and he always filled his house with pretty women. The Irish race meetings, especially Punchestown, we thoroughly enjoyed'.[52]

But the idyllic sporting nature of the county was tarnished a couple of years later when, in April 1882, for the first time in its history, the Punchestown Meeting had to be cancelled. The escalation of the 'land war' in 1880 had provided the English language with a new word when the English born Capt. Charles Boycott and resident of county Mayo, was ostracised by his neighbours on account of his opposition to the Land League. Further violence followed the imposition by the government of new coercive powers, and the Kildare Hunt Club found it necessary to cease hunting and to abandon the 1882 Punchestown Meeting. Col. J.R. Harvey who was serving with the 5th Royal Irish Lancers at Newbridge that year afterwards commented 'that the country people, even in Kildare, could hardly have been said by their best friends to have done anything this year to maintain that reputation for love of sport generally attributed to the Irish people; and in consequence it was resolved that there should be no Punchestown; a blow felt very much by the people, more especially about Naas, which it may be safely supposed was not without its good effects, though the present loss was ours of course, no hunting meant fewer horses in the regiment, but the card of the day was very much strengthened by a sporting match and a couple of open runs'.[53]

In addition to the Military Gold Cup there was a race for maiden hunters, the property of officers quartered at the Curragh, and in 1885 officers from the Royal Horse Artillery, 16th Lancers, 14th Hussars, Royal Dragoons and 5th Lancers participated. It was a major occasion as the Prince and Princess of Wales were again in attendance (Fig. 20).[54]

In the month of August the Grand Military Race Meeting on the Curragh was regarded as an annual reunion for the soldiers. Though the attendance was limited in 1885, due to bad weather, those who were there enjoyed 'lavish hospitality', and the music of a military band. The races included a hack race, polo pony flat race, chargers' plate and a brigade hurdle plate. Major General the Hon. Charles Thesiger, general commanding in the Curragh, was the chief steward.[55] (He was described by one of his subalterns as 'a jolly old fellow, whom we all liked'.) The following year, on a sunny day, General Prince Edward of Saxe-Weimar, Commander of the Forces, with his entourage, attended the Curragh Military Races. He was popularly known in the Curragh as 'Bally-Sax'. Then it was reported that 'various regiments dispensed boundless hospitality' to the music of the band of the 2nd Battalion Black Watch.[56] Earlier that year the 5th Dragoon Guards Regiment had held their races at Newbridge, and at the Kilcullen Steeplechase there was a military plate for officers' hunters.[57] An officer who served on the Curragh in 1886 described 'real point to points [being] then the order of the day, and I remember two in particular. One, when the [Royal Scots] Greys were simply told to fall in on their barrack square at Newbridge. They were then taken clear of the town, the round tower on top of the hill of Allen was pointed out to them and the only instruction they got was that they were to ride straight for it, and that they would find the winning-post at the foot of the hill. The 9th Lancers, later, were similarly taken up the slopes of the foothills of the Wicklow mountains above Punchestown, whence they could see the round tower of Kilcullen, and told to ride to it'.[58]

If the sporting life of the county appealed to the officers, they were again reminded of an undercurrent of unwelcomeness amongst some of the ordinary people in 1885. Three officers recently arrived in the Curragh found such an atmosphere one autumn afternoon as they walked in mufti through a village 'not a thousand miles from the Curragh'. The gentlemen were very surprised at the reception they got from a crowd of people who were gathered around a band playing 'the usual national airs'. The officers were 'hissed freely' by the crowd, a fact which seems to have surprised the reporter from the Unionist newspaper the *Kildare Observer*, as much as it did the officers. He wrote 'Ireland seems to be getting worse and worse. Military officers were always considered safe from insult, and indeed until very lately, they were treated with every mark of respect by Irishmen'.[59] That scribe must have been aware of, but not necessarily sympathetic with, the political atmosphere of the country, and of the aims of the Irish National League, and of the increasing influence of the Gaelic Athletic Association which had been founded just a year before. John Wyse-Power, editor of the *Leinster Leader*, was one of the secretaries to the Association.[60] The first football game played in county Kildare under the rules of the G.A.A. was at Sallins in February 1885, and sports under the rules of the Association were also held that year at Kildare, Monasterevan, Clane, Carbury and Enfield, all of which activity indicated the rapid growth of the organisation in the county.[61] That new interest in the national games and language was also to be found in the pages of the paper edited by Wyse-Power during the 18 months in which he held that seat.

The attraction of the military to a certain section of the population was observed by Maud Gonne in the 1880s when her father was serving in Dublin. She 'hated the long ceremonious feasts which the towns' people delighted in giving. They slavishly and lavishly entertained the British military. I could never understand why, for the younger officers hardly concealed their contempt for the natives. Tommy [her father] checked me when I too openly joined in laughing at our hosts and I heard him sharply rebuke a young captain who deliberately pretended to mistake the master of the house for the butler; "if you accept a man's hospitality, at least behave like a gentleman", I heard him say'.[62] In the eyes of the socially aspiring class the military and the landed gentry were the elite, and many of the former belonged to the landed families of Ireland and Britain. The gentlemen were much sought after for sports and entertainments of all sorts.

Consequently, the military gentlemen continued to enjoy their parties and their sport, and in the spring of 1887 the 11th Hussars from Newbridge barracks held steeplechases over a course selected by the Baron de Robeck at Beggar's End, not far from the baron's home at Gowran Grange near Naas. The Curragh Brigade Pony were also held during the summer of that year.[63]

In the early 1890s Punchestown was described as being 'then in the heyday of its fame. On the first day, tall hats were *de rigueur*, and the lord lieutenant drove up the course in state, escorted by the Kildare M.F.H. and hunt staff, their red coats and postilions' liveries making a brave show. There were to be seen all the best steeplechase horses and steeplechase jockeys of the day. The regiments quartered at the Curragh all had marquees in which they dispensed lavish hospitality . . . I think 'Roman Oak', the property of one of the officers of the 5th Dragoon Guards, who won at Punchestown about this time, was the best looking steeplechase horse I have ever seen'.[64]

The Curragh Military District held an annual point-to-point at Tully, near Kildare, and at the 1896 meeting the attendance was 'fashionable, and military'. Officers were also amongst the winners at the Kildare Hunt Sportsman's Races held near Ballymore Eustace that spring, as they were at the Hunt Horse Show in September when gentlemen from the 10th Hussars, Scots Guards and East Yorks took part in the jumping competitions. The officers also participated in the military chases at Ballymanny, and at Col. Crichton's private course at Mullaboden.[65] Also that year, with musketry and field training suspended at the Curragh as usual for the two day Punchestown meet, there was again a large attendance of military, of all ranks, and their families. Field Marshal Lord Roberts and the Lord Lieutenant and Lady Cadogan were the principal personalities present that April.[66]

Early in 1899 the Hussars held their Point-to-point at Barretstown Castle, the home of the Borrowes family near Ballymore Eustace, with the Earl of Mayo, Lieut. Col. de Robeck, Col. Crichton, Sir Kildare Borrowes and Mr T.J. de Burgh as stewards.[67] When the Army Point-to-point had been held a short time before it was at Cockeranstown, on the land of three other county Kildare gentlemen, Lord Cloncurry, Sir Algernon Aylmer and Sir Fenton Hort.[68] The races of the 14th Hussars at Kill in March had an unhappy ending when a soldier who had been working in the refreshment tent died of alcoholic poisoning in the back of a cart on the journey back to the Curragh.[69] However, the high-point of the year on the turf was once again Punchestown as it was graced with a visit from the duke and duchess of York, accompanied by Lord Roberts. As the guests of Col. Viscount Downe, who was in command of the 3rd Cavalry Brigade on the Curragh, they also visited the camp.[70]

The popularity of the Curragh to the officers as a station was amply displayed by Lieutenant (later General) Alexander Godley 1st Battalion Royal Dublin Fusiliers (Fig. 21), of the Killegar, county Leitrim family, when he was posted there in 1886. His battalion orders for the move concluded with the paragraph: 'Dogs will make their own arrangements.' Clearly it was a sporting unit![71] Godley later wrote that 'the Curragh in those days was a delightful station. For a season I ran the garrison cricket club, but then became obsessed by polo, and from that time on played little cricket . . . We also started a regimental polo team, and in combination with the Royal Munster Fusiliers, the Guards battalion in Dublin, and others, held an infantry regimental polo tournament . . . we hunted with the Kildare hounds and a pack of harriers kept by a civil veterinary surgeon, named Pallin, at Athgarvan Lodge on the edge of the Curragh'. The Kildare pack, in Godley's estimation, was then 'at the zenith of their fame, with Dick Moore [Major R. St Leger Moore late 9th Lancers, of Killashee] as master . . . sport was of the best. It would be hard to find a more delightful field or better sportsmen than hunted in Kildare in that time'.[72] Successive viceroys and their staffs came from Dublin to hunt with the Kildares, as did regiments stationed in the city.

In his reminiscences Godley recalled some of the Kildare families with whom he socialised and hunted, including the de Robecks (Fig. 22) at Gowran Grange

Fig. 21. *General Sir Alexander Godley G.C.B., K.C.M.C. From a landed county Leitrim family, he soldiered with the 1st Bn. Royal Dublin Fusiliers in the Curragh in 1886 and found it 'a delightful station'.* Photo: Life of an Irish Soldier, *(London), 1939.*

Fig. 22 (a) *The wedding of Dorothy Zoe de Robeck, daughter of the 5th Baron de Robeck, to Major Digby Robert Peel 60th Royal Artillery, of Alexandria, Egypt, at the de Robeck's temporary home, Osberstown, Naas, on 8 October 1910. The groomsman is Major Edward Conolly, 60th Royal Artillery, of Castletown, Celbridge. Courtesy Major John de Burgh.*

(b) *Sergeant Major Frank Kelly and his family outside his tar and felt roofed married quarters in the Curragh Camp in 1919. Mrs Kelly (neé Darling) left, with her son (kneeling) and her parents. Her father Charles Darling served with the 7th Queen's Own Hussars in Newbridge. Standing (back centre) is William Darling 12th Royal Lancers, home on leave from India. When he retired from service Charles Darling opened a barber's shop in the camp; it is now the only remaining pre-1922 commercial premises there.* Photo: *Courtesy Reg Darling.*

on the edge of Punchestown, a great rendezvous for that historic meeting'. The baron de Robeck of the day was 'father of the celebrated admiral, and of the late baron, who was a distinguished horse-gunner and subsequently master of fox hounds. His grandson, also a horse-gunner, carries on the sporting tradition of the family'. A daughter of the family had married Captain Bill Tremayne 4th Dragoon Guards, a hunting companion of Godley. Harry de Robeck, the baron's eldest son, was 'at Newbridge along with other hard-riding gunners . . .' Other county friends included Percy La Touche of Harristown, Willie Blacker of Castlemartin, Lord Mayo of Palmerstown, 'Cub' Kennedy of Straffan and Sir Anthony Weldon of Kilmorony, Athy. Lady hunting friends were Eva Beauman from Furness, Naas, who afterwards married Blacker, and Leila Crichton, daughter of Col. Hon. Charles Crichton, Grenadier Guards, of Mullaboden, and who married Sir John Milbanke V.C. 10th Hussars, and who was killed at Gallipoli; she later married General Sir Bryan Mahon.[73]

Invited to the viceregal balls in St Patrick's Hall in Dublin Castle, young Godley admired the ladies: 'to my mind the duchess of Leinster stood out among them', but other ladies connected with county Kildare whom he admired included Mrs Harry Greer, Catherine Conolly, later Lady Carew of Castletown, and Maud Gonne, who had lived on the Curragh when her father served there with the 17th Lancers. Lieutenant Godley was familiar with the numerous training stables at the Curragh, including that of Captain Harry Greer, Highland Light Infantry, who lived at Croatanstown Lodge, and Peter Purcell, who 'had just left the 5th Lancers and was living at Whiteleas'.[74]

Godley also hunted in county Carlow, where he had many landed friends, and he had happy memories of that time when he once managed 'to hunt thirteen days running, the Sunday with a Land League pack near Athy, being by no means the worst of them'. Posted to Newry in 1890, he continued to come to Kildare for part of his winter leave to continue hunting there, and when he some time later got a posting to the depot of his regiment at Naas he was overjoyed. There he enjoyed 'cheery dinner-parties' in the officers' mess, and the work was not too demanding.[75]

At Naas, finding himself short of cash he decided to try and make his polo pay for itself by purchasing raw ponies and training them. Some of the ponies 'came out of turf carts . . . I was fairly successful, and sold several for good prices'. He played polo locally, and for the All-Ireland Polo Club in Dublin, but his 'two years at Naas passed all too quickly. With a marvellous little pony, called "Duchess", which I bought for a fiver, an old governess-cart, with the bottom half out of it – an advantage for my long legs – and an old set of donkey harness, for the patching up of which I carried a liberal supply of string – the whole outfit costing less than a ten pound note – I travelled Kildare and Carlow from end to end, and penetrated also into Meath'.[76]

However, a decade after Godley's arrival in Kildare, there seems to have been some friction over the hunting between the farmers and the officers, as reported under the heading 'Hint to Military Sportsmen' in the local paper: 'In future military people who follow the Kildare Harriers should be careful as to where they ride. At last week's Newbridge Meeting they angered the farmers by riding over the new grass fields.'[77]

The annual ball of the Kildare Hunt, traditionally held in the Town Hall at Naas, was the main social event of the hunting season. It was always fully reported in the local press, and the attendance given. In 1896, for example, Lord and Lady Roberts, Miss Roberts, and officers from the 10th Hussars, 15th Hussars, Scots Guards and 19th Foot, as well as numerous other military gentlemen, attended. The gentry of the county and their families were, of course, widely represented there.[78]

The officers were regularly invited to the country houses in the county. When the marquis and marchioness of Drogheda gave a ball at Moore Abbey, Monasterevan, on 19 January 1876 there was 'a huge attendance . . . including Brig. Gen. and Mrs Seymour and *aide de camp* and officers from the Queen's Bays and Royal Artillery from Newbridge. From the Curragh came officers of the 17th Regiment and Depot, of the 45th and 62nd regiments; Col. Dunne and officers of the 99th Regiment and Depot. Mr D. O'Connor and officers of the Control Department, officers of the Royal Engineers, etc.' After the ball it was arranged that the train from Cork stopped at Monasterevan at 2.45 a.m. to bring guests back to Newbridge.[79] A further indication of the involvement of gentlemen from the county with the military establishment in 1880 was the fact that Capt. St Leger Moore 5th Lancashire Regiment, from Killashee, was aide de camp to Major General Seymour, while Capt. the Hon. H. G. F. Crichton 21st Hussars, who lived at Mullaboden, Ballymore Eustace, was brigade major.[80]

The continuity over the years of the hospitality of the Drogheda's to the military was evident a decade after that ball when in November 1886 they gave a *battue*; amongst those who shot over the woodlands at Moore Abbey were H.E. Prince Edward Saxe-Weimar, Commander of the Forces, his A.D.C. Capt. Darby and other gentlemen.[81] Other county families also entertained the officers. A couple of months before that outing gentlemen from the 10th and 11th Hussars and 16th Lancers had attended a garden party and tournament at Palmerstown, Naas, the home of the earl of Mayo.[82] Lieutenant Godley, also in 1886, enjoyed 'great bachelor parties that David Mahony of the Grange [Grange Con], who was secretary of the Kildare Hunt, used to have for the meets in Thursday country. About twenty of us would sit down in our red evening hunt coats to drink marvellous claret and to listen to Percy La Touche's inimitable stories'. While later stationed at Naas Godley enjoyed 'the great centre of hospitality at Mullaboden', and Col. Charlie Chrichton's private race-course where meetings and gymkhanas were held.[83]

The officers returned the hospitality of the Kildare families by inviting them to entertainments in the camp or barracks, and sometimes there was a grand ball, such as that held in the month of August 1887 in the gymnasium at the Curragh Camp. It was made a very grand affair by the attendance of the Prince and Princess Edward of Saxe-Weimar and numerous high-ranking officers with their ladies. The local paper reported that the 'neighbouring houses had been filled, [with guests attending the ball] the more the merrier', and amongst the county families who participated in the evening were the de Robecks, de Burghs, Borrowes, La Touches, Blackers, Moores, Humes, and Cosbys. Lieutenant Colonel J.A. Connolly V.C. 49th Regiment, a veteran of the Crimea, was also there. He was resident in the magistrates house in the Camp (and there he died on 23 December 1888).[84] The gentry mingled with officers from the Grenadier Guards, Royal Artillery, Cameronians, Suffolk and Shropshire Regiments who were then stationed in the camp. The brilliance of the gathering was remarked on by the man from the *Kildare Observer* as the success of the ball was such that 'the sun had risen before it was over'.[85]

During the drill period of 1891 the social life of the district was again stimulated by the presence of royalty, H.R.H. the duke of Clarence and Avondale who was with his regiment, 10th Royal Hussars. He was the 27 year old bachelor Prince Albert Victor, heir to Edward VII who had himself spent a memorable season in the camp 30 years before. The *Leinster Express* of 15 August reported the arrival of the regiment at the Curragh, having come by road from Dublin. Captain Wogan-Brown met the prince and his squadron at Moorfield and led them into their camping ground near the Harepark hospital. During Albert's six weeks under canvas on the plain the weather was bad, but he participated in the exercises and sporting fixtures with his fellow soldiers. Games of polo were played at some of the country houses, and when Thomas Conolly of Castletown came of age in September the prince joined the nobility and gentry of the county at the celebrations. When Albert succumbed to complications following influenza in January 1892, the *Leinster Express* noted that 'in Kildare and Dublin the Duke of Clarence was well know having many friends . . . He had visited Punchestown about six years ago with his father, and recently was stationed at the Curragh with his regiment where he won the respect and esteem of everyone with whom he came in contact'.

The year 1891 also saw the publication of the first newspaper to be produced in the Curragh camp; *The Curragh News* appeared on 7 February when its editorial stated that 'the want of a local journal has long been felt in the Curragh camp and the important districts which adjoin, the population of which hold intercourse, social and business intercourse, with the garrison at Newbridge and the Curragh'. It was the editor's intention that his paper would report on every event of interest in the camp and Newbridge, military, sporting and general; he promised that 'as the only military paper in the country, we will publish all matters of importance to the army in Ireland and which may serve to inform and instruct our readers whose interests are concerned'. The good intentions of the editor did not succeed in making the *News* viable, and after eight months he ceased publication, complaining that the paper had been published 'under conditions which did not permit us to develop it as we would like', but he expressed a wish that some time in the future he might be able to resume publication, but this was never to happen.

The unpleasant weather of that summer was not forgotten by General Aylmer Haldane some 60 years later when he was writing his autobiography. Stationed at the Curragh in 1890, he was adjutant of the second Battalion Gordon Highlanders. He afterwards recalled that 'it was excessively wet, and though the horses of the [Hussars] regiment stood in the open almost up to their girths in mud one could not help admiring the way they would turn out to take part in a field day as if the mud was of no consequence. Their colonel was Lord Downe, who had come to them from the Life Guards, and he was a good deal more than a "spit and polish" commander'.[86]

Prince Albert Victor sometimes came to dine in the Highlander's mess to hear the bagpipes played on guest nights, and he also visited the officers' married quarters. Haldane remembered that one evening, at a dinner party organised by the wife of a brother officer who professed to being a palmist, she examined the palms of the guests, including the prince. Some days later when the adjutant again called on that lady she told him in confidence that she had been shocked to note that the prince's life-line was short and ended abruptly, and she feared that some disaster would befall him. Not many months after her prevision came true.

The enlargement of the social scene in county Kildare due to the presence of the army was undoubted. If the officers and their families reaped most benefit from the hospitality of the many hospitable big houses, they reciprocated by providing eligible young men, lavish barrack entertainments and frequent professional displays on the training grounds. Almost all sections of the population were at some time or another entertained by the military, even it were but the playing of a military band.

Chapter 9

Social Clubs and Societies

There were also entertainments to which all ranks might bring guests. There was a constant round of theatricals and concerts hosted by the various regiments, such as in 1885 when the Curragh Brigade Amateur Dramatic Club gave a series of performances (Fig. 23). The reporter from the *Kildare Observer* sagely observed that 'the monotony of camp life in remote districts has been at all times proverbial in military circles. The routine of duties prescribed by the discipline of the service becomes invariably irksome at such a place as the Curragh Camp, where the population is confined almost exclusively to the military element, and were it not for the fact that entertainments are got up from time to time by members of the garrison and their friends, that interesting operation which has been so aptly termed "killing time" would – especially during the winter months – be tedious in the extreme.' The efforts of the club, he believed, 'earned the gratitude of almost every individual residing upon this dreary expanse of treeless plain. After sundown a truly Cimmerian darkness appears to envelope the apparently interminable rows of wooden huts which are laid out here with the rectilineal accuracy of the streets of an American city, and there is no inducement whatever to the inhabitants to venture out of doors except upon those welcome occasion, when balls, concerts, or private theatricals are held in the gymnasium, one of the best known buildings upon the Curragh'.

The performance of the dramatic club itself 'merited praise for the string band of the 1st Lincoln Regiment, a recitation by Lieutenant Gamble of the Northampton Regiment and a vocal contribution from Assistant Commissary-General Bridgeman'. Private Burbidge, on the violin, accompanied Miss Bolton who sang, and she was later joined in a duet by Mrs Crosse. There was a farce entitled 'Urgent Private Affairs, in which Mr Burgess played the part of Sally Nokins', and an impersonation of Major Polkington, a jealous spouse, by Lieutenant Campbell. The proceedings, 'which were warmly applauded, were brought to a close by a performance of the National Anthem'.[1] When the Brigade Dramatic Club resumed performances after the summer they were joined by the band of the 1st Lincoln Regiment for two evenings.[2] On another evening the 1st Battalion Liverpool Regiment (the King's) gave theatricals in a schoolroom at I lines, and it was patronised by senior officers and ladies of the brigade.[3]

The townsfolk of Newbridge welcomed invitations to the concerts held in the riding school at the barracks there. In August 1885 the band of the 5th Dragoon Guards gave two recitals,[4] and before Christmas of that year, under the patronage of Gen. Thesiger, the band again performed, the proceeds to go to the Protestant parochial room for the town.[5] At Naas barracks the Royal Dublin Fusiliers commenced a series of popular sing-songs in the library of the barracks in September. They were open to the public, and attracted large attendances on Wednesday, Thursday and Friday evenings.[6] Weekly smoking concerts were held in the Curragh Camp, and at those held in the recreation rooms of the Dublin Fusiliers there was an orchestra and songs; at the number 2 hospital smokers were given for the entertainment of the patients and their friends.[7]

An expected result of the extensive social intercourse between the military, line and militia, and the civilian population was matrimonial unions. As well as the society weddings of the officers and gentlewomen of the county there were regular unions between the other ranks and the local girls. A survey of 7,103 marriages between the years 1802 and 1899 in five Protestant and three Catholic churches in the neighbourhood of the Curragh, and the garrison church in the camp (named St Brigid's from 1893), shows that in that time there were 966 military weddings. The majority of the ceremonies (311) took place in the protestant church at Newbridge, and the women were mainly from the town or from the Curragh camp and locality. A total of 225 military weddings took place in the Roman Catholic churches in Kildare, Kilcullen and Newbridge, with 145 of those in the latter town.[8] In the Curragh Roman Catholic garrison church there were 274 marriage ceremonies between 1855 and 1896; of the 107 women whose addresses were recorded 53 were from county Kildare (with 24 from the Curragh camp), and another 52 from different parts of the country, and two from England.

In an analysis of the 966 military weddings (table. 1), using the strict criteria of local address and surname, and excluding those living in barracks, it was found that 387 of the brides could be regarded as 'local'. However, the exact number of Kildare born women is impossible to clarify as the registers do not give place of birth, and women who had followed their men to the county may have given their address of residence there. An estimate of 33% of the 966 unions as having been with local women is considered to be reasonable. The high proportion of brides with local addresses who married in the Church of Ireland may be explained by a high matrimonial rate between soldiers and women from military families. The Protestant population of the county Kildare in 1881 was 9,869; Roman Catholic 65,935. The church registers also indicate that of the 2,739 births registered in the Curragh Garrison Roman Catholic church from 1855 to 1899 between 3% and 4% of the fathers were recorded as being Protestant. Of a sample of 193 baptisms, 119 were of adult converts, predominantly soldiers, and 65 children of mixed marriages were baptised as Roman Catholics. In May 1881 Maria Loasby, her daughter and her infant son were all christened on the one day in the Curragh Catholic church. But in another instance a man of the 3rd Light Dragoons whose son was given conditional R.C. baptism in hospital, immediately afterwards took the child to the Protestant church for baptism there. From the records of the nine churches surveyed it may be deduced that there were frequent marriages between local Kildare girls and the soldiers, and that the children of such unions, if the mother was Roman Catholic, were baptised in that faith.

Table 1. *The number of marriages between 1802 and 1899 recorded in five Church of Ireland and four Roman Catholic churches in the Curragh area.*

	All marriages	Military marriages	Brides with local addresses
Roman Catholic	—	514	221
Church of Ireland	—	452	403
TOTAL	7,104	966	624

Many of the romances which led to the altar could have commenced in the social gatherings to which the local people were invited. But there were also barrack occasions to which men only were admitted, such as in the bars of the various messes.

Civilians were always happy to avail of the facilities of the sergeants' messes in the various barracks. In the messes the price of beer, spirits and tobacco was considerably lower than in civilian establishments, and while guests were not supposed to pay for their rounds in the messes, in reality they did, and in many messes it became usual for civilians to be admitted at the barrack gate if they mentioned the name of a member of the mess. As the messes were also open when the public houses might be closed, and even though the visitors paid a little more for their drinks than did the soldiers, the messes were still very attractive places. Of course the committee which oversaw the management of a mess welcomed the additional revenue as it increased mess profits, and enabled them to organise subsidised functions for their members, or to purchase effects for the mess rooms.[9]

By 1896 pantomimes had been introduced to the military entertainments, and performances were given in the Curragh camp gymnasium; local residents contributed to the shows with Mrs Greer of Crotanstown House making the costumes. The bands of the East Yorks and 1st Yorks took part, and 'the whole of the officers and ladies of the camp, and a very large number of the gentry of the neighbourhood' attended.[10]

To pass the winter evenings there was a variety of regular entertainments; the 1st Royal Munsters hosted fortnightly practice dances.[11] Magic lantern shows were given in the chaplain's room at D square,[12] and whist drives were held in the sergeants' messes.[13] Billiard competitions were organised, and following one in February 1896 the 1st Yorks gave a tripe supper, 'to which sixty sat down. It was followed by a smoker for all the sergeants of the garrison, 200 were present.' The fun there consisted of 'singing, reciting in the Yorkshire dialect, dulcimer, etc.'. It was generally agreed that billiards and whist tournaments should be held more frequently.[14]

The whist drives and billiard games held in the Garrison sergeants' mess also attracted participants from outside the barracks, and the travelling shows which came to the camp or to Newbridge found strong support from the soldiers. Sometimes the enthusiasm of the men got out of hand, as in Newbridge in the summer of 1887 during a visit of the German Prof. Louis' travelling boxing saloon to the town. A fracas developed into a military riot between soldiers from the Royal Artillery and the Scottish Rifles and the orderly officer had to be called from the barracks. He put an end to the affair by arresting ten of the men.[15]

In 1896 the Curragh Dramatic Club continued its performances, under the bandmaster of the York Regiment, and they gave a subscription dance in aid of the Church of England Soldiers' Home.[16] The Curragh Soldiers' Institute was to benefit from the theatricals given by the 19th Green Howards later in the year.[17] In April the Calder O'Beirne Opera Co. staged *Maritana* and the *Bohemian Girl* to packed houses at the Curragh gymnasium,[18] and in May the medical staff corps gave a quadrille party at the station hospital for 200 guests. Music was provided by the band of the 2nd Yorks.[19] With the coming of summer the 4th Battalion Dublin Fusiliers hosted a camp fire smoking concert 'at which nearly the whole of the troops in the camp attended'. The stage was lit by Chinese lanterns.[20] A more decorous occasion was the organ recital given in the west church one June afternoon. The organist was William Firth A.L.C.M.[21] With such a regular round of performances it was not surprising that in the summer of 1896 a site was chosen for the building of a camp theatre.[22]

The officers of the garrison held their own parties, such as the subscription dance in the officers' mess hut in I lines in July 1896,[23] or a regimental dinner given by the Munsters for their past and present officers some time before that.[24] In the lulls between such social occasions the gentlemen's favourite after dinner amusement was to 'draw the ash bins for a Curragh buck'. This, according to General Godley's reminiscences, was their 'name for the sheep who tried to supplement from the refuse the scanty fare provided by the short grass of the Curragh. These animals were very active and took a lot of stalking and catching. Once caught, they would be hurled in through a window, on the top of some unfortunate who had gone to bed. The wreckage of his room by the maddened beast can well be imagined'.[25]

A more public performance was that advertised in December 1896. It was 'An Assault at Arms', to be staged in the camp gymnasium, under the patronage of Gen. Combe. Featuring the staff of the gym, there was to be sword drill, fencing, Indian clubs, dumb-bells, horizontal and parallel bars and musical drills. It was due to start at 5 p.m., and carriages were to be ordered for 7 p.m. Reserved seats cost 2/–, others 1/– and 3d.[26]

At Naas the Dubs Depot Dramatic Club gave performances in the theatre hut, while in Newbridge Town Hall at a concert organised by Moorfield Gaelic Athletic Club soldiers figured on the programme: W. Tobin R.E. sang, Private Pittaway 10th Royal Hussars gave a song and a step-dance, while Messrs Entwhistle, Wheeldon and Wilde of the medical staff performed on banjo, mandolin and concertina.[27] The association of the garrison with the football club was an indication of the acceptance of the military by the local men.

Two performances were given in Naas barracks in May 1897 of an entertainment consisting of a concert and a farce. They merited a long notice in the *Leinster Leader* which praised the success of the performances, including those of Col. Legget, his wife and son. The item which received the most genuine appreciation was Colour Sergeant Southam's rendering of *The Soldier of the Queen*, although the colonel's lady's singing of a typical Irish song *Irish Reel* won great applause. The involvement of all ranks in the entertainment was an indication of the good morale of the unit, and the filling of the hall to 'its utmost capacity' proved the popularity of the occasion with both the military and civilian audience.[28]

The year 1899, and in which the Boer War commenced, was planned to be one of normal routine on the Curragh. During the season of Christmas and new year there had been parties, with as many as 200 children and 40 mothers attending that of the Army Service Corps. Concerts were given in the Soldiers' Institute, the 14th Hussars gave a dance, and at a sergeants' mess dinner and concert Lieut. C.J. Hartley, a famous dark blue cricketer, showed another skill 'in rendering *Matthew Hannigan's Aunt* and *Fr O' Flynn* (in which his vocal powers were invariably seen to such advantage) in a purest brogue'. In addition to a famous sportsman there was also a soldier poet then serving in the camp. He was Cpl. Harold Hanham, Royal Fusiliers, and he had published some verse in the *Daily Mail*, where he was termed 'The Soldier Poet'. Lord Roberts wrote to congratulate Hanham on his poetry, and the letter was read at a meeting in the camp of the Army Temperance Association.[29]

Sporting fixtures brought not only the military and the civilians together, but the cricket pitch also encouraged the officers and other ranks into a social mix. Cricket was a sport much identified with the military in county Kildare; a historian of the game has commented that 'for almost 100 years they had not only provided continuous top class opposition to local teams but also an ongoing direct supply of players. There were four bases in the county, at the Curragh, Newbridge, Naas and Kildare, with the first two being particularly active on the cricket front.'[30]

When, in 1841, the duke of Wellington had ordered that cricket grounds should be laid out at each end of barrack stations throughout the Kingdom for the use of officers and other ranks he also directed that the colonel of each regiment should encourage the game.[31] The first organised cricket in county Kildare is believed to have been played at Newbridge, and there is a report of a game there in September 1838 between a team from Carlow and the Newbridge Garrison Club.[32] By 1866 two cricket pitches had been laid down at the Curragh and where 'the sod was continually renewed where necessary'. At Naas cricketers from the barracks enjoyed the facilities of the County Kildare Cricket Club from 1881, a club that had evolved from Naas Shamrock Cricket Club and which had been founded in 1860.[33]

In the season of 1887, for example, the Curragh Brigade played the County Kildare Cricket Club at the County Club in Naas, and the Cameronians met the latter club on the Curragh pitch.[34] When the Curragh Brigade defeated Dublin University on the Curragh in May of 1886 the winning team consisted of officers, non commissioned officers and men of the 5th Dragoon Guards, 24th, 26th and 73rd Regiments.[35] Other matches that year were played between the Royal Dublin Fusiliers and Baltinglass and the R.D.F. and the County Kildare Club.[36] Lieutenant Godley 1st Bn R.D.F., then stationed there, commented that 'we had a very good regimental team. Two of my brother officers, Captain Sheppard and Price, and three of the bandsmen, Ashcroft, a wicket keeper, and Maylam and Mayhew, fast and slow bowlers, were almost first class'.[37]

Cricket also featured during what the *Kildare Observer* described as 'a gay week' on the Curragh in August 1887. The gaiety was to entertain the commander in chief, Prince Edward of Saxe Weimar. The programme also included tennis tournaments, a race meeting and a ball, all of which must have caused considerable excitement amongst those fortunate enough to be invited.[38] Officers from the Naas barracks were members of the County Club at Oldtown, Naas, and in 1886 Col. Colville Frankland, Commanding the 102nd Regimental District, was on the committee of the club.[39] Lawn tennis parties were sometimes held in the Naas barracks for which invitations were issued to the local gentry.[40] Some of the county gentlemen would meet the officers on the camp polo grounds, but players from further afield also came to the Curragh grounds. Sub-lieutenant Ian Hamilton 12th Suffolk Regiment, brought two ponies with him when he reported to the Curragh camp in July 1873, and there he was introduced to polo.[41] The Castlecomer Club, which was but two years established, met the 3rd Dragoon Guards on the Curragh in August 1876;[42] twenty years later when the county Kildare team met county Derry at Derry three of the Kildare team were officers.[43]

When the new artillery barracks was built at Kildare in 1902 it had a cricket ground, and the first match was played there in June 1906 between the 32nd Brigade from the Curragh and the 33rd Brigade Royal Field Artillery.[44] The withdrawal of the British Army in 1922 was to have a demoralizing effect on the county players, and the game was never again to have the same popularity.[45]

Annual sports days in the camp were a major attraction for the local people of all classes, but when they were held near the Stand House and enclosure those areas were reserved for the officers and their guests. The sports which attracted the people of the neighbourhood to the Curragh were the various regimental football, athletic, gymnastic and boxing fixtures. Sometimes there was betting on an event, and it was evens on the three miles foot race for £10 a side, between Pte. Frazer R.E. and Sgt. Ashcroft 99th Regt in April 1876.[46] When the 2nd Bn West Kent Regiment held its sports 500 yards in front of C lines in August 1885 visitors came from Newbridge and elsewhere for the sport, and to be entertained by the band of the 1st Bn The King's Liverpool Regiment.[47] The sports organised by the 11th Hussars at Newbridge in 1886 attracted a big attendance of 'military and friends of the regiment . . . there was profuse hospitality, for which the gallant 11th was famous'. The programme that afternoon included a sword exercise competition, a musical ride, a lance exercise exhibition, wrestling on horseback, a mounted quadrille, tent-pegging and many other exciting events.[48] At Naas in September 1896 the Dubs sports attracted 'the gentry and public to the pretty grounds of the barracks',[49] and a couple of months earlier 'an immense assemblage of military and civilian on-lookers' attended the 28th annual All-Ireland Athletic Meeting at the Curragh.[50] Indeed, it would seem from the reports in the county press that, without the constant amusements provided by the military, social life in county Kildare would have been much less varied and exciting.

The possibility of going skating on the Curragh must have been attractive, but it is not known if the Curragh Skating Rink Co. which was wound up voluntarily in October 1876, ever realised its intention of 'reorganisation and amalgamation'.[51] It is not mentioned amongst the variety of entertainments which were held in camp in 1896; they included boxing competitions, a bicycle gymkhana on the garrison cricket ground (in aid of the Soldiers' and Sailors' Families Association), cricket, rackets, monthly golf handicap competitions, tennis tournaments and football matches.[52] When the team of the 1st Lancashire Fusiliers came from Athlone to play the 1st Derbys at the grounds to the north of the magazine on the Curragh in December they were accompanied by some 200 comrades, all of whom were entertained to dinner by the Derbys after the match.[53] Other matches that year saw the Royal Engineers from the Curragh meet the band of the Sherwood Foresters, (the latter team had been victorious over Bohemians in the 4th round of the Irish Senior Cup a week earlier),[54] and the Scots Greys beat the Derbys [Nottingham and Derbyshire Regiment] by four goals.[55] All of the

Fig. 23. *The cast of an entertainment photographed beside the gymnasium in the Gough Barrack area, Curragh Camp, 1919. Sergeant Major Frank Kelly (see* Fig. 22b*) is on the extreme left of the group. Courtesy Reg Darling.*

Fig. 24. *Group taken at the inaugural ceremonies of the Curragh Camp Masonic Lodge in Newbridge on 28 March 1906. Some of the N.C.O.s and men are from the 1st Bn. South Staffordshire Regiment which was stationed in the camp from January 1904 to November 1906. Courtesy Lodge No. 215 Newbridge.*

regimental teams had their own following, and the progress of the favourites was a matter of some interest to both military and civilian football enthusiasts.

Of less interest to the people of the neighbourhood would have been the official formation of the garrison golf club, in connection with the officers' recreation club, in March 1883. Major Gen. Fraser, Commanding Curragh Brigade, had sanctioned the institution of the club, and Lieut. A. G. Balfour 1st Battalion Highland Light Infantry was appointed secretary. Civilians were not admitted to membership until late in the 1900s, and then membership was confined to the local gentry.[56]

The popularity of the Curragh as a military station was clearly declared in a military magazine in May 1896. The Curragh was described as 'the most delightful of camps' in summer and autumn, if the season was not exceptionally harsh: 'the air, scented with the odour of fresh grass, and the perfume of wild flowers, exhilarates and cheers'. It was known that the soldier of every rank was 'indeed at home in sweet Kildare'. The angler could fish in the canal or the local streams without let or hindrance, many of the regiments kept packs of beagles, and for those who could afford it there was 'the flying Kildares, a pack of hounds second to none, mastered by an old hand St Leger Moore'. And of course there were the Curragh races: 'it is then that the camp is arrayed in all its festive glory, for who would be absent when "On the Curragh of Kildare, all Ireland will be there?"'[57] From the description of the Curragh by that journalist it would seem that the idyllic picture of the place was a reflection of the soldiers' experiences, rather than just a flight of the writer's imagination.

The public interest in army sport resulted in weekly columns in the loyalist *Kildare Observer* on cricket, polo, and Association Football. As well as reporting on inter-regiment games, matches between other clubs, such as with Clongowes Wood College in cricket, or the county teams in polo and golf, were reported.

Matches of another sort also appeared in print when the marriages of soldiers to local women were noticed. Such was that of the Colour Sergeant B.F. Bruen R.D.F. to Catherine Masterson, daughter of the owner of the Temperance Hotel in Naas. It was a full-dress affair with an honour-guard.[58] What was described as a 'Fashionable Marriage' was that between Lieut. John Poé R.A.M.C. from Kilkenny to the eldest daughter of R.J. Goff J.P., the Newbridge auctioneer.[59] The established status of some of the county ladies was confirmed with their inclusion in a list of those who attended at the viceregal court in Dublin; they included members of the Clements, Cane, de Burgh, Crichton and More-O'Ferrall families, who also had strong military connections.[60]

The status of the officers serving in Ireland was that of a country gentleman, and indeed, as has been noted, many of them came from landed Anglo-Irish families. Nora Robertson, who herself was the daughter of such an officer, believed that 'the Anglo-Irish country gentleman of my day took their colour absolutely from the garrison, not only the patriotic orientation of the latter but their social and mental angle'. She saw that 'most younger sons, and eldest sons until they inherited their patrimony, joined the fighting services',[61] but the popular belief that the Anglo-Irish provided a monopoly of the officer corps has been disproven.[62]

Of course the officers, both Irish born and English, were welcomed by their social equals, as was the army in general in areas where it contributed to the economy and social life. But the officers, whose duties did not usually take up much of their time, were enraptured with the splendid sporting facilities of the countryside and had little involvement in ordinary Irish life. In the opinion of Nora Robertson, only child of Lieut. Gen. Sir Lawrence Worthington Parsons (of the family of the earls of Rosse), and herself a sympathetic observer of the changing political scene, the understanding of the English garrison officers of the people and their politics was superficial: 'In the Kildare Street Club it was impossible for them not to compare the Irish with the Indian situation. The Irish too should be treated kindly, fairly, led, not bullied. Agitators were the curse of both countries, Home Rule was the invention of disloyal fanatics or paid agitators who only wanted to pull the Empire down. It was all quite simple'.[63] Or, as another commentator has put it: 'Throughout the nineteenth century . . . toward Ireland and the Irish the officer was likely to maintain the same combination of prejudice and condescension that characterised his response to African and Asian colonial natives'.[64] While many officers would have had a more enlightened view of the Irish and their problems, it was despite the fact that 'their social and professional world encompassed only a small portion of the actual country in which they served'.[65] The commentator just quoted is the American historian Elizabeth Muenger, and even if her views reflect those of Nora Robertson they must be appreciated also as the fruit of the research of a disinterested foreign scholar.

The evidence from this county, however, would show that the English officers in general fitted in well with their social peers, and had a kindly view of the people in general. Many of them settled here on retirement, though after 1922 some of them returned to England when the social scene changed, and there was political uncertainty. Irish born officers, normally being of a different social class and generally holding opposite political views to the majority population, had the same relationship with the country and the people as did their own families. Some of them, such as General Sir Bryan Mahon and Major Bryan Cooper, endeavoured to adjust to the post 1922 new Ireland and entered politics. When Cooper died in 1930 he was a member of the *Cumann na nGaedheal* party. His coffin was draped with both tricolour and union jack. Captain Derick Barton served in the 17th–21st Lancers in the 1920s, in the Irish army during the Emergency, and again in the British army for the last years of World War II. The careers of those men support the popular Irish saying that: 'there are many shades of green'.[66]

That 'Indian summer' of the old order was recalled by Hector Legge, a former editor of the *Sunday Independent*, when reminiscing on his boyhood in Naas: 'In the years before the 1914–18 war the County Kildare Club [at Naas] was one of the outstanding sports centres in Ireland. In cricket, lawn tennis and hockey, its members were to the fore . . . At that time the British Army was strong in Ireland. There were barracks in Naas itself, Royal Dublin Fusiliers, in Newbridge, Royal Field Artillery, and the many cavalry and infantry regiments on the Curragh. For the officers the County Kildare Club offered golden opportunities for them to play cricket and tennis. It was the age of what was referred to as "the gentry". Young ladies of those families made many a marriage with army officers through the tennis court.[67] Colonel St Leger Moore lived nearby at Killashee. He was well known for the big dinner parties he gave, opportunities, no doubt, for fair young ladies to meet exciting young military men'. An indication of the effect of the Great War on the club was the fact that no annual general meeting was held from May 1914 until June 1919.[68]

Derick Barton of Straffan, in his autobiography which looked back over 90 years, also reminisced on those halcyon days; before the Great War, he recalled, 'with the army still in occupation on the Curragh, in Kildare and in Newbridge, there was an enormous following [of the Kildare hunt] and it has always remained a vivid picture in my mind, these 70 years later . . . there was generally a guest or two from the garrisons at the Curragh or Newbridge [staying at Straffan]. I am not sure if their horses came with them, there was ample room in the stables and for the soldier grooms in the rooms overhead, or they may have come by the train to Straffan Station next morning'.[69]

If the idyllic life of the gentlemen may have seemed even more idyllic in retrospect than in reality, that of the rank and file was also sometimes perceived as having been more leisurely than it was. A postcard of 1908 entitled 'Camp Life in Newbridge', and sub-titled 'On Guard?', depicts a soldier lying snoring on his back, with his rifle at his feet.[70] Since the days of aid of the civil power in 1881 when a posse lost its arms at Hackettstown, and after the field experience of the Boer War, it was unlikely that discipline was such that sentries on duty could so relax. But even in its comic intention the card suggested that garrison life in county Kildare for the other ranks might also be pleasurable, a supposition supported by the evidence presented in this chapter.

Apart from the cricket pitch, another rank-less meeting place for the officers and other ranks with each other, and with men of a district, was the masonic lodge (Fig. 24). From the granting of the first travelling warrants to the 1st Battalion Royal Scots Regiment in 1732 by the Grand Lodge of Ireland, there were regimental warrants. In all, over 200 such warrants were issued in Dublin.[71] Writing on the travelling warrants Brother R.E. Parkinson remarked that there was 'hardly a unit of the British army which has not served in Ireland',[72] and where they came in contact with local lodges, or formed their own. He believed that next to the Colours the soldier mason loved his lodge. In periods of boredom the practice of his craft 'solaced the soldier in out-of-way stations, and the principles lifted him above the weariness, and even sordidness of his lot'.[73]

Brother Parkinson even went so far as to claim that the work of the military lodges 'played some part in tempering the regiments of the British army into the incomparable weapon they were, and are. Once within the portals of the regimental lodge, where all external distinctions of ranks were laid aside, a man's self respect was fostered, and mutual regard and even affection was encouraged between the officer and man which built up the *esprit de corps*, and pride of every regiment'.[74]

Without a comparative table by ranks of the membership of lodges it is impossible to say to what extent the masonic code influenced the morale of a regiment. Undoubtedly but it must have had some bearing in the relationship between the officers and other ranks who assembled together in the lodges, but the vast majority of all ranks could not have been influenced by the brotherhood.

In the vicinity of the Curragh there were masonic lodges at Athy (no. 167) from 1740,[75] Naas (no. 205) until 1870,[76] the Curragh (lodge 397) until 1937,[77] and at Newbridge from 1859. The latter was the United Services Lodge (no. 215), and it still flourishes.[78] Early in the nineteenth century there was a lodge in the Kildare militia, but it had been closed by 1816,[79] and there were lodges in the 1st and 2nd squadrons of Lord Drogheda's Light Dragoons until they were disbanded in 1821.[80]

Records of the Curragh and the United Services Lodges at Newbridge give some impression of their membership. The minute book of the Curragh Lodge (no. 397) for the years 1917–1920 shows that the members were uniformly addressed as brother,[81] but in the attendance list individuals were sometimes given their army rank also, as for example, Brother Captain Gillies. In the Grand Lodge registers for the same lodge for the years 1906–1908, 49 out of the 92 entries of members were designated as soldier, and the remainder as non-commissioned officers, such as 'colour sergeant and sergeant of the Black Watch', and civilian employees of the military, including a canteen manager and a foreman of works. Only one officer is listed, Lieutenant Quartermaster John Barrow 19th Hussars.[82]

The register for the Newbridge United Services Lodge (no. 215) for the years 1906–1914 shows that the membership again consisted of a mixture of non-commissioned officers, soldiers and civilians.[83] However, it is the understanding of a current member of the lodge, whose grandfather was also a member, that there were many cavalry officer members, and that the Curragh lodge was mainly for other ranks.[84] Regiments with travelling warrants and which served in the Curragh would have held their own meetings. One such regiment was the 4th Dragoon Guards, and its brethren of St Patrick's Lodge (no. 295) included officers, N.C.O.s and men. The 4th Royal Irish Dragoon Guards garrisoned Newbridge regularly from 1823, and they were in the Curragh camp for training in 1863 and

1864, and again in the drill seasons of 1873, 1874, and for eight months in 1877/78. The lodge of the Worcestershire Regiment was named 'Glittering Star' (no. 322), and its membership also included all ranks. Battalions of the regiment were on the Curragh in the 1860s and 1880s, and for almost three years in 1891–93.[85]

At Newbridge the United Services Masonic Lodge (No. 215) building fund was to benefit from a concert held in the Masonic Hall on 9 June 1899. Despite the fact that the Master of the Lodge, Mr Church, a jeweller from the Curragh Camp, was unable to be present the concert was very successful. A packed hall enjoyed Miss Gray's pianoforte solo, Miss Gibson's song, Mr Walker's comic song and numerous other performances. In addition to the presence of almost all of the members of the lodge, and of Major O'Brien and the officers of the 14th Hussars, there was good attendance of the people of the town.[86] The cavalry were strong supporters of the lodge (which also served as the Curragh Lodge No. 397 until 1937) which was originally in Eyre street, and of the building of the new hall in George's street in 1897.[87]

The premier gentleman in county Kildare, Augustus-Frederick (1791–1874) 3rd Duke of Leinster, was Grand Master of the Craft, and Sovereign Grand Commander in Ireland.[88] The acceptance and social standing of the masonic order in Kildare was exemplified in July 1899 when a masonic choral service of the Provincial Grand Lodge of the midland counties was held in St Brigid's cathedral, Kildare. It was attended by people from the Curragh camp, and those from further afield could avail of specially reduced train tickets.[89]

That the Masonic Order was held in good regard in the Curragh district is not surprising. The strong association of the craft with the military encouraged support from the business and trade community whose livelihood largely depended on the army, and those of the Protestant civilian population which socialised with the military could also have sought membership of the brotherhood. The fact that the principal aristocrat and landowner in the county held for a time the highest office in the order must also have enhanced its position amongst all classes. Nevertheless the lodges would not have facilitated social intercourse with the local people to the extent that the hunt, the sports meetings, and the barrack entertainments did. But the concerts and other public occasions organised by the Masons attracted not only the families of the brethren themselves but also persons who appreciated such events, and who generally would have been associated with the ethos of both the Masons and the military.

The view expressed by the duke of Cambridge in 1857 had indeed come to fruition. That 'the friendly intercourse that is carried on between the armed forces in the country and its population in general is of great benefit to the army and of advantage to the country, and feeling favourable to the army is thereby encouraged', was certainly true of the relationship between the civilians and the soldiers in county Kildare.[90]

However, despite the obvious social benefits to the county from the presence of the soldiers, the most significant gain was a financial one. As with the social benefits, all classes prospered on the flow of money from the War Department and the regiments, and the importance of that contribution to the economy was best illustrated by the widespread dismay displayed with its withdrawal in 1922.

Chapter 10

TRADE IN AND WITH THE CAMP

Commercial activity generated by the creation of the camp on the Curragh could be divided into two areas, that necessary within the confines of the encampment, and trade conducted throughout the district and further afield.

The size and self-contained nature of the camp were amongst the features which never failed to impress visitors. Large numbers of civilians were employed there in various capacities, as manual, skilled, or unskilled workers in the barrack workshops, and as employees of the Board of Works which provided maintenance services to the buildings and installations. Civilian managers with their staffs operated the canteens, and a certain number of entrepreneurs were allowed to open shops intended to cater for some of the necessaries of the military and their families. Licensed sutlers were also permitted to regularly enter the confines of the camp to peddle certain commodities to the men. As late as the 1940s an old woman in a donkey and cart frequented the lines in the barracks to sell fruit and vegetables on a licence which had been originally granted by the British army. Domestic servants might also be employed by the married officers and the warrant officers quartered in the camp. Some of the servants might be recruited from the families of the camp rank and file, while others could be from the neighbourhood of the Curragh.

The weekly market in the camp attracted vendors of a wide variety of provisions. Smallholders from the hinterland of the plain sold their fruit, vegetables, eggs, milk, meat, poultry and other produce. As the *Leinster Express* reported on 22 September 1855, 'Markets were held on the west side of the camp on Wednesdays and Saturdays giving the place the appearance of Donnybrook Fair. Lines of knock-kneed tables and donkey carts fixed in a line displayed perfumery, soap, razors, blackening brushes, shoes, boots, clothing, beef, mutton, fish, turnips, cabbage, cheap ware and delph, etc. The donkeys wandered around amongst the crowds, while the traders, who had come from all parts of the country, sold to those who had no alternative at usurious prices.' The market also provided a social outlet for the occupants of married quarters, a neutral venue where they would meet residents of all the barracks in the camp, as well as civilians from the countryside and neighbouring towns.

Newbridge was the town which most benefited from the presence of the military as not alone did it cater for the considerable garrison in the town itself, but also for the men and women of the Curragh camp. There was a choice of commercial establishments in the town which catered for the requirements of officers and their ladies as well as those of the other ranks and the civilian community. The population of the town was increased by the residence there of personnel employed in the town barracks and in the Curragh camp, as well as the families of soldiers who were not on the married establishment, and a variety of camp followers.

To a lesser extent the towns of Kildare, Kilcullen and Naas also profited from the military. Naas was the county town, with a barracks which was generally occupied by up to a couple of hundred men, the seat of the county administration, a court house and jail, and a link to the Grand Canal. Kilcullen and Kildare were recreational and residential towns for the men from the Curragh camp. As a railway junction, Kildare further attracted constant trade from the camp.

Contractors of victuals, fuel, fodder, cattle, or horses to the military prospered on the trade. The movement of supplies to the barracks, by rail, road and canal, stimulated the transport business, while ferrying the passengers from the camp to the nearby towns and railway stations was a minor industry in itself. Professional men, such as doctors, dentists and lawyers, benefited from the increased population in the county, as did the churches of various denominations. In an era when church attendance was high (and compulsory for the men living in barracks), the clergymen's stipends were inflated by War Office payments if they officiated to the troops, by increased donations from the enlarged congregation, and by an increase in baptism, marriage, and funeral offerings.

During the construction of the buildings in the Curragh camp the contractors had opened canteens to cater for their workers, and subsequently a canteen for civilians remained in operation. The original plans for the camp included the provision of a sutler's hut in each regimental area, to a total of ten within the lines. Each hut was to be a 38 feet by 20 feet structure, half of which was to accommodate the canteen area. There was a small bar, with a counter and two stores. The remainder of the hut was living accommodation for the supervising non-commissioned officer, and the managers were to be appointed by the commanding officers. The sale of spirits was forbidden in the canteens. That all of the designated huts were not required as canteens is suggested by a schedule of 1858 in which one of the huts was listed as being in use as a stable for 40 horses.[1]

Camp Regulations specified that 'no strangers or civilians, except the wives and friends of soldiers, are to be allowed to visit the sutlers' booths, licensed to sell articles to the soldiers, and their huts must be cleared at tattoo'. The provost-sergeant, aided by an assistant provost-sergeant in each brigade, and a non-commissioned officer in each regiment, was held responsible that the sutlers complied with any regulations which might from time to time be issued for the preservation of order in the camp.[2]

The control of the canteens was one of the matters dealt with in an official report on barrack conditions which was published early in 1855. It saw the canteen system [throughout the army] as being run by 'capitalists' who paid a fixed rent for the premises (if provided), and 'head money', based on the number of men in the barrack. The manager was seen as 'merely a shopkeeper who speculates on his profit'. Spirits could not be sold in the canteens, nor could civilians from outside the barrack avail of the facility. The committee recommended that pensioned non-commissioned officers should be employed as canteen managers, and that the prices charged would be fixed by an officers' committee. It was recommended that newspapers and periodicals should be sold in the canteens.[3] Such a retired army man who sought employment in the camp, was Barrack Sgt. Boleshaw, late Troop Sgt. Major 17th Lancers, whose application to be a sutler was approved in January 1860.[4]

While it is not clear as to what extent the stipulations of that official report were adhered to in the Curragh camp, by September 1855 Messrs Duffy, Walsh, M'Donnell and Cleary managed establishments which were described as 'admirably kept . . . they close at 7 p.m. They can supply all necessaries, but only to their own battalions'. Soldiers who were found in canteens outside their own lines were taken into custody.[5]

However, the high standards of management seem to have slipped fairly rapidly as within a few months the Weights and Measures Sergeant had three of the managers fined at Kildare Petty Sessions.[6]

Subsequently there were changes of managers from time to time, as in October 1858 when Daniel Guidera, a Wine and Spirit Merchant of Barrack street, Dublin, applied for the use of the vacant civilian canteen at C barrack. In September 1859 a drink licence was approved for Peter Doyle, keeper of the civilian canteen, as he claimed that he was not getting any custom due to being unable to sell wine and spirits. It was ordered that, if necessary, a non-commissioned officer should be on duty at the canteen to prevent soldiers from going there. The Officer Commanding the Forces in Ireland, Lord Seaton, took a personal interest in the matter, and finally he recommended that in the next year arrangements were to be made to prevent civilians and soldiers from using the same canteen at C encampment.[7]

Difficulties with the canteens were to continue and in 1869 instructions were given that Mr McDonnell and his establishment might be ejected from the camp if he continued to ignore the regulations. In April of the following year the enlargement of the ten canteens at the camp was included in the Estimates, and the necessity for McDonnell's canteen was being questioned.[8]

Outside the lines approved sutlers might construct their own premises on a rent free site, and on the map of the Curragh dated 1855 (and included with

Lugard's narrative), four such huts are shown.[9] A painting of the camp as it was in 1861 depicts one of them as being thatched.[10] Private competition was the system devised by the military for the positions of brigade sutlers, and their testimonials were examined for the selection of honest traders. Agreements were entered into 'embracing such regulations as were deemed requisite to preserve due command over their occupation, and the discipline of the troops', who could conduct their business with whichever sutler supplied 'the best article at the most reasonable price'.[11]

After a few years in operation the trading system adopted for the camp was seen by the military to be successful. Lieutenant Col. Lugard held that 'the system adopted for the canteens and sutlers for this camp, by bringing competition into the market, is good in principle, but the regimental canteens, if viewed as a permanent measure, are too small, and the comfort of the non-commissioned officers and men is in consequence curtailed'.[12]

The importance of trade with the military was evident in 1872 when, to sustain its contract with the canteen of the 14th Hussars in Newbridge, the Phoenix Brewery 'delivered porter to them at a low price of 43/3d per hogshead, and gave sawdust, pint measures and spittoons free'.[13] No doubt, but the competition from other breweries, including Cassidy's in Monasterevan, encouraged all of the suppliers to be keen in their offers.

Casual traders entering the camp were also subject to camp regulations, rule nine of which was that 'the sutlers will be furnished with passes signed by the assistant quartermaster-general, which will admit them into the lines', while rule 11 required that 'all carts bringing baggage or provisions into camp will pass by the rear road entrance'.[14] But local people who had traditionally crossed the sward with their wares continued to do so, such as Mary Ryan who sold vegetables in the camp. She was acquitted of a charge of trespass in June 1865.[15] Three years later James Doolan and his wife from Brownstown were described as 'persistent offenders' by the ranger when he had them charged for trespass with their ass and cart between Knox's corner and the camp market, but that case was also abandoned.[16]

An insight into the importance of trading with the soldiers is found in a letter from Dr Samuel Chaplin of Kildare to the Mansion House Relief Committee in February 1880. Having explained that some 100 labourers, who normally worked for the local farmers, were in poor circumstances due to the bad harvest of the previous year, and terrified of the prospects of having to enter the workhouse, Dr Chaplain went on to give information on a further 24 labourers who:

> had passes for the Curragh camp, small pedlars, etc., who had been attracted hither by its proximity. The camp followers live by selling small wares, vegetables, watercresses, etc. to the soldiers or exchanging them for broken bread and broth meat. When the camp is full they are enabled to support themselves and families but when the number of

> men there are much reduced, as at present, they become very poor. Both classes (the unemployed labourers and camp followers) have suffered severely during the last winter and their sufferings have been intensified by the severity of the weather and the want of fuel . . . they cling to their cabins, fearful that if once they give them up others will take their place, or that they may be levelled and that when they are enabled, if ever, to leave the union house they will find themselves tramps and looked upon as degraded.[17]

That the facilities of the encampment were constantly improving was shown by the intentions of two Dublin military tailors to open workshops there. Mr Kohler, 7 Lower Sackville Street, Merchant Tailor, and Mr Robert Sexton, of J.B. Johnstone & Co., Military Tailors, Dawson St, sought exclusive priority for their establishments. The general commanding had no objections to their plans, if the Curragh ranger sanctioned the applications, as he believed 'the facility would be of service to the troops'. Another trader, Mr J.H. McCormick of Christ Church Place in Dublin, wanted permission to erect a temporary hut to sell stationery; again the General Commanding recommended the request, as he thought it would be useful to the troops, but first he ordered that the character and manner of business of the applicant was to be checked by the police.[18] Considerable trade was also generated by the provision of forage for the large numbers of horses required by the military. It was procured through the Army Agents for forage, listed on 18 April 1856, as Sir E.R. Borough; Armit & Co.; Messrs R. Cane & Sons; J. Atkinson, Esq.[19]

Sometimes the lives of the gentlemen in the camp were disturbed by complaints from outside traders: in May 1859 there was a letter from a shoemaker in Newry that officers of the 2 Bn 76 Foot, then in the Curragh, had not paid for their boots; he had been given a dud cheque and since then there had been no payment from the three officers concerned. The shoemaker based at the 2nd lock on the Grand Canal was also seeking 11/–, owed to him by 5 Coy 2 Bn 14 Regt.[20] In March 1860 Newbridge grocer Edward Mathews made a claim against the mess sergeant of the officers' mess, Military Train for £35.11s.5d for goods supplied. He had not seen the caution concerning default from military contracts in *Saunder's News Letter*, and he felt that the officers of the unit were bound in honour to pay.[21] It was not unusual for young officers to be short of cash or to be in trouble with their mess or trade bills. The provision of feed for his horse was a big expense for a gentleman, and occasionally there were difficulties, for example in July 1866 when forage contractor John O'Brien made a complaint to the officer commanding. 5th Foot that Capt. Chapman owed him money.[22]

Hawkers in the camp might also cause problems, such as those expressed by 'a father' writing to Lord Seaton in 1857. He claimed that they were selling 'immoral pictures and jewellery to allure officers into debt'.[23] Seaton's reaction to that complaint is not known, but he was not deterred from allowing hawkers into the camp, and in August 1859 he gave permission to a colporteur to enter camp to sell religious tracts.[24]

The apparent plenty in the Curragh Camp attracted not only colporteurs, but also other characters such as Denis Barrington O'Sullivan from Armagh, otherwise known as 'The Wandering Star'. He travelled from place to place peddling almanacs, song books and holy pictures, and was a frequent visitor to the camp. He was very popular with the soldiers as he would entertain them by composing humorous verse, or giving the solutions to the puzzles in the almanacs which were then popular reading material. He died in the Naas Workhouse in the winter of 1876, and was buried in a pauper's grave which remained unmarked for 20 years when a monument was erected.[25]

The carmen, who provided transport to the outside world, frequently caused annoyance to the military authorities as they trespassed on the camp roads. The general officer commanding sought support from Lieut. Col. Shaw R.M. in July 1861, but he said that the magistrates at the Newbridge Petty Sessions had decided that they had no jurisdiction over the carmen for such trespass, and that the opinion of the Law Advisers should be got.[26]

To cater for the accommodation of visitors and officers' families to the Curragh several hotels were opened. William Brereton of Kildare announced the opening of a commodious and comfortable hotel in November 1855, and following the queen's visit in 1861 the Victoria Hotel was functioning at the Stand House.[27] In Newbridge there was the Albert Hotel, advertised as being in the immediate vicinity of the Curragh Camp.[28] The Lion Hotel, at Kilcullen was owned by a well known cricketer, Peter Doyle, and on its opening day in July 1865 a cricket match and other sports were held in the spacious grounds.[29]

It was accepted that 'the large flow of money which necessarily takes place in consequence of the extensive military depot' benefited the strong little garrison town of Newbridge,[30] which had improved more than any town in Ireland in a few years on account of its proximity to the Curragh Camp. It was recommended as a base to industrialists.[31]

Local enterprise was also shown with the opening by Mrs Hilton of the Camp Inn at Ballymanny in January 1857. The weekly tenancy cost her two shillings and six pence, payable to the Crown.[32] However, another enterprising venture planned for the old Standhouse at the race-course failed to materialise. In 1860 a doctor from Blarney sought to lease the building with the intention of opening there a Turkish or Roman Baths. He emphasised 'the great utility of such baths . . . to the public as well as to the troops stationed at the Curragh and the barracks at Newbridge'. A lease of the premises was refused as it was not official policy to grant leases for private enterprise.[33]

An indication of the benefit of the presence of the military to the county may be deduced from the variety and quality of goods and services advertised in the county press, and from the advertisements placed in the papers for the supply of the camp and barracks. The *Kildare Observer*, printed and published at Naas, regularly carried this notice: 'The *Kildare Observer* is published in Naas, an Assize Town and important market centre, having railway and canal communication with all parts of Ireland, and is in the vicinity of the great Curragh Camp and Newbridge barracks, and circulates extensively amongst all ranks and classes in this and the adjoining counties.' Then there follows a list of the places in which the newspaper circulated, including the Curragh Camp and Newbridge.[34] But that the town and district did not get the expected maximum benefit from the camp was mentioned in a directory to the county in 1910. It said that 'it is a well-known fact that many thousands of tons of agricultural and garden produce pass through the town to the Curragh and district that could be easily produced in the neighbourhood', if less of the fertile country was used for grazing, and more for mixed farming.[35]

It was decided in April 1858 that new military purveying districts were to to be formed in Ireland, including the Camp Curragh, Newbridge, Naas, Kilkenny, Carlow, Duncannon Fort, Arklow and Wexford. Lord Seaton sought information on which would be the most suitable district for a headquarters, and the recommendation was that 'from the large force stationed at the Curragh Camp and Newbridge, one of these posts appear to be desirable'. The War Office approved of headquarters for the new purveying district at the Curragh Camp, and 1st Class Purveyor's Clerk Bush was appointed to take charge of the purveying district there.[36] As with other government appointees such officials could be moved to other stations, as when the assistant commissariat general at the Curragh was posted to St Helena in February 1859, and a new man was to be appointed in his place.[37]

A couple of years later changes were made in paying and accounting for army expenditure in the Dublin and Curragh Districts. All public monies hitherto dealt with by the Barrack Masters in the Royal, Richmond and Portobello barracks in Dublin were to devolve from 1 October 1861 on Deputy Assistant Commissary-General Long, and those dealt with by the purveyor and barrack masters at the Curragh and Newbridge were to go to Assistant Commissary-General Walker. Individuals who had accounts with the commissariats were expected to pay them promptly. If they did not they got reminders, like that issued on 31 October 1866 to Mr Matthews, deputy adjutant generals department, for £1.9s.8¼d. for bread at 1¼d per lb; 52¾ lbs of meat at 6¾d per lb, to be paid as soon as possible.[38] Advertisements for contracts were placed in the newspapers including the provincial ones. Contracts for coal, coke, turf and candles were sought in March 1865, and for palliasse straw in November, the tenders were to be submitted to the War Office, Pall Mall, London,[39] but when tenders were invited for forage, bread, meat, cattle, sheep and flour for the half year ending in November of the same year they were to be sent to the Commissariat Office, 4 Palace St, Dublin.[40]

It was not surprising that such extensive local purchase as that of forage could lead to unscrupulous dealing. There were stories in the middle of October 1865 that a number of quartermasters and sergeants in the camp had, over a number of years, accepted bribes from contractors. Quartermaster Malcom Kier, Military Train, was arrested and charged with taking bribes from the late forage contractors to the camp, Messrs Wright and Ryan of Clonmel. Mr Wright, in evidence before the court martial at the Royal Barracks in Dublin, presided over by Maj. Gen. Napier, said that his firm supplied hay to the Curragh in 1863 and 1864, and up to the present year. During that time he had given sums of from £5 to £50 to Kier 'so that he would always be friendly towards him, and endeavour to remove any objections that might be raised to the forage he sent in'. The firm had decided to discontinue the practice of awarding Kier, and the hay had been rejected when delivered. While the verdict of that trial is not known, a sequel to it may have been the failure of Francis Noonan, the representative of Messrs Wright and Ryan at Newbridge, to get the local appointment of inspector of nuisances in December 1865, even though he was recommended as being very connected with government contracts.[41]

The farmers sold beef as well as hay and oats to the camp, and the local small-holders also benefited from the proximity of the camp as they found a good market there for their milk, butter, fowl, eggs and vegetables. A farming family from South Green, Kildare, can recall the substantial business which their grandfather did with the army. James E. Dunne, who died in 1934, grew hay, alfalfa (esparto grass for the horses) and vegetables especially for sale to the army, which he also supplied with milk. The old man used to tell a story that when some contractors brought hay to the barracks in Kildare they were checked in at the front gate; then, in collusion with some of the soldiers, they drove the load out the back gate, and soon presented it again at the main gate.[42] That some of the contractors to the camp benefited illegally from their trade is suggested in local tradition where it is told that 'the carts of certain contractors never returned empty from the camp'.[43]

A notice in the first issue of the *Kildare Observer* for 1885 described John Maguire, Main street, Naas, as forage contractor to the garrison at the Curragh Camp, and the military barracks and R.I.C. at Naas. A year later he still held the contract, and offered for sale 'a plentiful supply of good old hay and oats suitable for gentlemen's horses for the coming hunting season, always on hand at moderate rates'.[44] A decade later contracts were invited for the supply of oats, hay and straw for six months from 1 May 1896, quotations to be sent to Louis Butler, Compor Farm, Portsmouth.

Contracts for one year were being offered in February 1885 for the purchase and removal of stable manure from the barracks in Dublin and in the Curragh and Newbridge,[45] and tenders were invited for the supply of coal, coke, turf, bogwood and kindling wood, as required by H.M. Land Forces in Ireland. Tender forms were available from the Commissariat Office, I House, Palatine Square, Royal Barracks, Dublin.[46] Other tenders were available for the winding of clocks at Dublin, Birr, Athlone, Curragh Camp and Newbridge for a three year period;[47] for the painting, papering and whitewashing of buildings at the latter two stations,[48] and for the supply of potatoes and mixed vegetables to the canteen of the Depot R.D.F., Naas.[49] Six month contracts were available for the supply of bread, meal, flour,[50] forage and palliasse straw, live cattle and sheep in March 1886,[51] while three year contracts were offered for the services of artificers at Newbridge barracks, the Curragh Camp and the Lock Hospital, Kildare.[52] In 1868 the acting commissary general at the commissariat office in the Curragh was Mr W. C. Ball, with a staff of two.[53]

Recorded evidence of the trade by local people with the military is found in the accounts of the Brophy family of Herbertstown, Newbridge, with an office at the rear of Flag Staff in the camp, and at Newbridge and at Corbally harbour.[54] They supplied several varieties of coal to the camp from at least 1871, using a team of ten horses stabled at Herbertstown to draw the coal from the canal boats at Corbally. From 1882 Richard McGlynn, Brownstown, also supplied coal to the military, and surviving account books indicate the transactions:

Regimental Canteen 73rd Black Watch: 1886: £24.1s.6d. (twelve months supply of coal).
Sergeants' Mess ASC: 1887: £6. 8s.4d. (Feb.–Nov).
H.R.H. Duke of Clarence: 1891: £2.14s.6d. (Aug.–Sept).
Officers' Mess A.S.C.: 1894: £15.12s.6d. (Jan.–Sept).[55]

The yearly total sales by McGlynn for 1883 amounted to £1,916.
In 1885: £1,219. In 1893: £1,344.[56]

In the neighbouring townland of Suncroft the Molloys of Newtown recall a family tradition of lucrative business with the army; they sold potatoes to the messes, and rented land for exercising war horses to the cavalry. A family connection of the Molloys, the O'Gradys of Liffey View, Rosberry, Newbridge, also rented land to the military, and they sold eggs in the camp. Ledgers from the premises of John Cummins, Albert House, directly across the main street from the barracks in Newbridge, record the sale of groceries and porter to the regiments there. The patronage of the military in the bar was also appreciated.[57]

The ferrying of passengers from the camp to the neighbouring towns and railway stations provided a living for numerous jarvies (Fig. 25),[58] but sometimes

Fig. 25. *Carmen (jarvies) wait at their stand in the main street at Newbridge. The ferrying of passengers from the barracks and the Curragh Camp was a major business. Mr E. Harrigan, Newbridge, was at the end of the nineteenth century regarded as 'first carman in county Kildare . . . his cars having the best pneumatic tyres'.* Photo; *National Library of Ireland.*

they felt that they had unfair competition from the military themselves. The Under-Secretary for War was asked in the House of Commons in June 1896 'if he was aware that military horses from the Curragh had for some years been used for private purposes, contrary to regulations, and injuring the hackney trade in Newbridge and Kildare?' He was requested to have the practice stopped. In his reply the under-secretary said that sometimes the commanding officers sanctioned such an arrangement, but now the complaint had been noted and a report sent to the commander in chief in Ireland. How he dealt with the matter is not known, but if there was a tradition of permitting army transport to be used for such purposes in isolated stations it is most likely that the complaint was ignored as the military never welcomed interference from the civil side.[59]

Sometimes a carman might receive a bonus from a passenger, as this story recalled by Henry Hughes, an old resident of Newbridge illustrates: 'at one time a certain jarvey who was hired to drive an officer to the Crown Hotel in Newbridge was told by his passenger as he was alighting from the jaunting car "I am sorry, jarvey, but I don't seem to have any change to pay you; will you take an image of my mother's head and face?"' The jarvey's reply was both crude and unprintable as to what the officer could do with the image of his mother. When the officer had walked away, the jarvey opened his hand to see what had been pressed into it; he found to his amazement a gold sovereign. The officer was a member of Queen Victoria's family, possibly the duke of Connaught, as his journey was from the duke of Connaught's residence.

Within the Curragh Camp itself private enterprise also flourished as demonstrated in a notice of October 1896 concerning the establishment of Sir Richard Dickeson and Co. They had moved premises from B Lines to a larger building at the south side of Hare Park, and the wet and dry goods canteen was under the management of Mr E.D. Woods; it was intended for the use of the workmen employed on the new buildings then in progress in the camp, and it was described as 'a great comfort as hot meals are provided at very ordinary rates'.[60] Another trader in the camp at that period was C. Lawless, grocer and provision merchant at the Market Square.[61]

The auctioneers Robert J. Goff of Newbridge thrived on their business with the military, the zenith of their trade coming with the long series of auctions prior to the departure of the British army in 1922. In the summer of 1885 they auctioned a choice selection of plants at the Market Square in the camp, on the instructions of General Frazer V.C.[62] Later in the year they disposed of the household goods, camp furniture, a horse, vehicles and a Victoria phaeton at H Lines in the camp for Col. Stokes, Commanding Royal Inniskilling Fusiliers, and other officers proceeding abroad.[63] When Goffs were advertising a licensed business house for sale in the Market Square, Kildare in 1889 they gave amongst its many advantages 'its proximity to the Curragh Encampment, it offers every inducement to add the hotel business to the above lucrative trade'.[64]

Included in the many advertisements in the local papers directed at the military were those of Callaghan and Co., Dublin, Military Outfitters;[65] Geo. Searight, Newbridge and Curragh Camp, agent for Dunlop's Steam Dye Works[66]; F. Blowney, Newbridge, Cricket and Tennis gear, Furniture for hire.[67] Edward Harrigan of Newbridge, carriage and undertaker, requested the attention of the military and general public to the fact that he had extended his undertaking business and offered coffins of every description; funerals could be arranged to any part of England, Ireland or Scotland.[68]

At Newbridge in 1887 the military sought to accommodate the Town Commissioners by permitting an encroachment along the main street outside the barrack wall for the holding of monthly fairs. The agreement required the Commissioners to erect a paling at 20 feet each side of the barrack gate, and a yearly rent of 12/– was to be paid.[69]

Just before the Christmas of that year there was a less agreeable contact between the Commissioners and the military when Commissioner Michael Clare resigned from office. It would seem that he had attended a meeting in the county at which, it was claimed, 'the government was maligned and condemned and criticised for their action against William O'Brien'. The fact that as a government employee Clare had not protested at the anti-government remarks annoyed the military and they told him to 'clear himself'. But they had not threatened to hand him over to the Provost Marshal, as had been rumoured in the town.[70] The sensitivity of the officers to any support for the Nationalist politician was no doubt aggravated by his presence at the Mitchelstown 'massacre' in the previous September, and the fact that in December he was in prison.[71]

The availability of contracts for the Curragh District, and for the canteens and messes stimulated trade in the county, and further afield. Tenders for the supply for a year of soap, candles, firewood, bottled ale, stout, wines, spirits, tobacco, stationery, provisions, coal, turf[72] and the delivery of oil on a house to house basis were invited early in 1899, and for flour and meat a few months later.[73] In the autumn the seasonal advertisements for forage and palliasse straw were published,[74] but as such notices were usual it did not necessarily mean that the contracts were awarded locally. That such was the case was to be publicly aired within a couple of years.[75]

However, the day-to-day spending of the soldiers in the camp and its environs kept a sizable amount of cash in circulation. For example, in January 1899 the pay of a private soldier was 1/– a day. From that monthly sum of £1.11s deductions of 2/2½d were made. These included 3d each for subscriptions to the cricket or football club, the shooting club and the library. One penny was deducted for hair-cutting, and the same for clothing. A man's laundry cost 1/3½ a month. In 1898 a stoppage of 3d per day for messing had been discontinued, but in reality it meant but a gain of one penny a day to the soldier as he lost 2d a day when his 'deferred pay' of that amount was discontinued. There could also be deductions for repairs to boots or uniform, for the purchase of articles of kit, or for barrack damages. The result was that the soldier was generally left with, on average, about £1.7.0 each month. The daily rate of pay then for a sergeant was 2/4d, and for a corporal 1/8d.[76]

In that year the average strength at the Curragh camp was about 4,000 men, and the disposable income of a private soldier from his pay of one shilling a day was slightly over eleven pence, the total amount of money which they might spend locally in a 31 day month was about £5,800. The then permitted 4% of private soldiers on the married establishment would have had less surplus cash available than their single comrades.[77] Non commissioned officers had more to spend, while the officers' affluence depended to a large extent on their family circumstances. Many of the single gentlemen might have had an allowance from their fathers, while those in wedlock could have the support of their wives' dowries, and they also had access to credit facilities.

The spending power of the military on the Curragh varied with the numbers which were there, and in the summer drill months the massing of troops brought substantial benefit to the neighbourhood (Figs. 26, 27). In the summer of 1899 the estimated strength in the camp was 9,000, including 1,500 militia.[78]

That the population of Newbridge and the other towns within reach of the Curragh camp, and of a wider farming area, undoubtedly prospered on the

18 ADVERTISEMENTS.

By Royal Warrant — To H. M. The King.

McCABE'S

FOR

Fish, Poultry, Game and Ice.

FIRST-CLASS QUALITY ONLY. MODERATE PRICES. — ALL ORDERS RECEIVE PERSONAL ATTENTION

South City Markets,
11 Rathgar Road,
48 Upper Baggot Street, } **DUBLIN**

85 Main Street, BRAY.

Central Road. CURRAGH CAMP.

Telephones:		*Telegrams:*	
DUBLIN (S. C. M.)	*2481 (3 lines)*	*DUBLIN (S. C. M.).*	*"POULTRY."*
BRAY	*22*	*BRAY,*	*"McCABES."*
CURRAGH	*7*	*CURRAGH*	*"McCABES."*

T. J. CUNNINGHAM,

Family Grocer,

Tea, Wine and Spirit Merchant,

ALE AND PORTER BOTTLER.

Camps and Families supplied with Jameson's and Power's Pot Still Whiskey, Dewar's Celebrated Scotch Whiskey, Guinness's Stout in Cask or Bottle. A Trial Solicited.

Telegrams: CUNNINGHAM, SUNCROFT, CURRAGH.

JOSEPH McCANN,

Practical Coach Builder, Coach and Motor Body Maker.

Every class of Carriage Work Made, Repaired, Painted, and Trimmed, on Most Moderate Terms.

Fig. 26. *Trade flourished both in the camp and in the district during the years of British occupation, as is evident in the pages of* Porter's Post Office Guide & Directory for Counties Carlow and Kildare. *1910.*

Fig. 27 The hut of Eason & Son Ltd. in the Curragh Camp. This branch of the Dublin firm invited inspection of its 'large and varied stock of pictorial postcards, which comprised Curragh Camp views, many specially taken for us, Beauty Spots of Ireland; Actresses, Comics, Manoeuvres. New stock of latest productions received almost daily'. Photo: *Curragh Camp and District. Illustrated and Described.* (Dublin) *c.* 1908.

presence of the army is certain, but to quantify that financial benefit is difficult. Newbridge had developed from being but a river crossing in 1811[79] to the status of a one side shopping street opposite the barracks in 1858.[80] From then on the town prospered, and the increase in the value of the various business premises is reflected in corresponding increase in rateable valuation. For example lot 1 in main street which had a valuation of £3.10s. in 1852, was valued at £16 in 1899 and £23 in 1909, while over the same period lot 37 went from a valuation of £8 to £19. Even the lesser streets profited from the military presence with lot no. 1 in Friary road going from 5/– in 1852 to £6.10s. in 1899. By the latter date the town consisted of nine streets and four lesser designations.[81]

The income to the Curragh area came from a variety of sources: from the War Department in contracts for victuals, fodder, fuel, etc.; in the wages of local civilian employees in the barracks; in the salaries of the officers and the pay of the other ranks and the allowances to their families. In 1885–6 it was claimed in Limerick that the presence of a battalion was worth £60,000 a year to the city.[82] In 1887 there were four infantry battalions stationed at the Curragh camp, as well as an attachment of engineers, commissariat and transport staff, medical staff, chaplains, etc. At Newbridge there were the 11th Hussars and two batteries of artillery. On the estimate from Limerick the military in the Curragh and Newbridge in 1887 would have been worth over £300,000 per annum.

In 1911 three infantry battalions, two cavalry regiments, and supporting divisional troops (a total of about 4,300 men) were stationed in the Curragh camp. It is estimated that their annual income was in the order of £127,000.[83] The daily pay of an infantry lieutenant colonel that same year was 23/–, of a lieutenant 5/3, a sergeant major 5/–, a sergeant 2/4 and a private 1/–. It can be assumed that all of the weekly pay of the other ranks was spent locally, and a proportion of the monthly salaries of the officers.[84] Then a pound of beef cost from 6½d to 10d, a pound of mutton from 4d to 10½d, 6 lb of bread 5d, 7 lb flour 10½d, tea 1/6½d lb, milk 3½d pt, butter 1/2½d lb, and a pint of Guinness 3d.[85] In addition to the military personnel in the camp in 1911 there were the families of the married men, and civilian workers and their families who resided there, bringing the total population to 6,055 persons.[86]

That same year in Kildare town the total population of the barracks was 1,020 persons, of whom 800 were soldiers; the balance being their families, civilian workers and their dependants. Newbridge barracks had a resident population of 823 persons, of whom 673 were soldiers, and at Naas the barracks was the home of 130 soldiers, but the total population was 230 (Fig. 28).[87] Civilians who were employed in all of the barracks in the county, and who resided outside the barracks, were not included in the barrack populations, but they amounted to a considerable number of persons, estimated in 1922 to number 302 Board of Works employees, and between 500 and 900 in direct army employment.[88] The

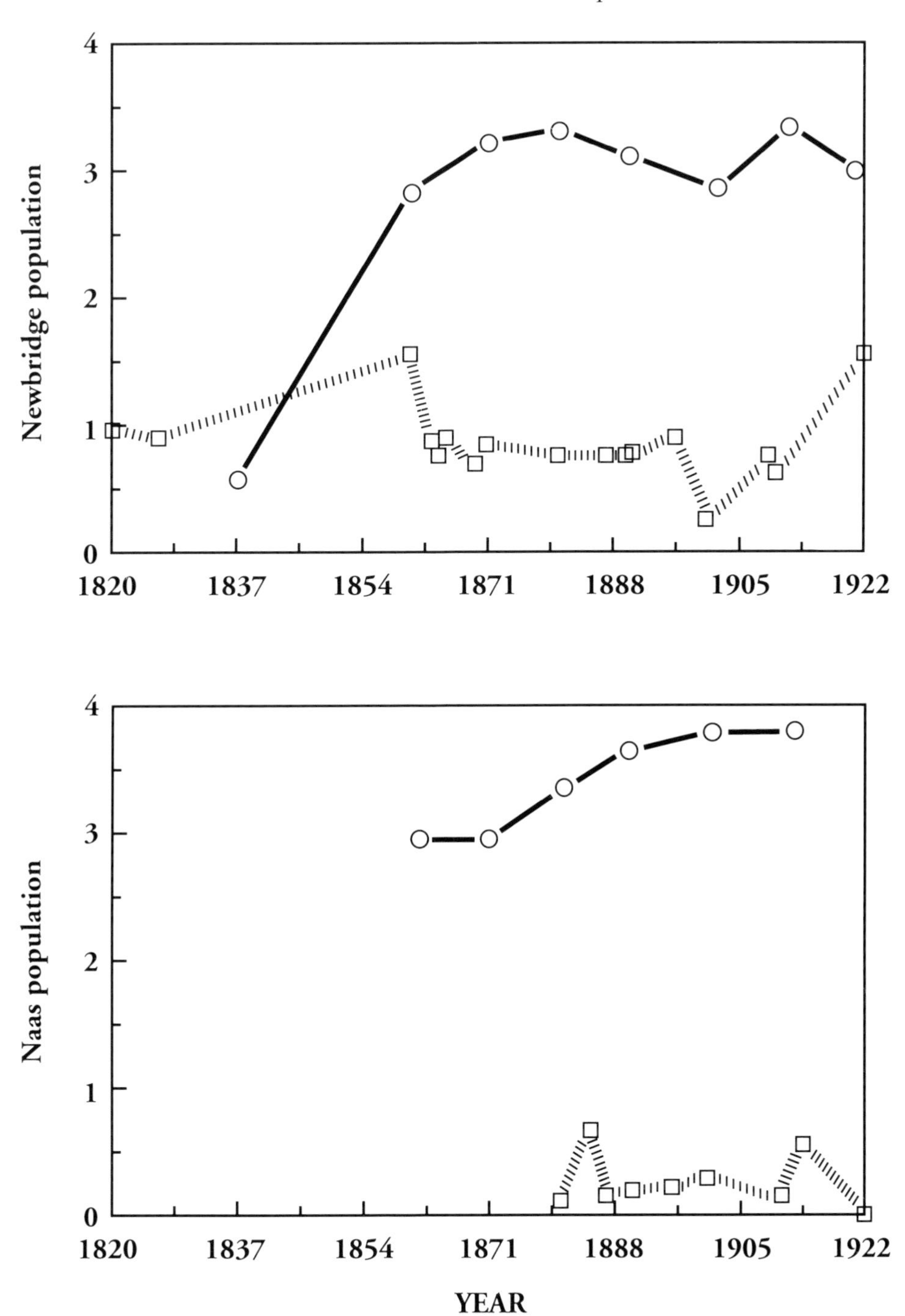

Fig. 28. *Population in thousands recorded for the towns (circles) and army barracks (squares) of Newbridge and Naas.* Source: *Census reports*

total military population of the county in April 1911 was 5,923, with 2,202 other residents in barracks, and non-resident employees averaging 1,000 (plus their dependants, not estimated), the total of which amounted to over 9,000 persons directly connected with the military resident in the area, or 7.5% of the total county population of 66,623 (Fig. 29).[89]

In the month of March 1914 life seemed to be proceeding as usual in the Curragh camp and some progress was being made in dealing with the perennial problem of traders there. It was decided that leases of plots within the camp were to be sold by auction for the building of shops, but the proposal alarmed the pass-holders who normally traded within the camp. They were assured by the garrison adjutant that they would not be excluded from the camp, a decision welcomed in the county press. One journalist expressed local feelings: 'These people have been for many years so mixed up and concerned with trading on the camp that they looked to it as the chief means of maintaining and sustaining their families and the decision of the authorities will do much to promote the continued popularity in favour of the sporting facilities which have always been at the disposal of the officers of the district.'[90]

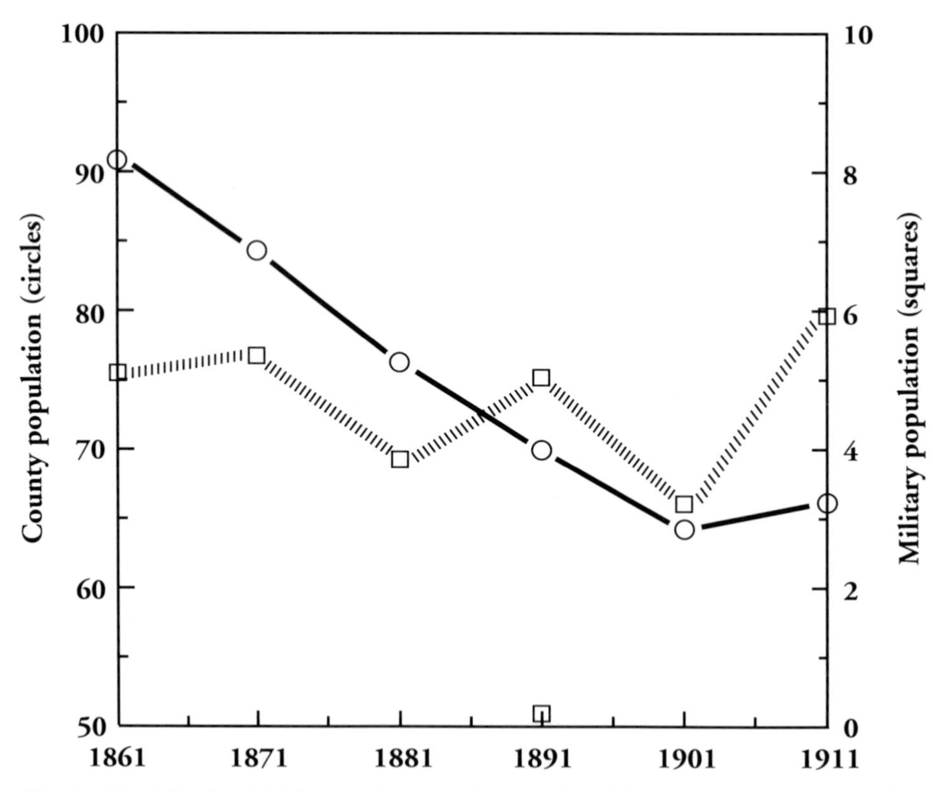

Fig. 29. *Total (Circles, Solid line) and military (squares, dotted line) populations in thousands, of county Kildare from 1861 to 1911.* Source: *Census reports.*

But the shop keepers in Newbridge were worried that their trade might be damaged. They were reassured by Capt. F.P. Dunlop that it was not the intention of the military to 'form any protection for shopkeepers in the camp, competition was welcomed'. The traders were described as deserving persons who had the privileges for many years, and their trade was practically their sole means of livelihood. Since the camp had been built the traders had given satisfaction, with quick sale and light profit.[91] Indeed, the importance of the camp in the local economy was noticed in a local paper in April in a report concerning the annual military estimates. The building of a 'large set of buildings in conjunction with the military hospital [in the Curragh camp] was giving much needed employment to tradesmen and others in the neighbourhood of the camp'.[92]

The declaration of war on 4 August 1914 by Britain on Germany was to create a completely new economic situation in not only Newbridge, but the entire district. The provision of food, fuel and other services and requirements to the ever increasing forces gave a considerable boost to trade. Before the war started tenders were sought for bread, meat, palliasse straw, fuel and wood, and for the purchase of old straw, scavenging, the removal of privy soil, and the supply and conveyance of water for the training season to the troops be encamped on Maryborough Heath.[93] From the month of August onwards advertisements appeared seeking quotations for the supplying of meat, forage and palliasse straw, bread (or baking from flour supplied by the War Department), potatoes, oils and wood, live oxen, and for the purchase of old straw, scavenging, and the removal of soil from the dry earth closets, as well as for the conveyance of stores at the Curragh, Newbridge, Kildare, Naas, Coolmoney and the Glen of Imaal, and other stations. Another service for which tenders were invited was the performance of military funerals.[94]

The Newbridge Sanitary Steam Laundry Co. Ltd held contracts for the barracks in Newbridge and the Curragh in 1914 when some financial difficulties arose. These were resolved, but a couple of years later, despite the general wartime boom in trade, the firm was in liquidation, with £1,900 of military contract money not accounted for.[95]

The wider commercial interests were also conscious of the opportunities in the Curragh area was demonstrated in a series of advertisements in the local papers for Tinori, a cure for corns, which was available in the chemists' shops in Kildare and Newbridge. It was illustrated with a picture of the inspection by a non-commissioned officer of soldiers' feet. The message was that newcomers to the barrack square might appreciate Tinore.[96] However, while recruits might indeed suffer from ill-fitting boots and hours of foot-bashing on the barrack square, their contribution to the local economy was more likely to be to the coffers of the regimental canteen or the publicans in the vicinity of the barracks than to pharmacists.

The figure estimated in 1921 as circulated by the military in the Curragh district was from £90,000 to £100,000 a month. Another estimated £28,000 monthly (£7,000 a week) was paid in wages to civilian workers employed in the camp.[97] The extent to which the military benefited the economy as a whole was not to be fully appreciated until they evacuated their stations in 1922.

Chapter 11

REBUILDING THE CAMP, 1870–1922

Further economic benefit was to accrue to the Curragh district with the rebuilding of the camp from 1870. As with the initial construction of the encampment a large workforce was employed, and the extra money circulated in the area increased the income of the already prosperous local community. Such manufactories as the brick works at Athy, Newbridge and Ballysax were at full output producing the bricks for the new barracks.

During the last three decades of the nineteenth century a building programme was undertaken at the Curragh which saw the transformation of the camp from one of wooden hutments to that of permanent, mainly red brick, buildings. The construction of concrete huts in the camp was under way by 1871, and three of them were occupied by the troops in May of that year.[1] In October 1878 and April 1879 War Office instructions were given for the handing-over of more concrete buildings in the camp for the use of the troops.[2] The huts, each to hold 18 men, and estimated to cost £5,000 were built by military labour in C Square,[3] and ten years later the barracks was given a cook house, drill shed and reading rooms.[4] However, in 1885, the concrete huts were found to be damp and unsuitable for use as billets by the troops.[5]

The large numbers of horses on the camp establishment necessitated the provision and maintenance of a considerable amount of stabling, and in 1873 some improvement was made to the stables at Donnelly's Hollow,[6] and work was undertaken on more stables in 1877.[7] In February of the following year proposals were made for the erection of sheds for 12 waggons and harness stores for the transport of three regiments in the Curragh;[8] that same month 46 waggons and 105 horses were allotted to the camp, and 10 waggons and 24 horses to Newbridge barracks. This transport was independent of that of the A.S.C., and was intended to be employed upon general service in the District.[9] The estimates for the year 1879/80 included the cost of building stables for 12 officers' horses of the Royal Artillery at A Square in the camp.[10]

Nor were the commercial amenities of the camp being neglected. By the end of 1877 a building had been erected in the market place and it was available for letting to a suitable tenant for use as a draper's shop.[11]

Improvements were also being made to the sanitary arrangements of the camp in 1878 with two-thirds of the cess pits having been filled in and drains being laid to the mains sewerage to the sewerage farm. Dry earth stores were being planned. Surgeon General J. Lamprey's Medical Report on the camp for 1882 expressed regret that 'during the summer months a bad smell from the sewerage farm, to the south of the camp, was complained of. It appears that the sewerage poured on the land is not sufficiently absorbed, and a bad smell, therefore, results. The subject is under consideration, but as there is no fall for the fluid it seems difficult to find an outlet for its discharge'.[12]

Early in 1880 preparations were under way for the building of a new Ordnance Reserve Store in the barrack compound in the camp,[13] and sites for dry earth stores approved.[14] At that time the maintenance of the roads and parades in the camp was one of the duties of the troops stationed there.[15] Plans were made for the building of a prison with 52 cells in 1882,[16] and three years later it was completed but unoccupied,[17] and in 1883 the construction of a garrison bakery was planned. Plans to appropriate Naas jail as a military prison were not proceeded with.[18]

Conditions on the Curragh in winter-time were graphically described by a journalist who went to visit there early in December 1885. He found it was not a pleasant experience, even when approaching there by car from Newbridge as neither the horses, cars nor roads were good. By that time many of the wooden huts in the camp had been replaced by concrete ones, but even most of those had been condemned, due to dampness. However, to the visitor's surprise, buildings of the same material were still being constructed, as well as bunks for sergeants with connecting links to the huts. He surmised that when the 'concrete tents' were completed they would be immediately condemned! The few buildings which had been built of brick were found to be very satisfactory. The roads within the camp were described as 'simply execrable, abounding in deep ruts, and every description of mantrap'. While the observer pitied the officers and men who had to traverse them when on duty, and very often in the dark as the oil lamps which illuminated the camp were frequently quenched by the wind, he could not understand why the 'considerable number of prisoners and defaulters ever to be found in a large camp' were not employed in repairing them, rather than on time-wasting punishment drills.

What really annoyed the pressman was the fact that the newly built military prison, 'an imposing structure . . . with every modern requisite for prisoners, heated throughout with hot air, altogether the most luxurious building in the camp', was unoccupied, even though it had been completed for 14 months, and certified for 56 prisoners. Prisoners on sentences of over 42 days imprisonment were still being sent to Dublin or Portarlington 'at a considerable cost in railway fares . . . and the loss of service of the escorts'. Men on lesser sentences were confined in the guard-room cells in the various barracks.

It was noticed by the journalist that six of the eight hospital blocks were completed, and that a verandah 'similar to that of St Thomas' Hospital, London', was to be erected. At Donnelly's Hollow the stabling which could house two cavalry regiments was 'in utter ruin simply for want of looking after', while stabling for 92 horses for the Army Service Corps was provided in the current estimates. The new bakery he considered to be the one successful building in the camp, and bread of excellent quality was baked there. As for the officers' messes, with the exception of that in C lines, they were poor and small. Even that in C lines had a bad mess-room 'due to the wretched quality of glass put in by the Royal Engineers'. G square had the most accommodation with quarters for 40 officers, 12 staff sergeants, 14 married soldiers and 700 men. The conclusion reached by the *Observer's* reporter was that the condition of the camp afforded 'ample reflection to the unfortunate taxpayer, who there clearly sees how the public money is misspent'.[19]

The gloom of the encampment, observed by the journalist, is also captured by Walter Osbourne R.H.A. (1859–1903) in his painting in muted greys of the camp 'from Gibbets Rath' [sic]. Apart from the clock tower and a church the only high buildings that are visible from the artist's viewpoint are the officers' mess in 'C' lines and the prison, which suggests that Osbourne's visit to the plain was at about the same period as that of the man from the *Kildare Observer*.[20]

However, the contents of the *Monthly Official Directory of the Curragh Camp and Newbridge* for December 1887 indicate that even if the structural state of the camp was not perfect, the social aspects were numerous, for the officers at any rate. The Officers' Recreation Club catered for the mental and physical relaxation of the gentlemen, while the many facilities of the camp gymnasium could be enjoyed by all ranks (Fig. 30). A regular fleet of hackney cars, the hiring of which was covered by the bye-laws made by the Newbridge Town Commissioners, plied between the camp and the outside world (a footnote recommended that a bargain should be made with the driver), while nine trains a day left Kildare, or seven from Newbridge, for Dublin.[21]

The quality of barracks both in England and Ireland had been adversely commented on for some time past, and an especially condemnatory report had been made on the Royal Barracks in Dublin in the 1880s. No doubt, but the publication of critical reports on barracks, such as that in the *Kildare Observer* in 1885, annoyed the government, and, finally, in 1890 a barrack renovation programme was commenced.[22]

Fusilier Horace Wyndham, a 'gentleman soldier' who underwent a couple of month's musketry training in the Curragh camp from February 1891, gave a description of conditions there when he published his experiences of army service in 1899. With his platoon from Wellington barracks in Dublin he had travelled by train to Newbridge under not very comfortable conditions as 'sixteen warriors'

Fig. 30. *A gymnastic class posing on the roof of the Curragh Gymnasium in 1897. Teams from this camp took part in the Royal Irish Military Tournaments held at the grounds of the Royal Dublin Society in Ballsbridge, Dublin.* Photo: Curragh Camp and District. Illustrated and Described. *(Dublin)* c. *1908.*

were crammed into a compartment intended for ten. From Newbridge the party marched to the Curragh and, though they were in 'marching order' (with packs and rifles), Wyndham preferred it to the train journey.

The accommodation in the camp he found was either in huts or under canvas as only enough stone barracks for a single battalion had then been constructed. He was billeted in a hut in 'M' lines, and his opinion of the structures was that 'they were not things of beauty, and internally they were about as unsuitable for living in as could well be imagined. From their great age, they dated from the Crimea, or from some even more remote period, the wooden planks forming the walls fitted so badly that at night the unfortunate occupants were half frozen. Winter in one of these huts is a terrible trial for any but troops who have just returned from a tour of duty in Halifax. They had no fire-places, but were provided instead with rickety stoves. Peat was the only fuel issued, and the allowance of this was very sparing. At night the rooms were filled with an evil-smelling smoke, which the zephyrs that whistled through the numerous holes in the walls that answered for chimneys distributed in all directions. Everything was in a most dilapidated condition, and I often wondered that our bungalows were not blown over altogether during the furious gales that prevailed during the winter'.[23]

The plans for the rebuilding of the Curragh camp were made in the offices of the Director of Fortifications and Works or of the Royal Engineers in England, and in general the buildings erected in the Curragh were adaptations of master plans which had already been used in England. An example is 'Standard Plan No. 23 Serjeants' Mess for One Battalion of Infantry' to accommodate 58 members; and 'Serjeants Mess for One Brigade Division of R.F.A.' to accommodate 28 members, prepared by the Director of Fortifications and Works in 1905. Buildings based on those plans were built at Gough barracks in the Curragh.[24] In recent years when a retired brigadier general and a lieutenant colonel of the British army met at a function in the former Beresford barracks officers' mess in the Curragh they immediately recognised it as a copy of their own former regimental mess at Colchester.[25]

At the Curragh major building programmes were under way by 1893 with the cavalry barracks getting 'superior married quarters', barrack blocks and stables, and a recreation establishment.[26] Work was in progress on the hospital, and it continued until withdrawal in 1922.[27] Officers' and non-commissioned officers' messes were built in Gough and Keane barracks, and the officers' mess at Beresford barracks was enlarged.[28] A cook house was built at C Square, and a guard house for two units of the Royal Engineers. A women's wash house, warrant officers' quarters and a mobilisation stores were completed.[29] In 1896 a site near the Flag Staff was selected for the proposed water tower and fire station,[30] for which a series of designs had been made.[31] The tower, which was to dominate the camp until the 1950s (when the campanile of the new Roman Catholic church was built), was completed in 1908.[32]

By the month of January 1896 eight of the ten blocks of the cavalry barracks had been completed, handed over and partly occupied by the 10th Hussars. Eight other blocks were under construction.[33] The engine driver at the Hare Park pumping station occupied newly built quarters there in March,[34] and by September the site for a barracks for two infantry battalions was selected at D and E Lines.[35]

To facilitate the many civilian workers employed in the construction of the new camp buildings the firm of Sir Richard Dickeson and Co. in 1896 opened a wet and dry canteen near Hare Park,[36] and the soldiers were forbidden to enter the squares under reconstruction. The quarters and canteens of the civilians were out of bounds to the military, and their barrack rooms, quarters and squares were forbidden to the civilian employees.[37] One of the contractors employed in the re-building of the camp was Patrick Sheridan from Birr. He was engaged in the construction of the Catholic Soldiers' Institute[38] and of the hospital. For the convenience of proximity to the extensive contracts Sheridan had moved his family from Birr to upper Henry Street in Newbridge, where he was joined by two of his brothers. They established a very successful building firm in the town and eventually owned a large area of ground there. Other employment then generated by the rebuilding in the Newbridge area was at Greatconnell where bricks were manufactured for the camp. The area is still known as the brickfield.[39] By an agreement dated 23 January 1894 the field for the bricklaying had been leased from James Kelly. A royalty of 2s/9½d per 1,000 bricks was paid.[40] Pallin's had a brick manufactory at Ballysax, and at Athy there were no less than 13 brickyards at Churchtown which manufactured bricks for the camp.[41]

The Army Medical report for 1896 observed on the fact that the original buildings in the camp were becoming outdated. The old wooden huts in the right wing of the camp were criticised and their cubic space was found to be insufficient for the number of occupants. The camp latrines, with the exception of those in the hospital, were of the old dry earth system, and in need of constant repair. The camp water supply was described as 'pure but hard'. It remarked on the fact that new barracks were under construction in the left wing of the camp, and part of the cavalry barracks had been occupied in February of that year. Again, the old huts there were described as unsatisfactory. The ablution rooms were damp and cold, and there was no hot water laid on. Married quarters had been inspected and the huts were found to be draughty and overcrowded, but the brick buildings occupied by the married soldiers were good. The drainage system of the camp was found to be bad. It was recommended that drying rooms should be provided in all barracks. However, the general health of the residents was considered to be satisfactory.

At Newbridge barracks it was noted in the 1896 report that the surface drainage of the roads needed improvement, and that the supply of fuel for the hospital was not enough for bath water to be heated more than two days a week. Nor was there hot water in the ablution rooms. The married quarters there were found to be insufficient, and some families were on a lodgings list.[42]

That the building of the new barracks in the camp was to be continued was known in July 1899 when a Parliamentary Return Schedule of Military Works listed the sum of £50,000 for the reconstruction of the large camp in the Curragh.[43] A couple of months later visible evidence of the work was the laying of the foundation stone of the Alexandra Nurses' Home by Miss Combe, the daughter of the commanding general. Mr Sheridan of Newbridge was the contractor.[44] A couple of years later the Home was to be the subject of an illustrated feature in *The Irish Builder*.[45]

By 1902 the rebuilding of the west end of the camp was largely completed and the five new barracks had been named, from the west to the east they were: Ponsonby, Stewart, Beresford, Army Service Corps and Royal Engineers.[46] Gough and Keane barracks on the east side were to follow. Each of these barracks had a range of married quarters; the finished Keane barracks had 43 quarters, 15 of which were for officers and 18 for civilians; Gough had 48 quarters for other ranks, Engineers had 44 including 3 for officers, 4 for civilians and one for the verger. At the hospital there were two quarters for officers and two for civilians. There was a total of 58 quarters at the A.S.C. barracks, including one for an officer and one for the Master Baker. In Beresford barracks there were 65 quarters, including 14 for civilians and 3 for officers, and in Stewart barracks there were 62 quarters, with 4 each for officers and civilians. Ponsonby barracks had 65 quarters, including 6 for civilians, and five for officers. The post office, with staff residence, was completed in 1900, and the water tower in 1908. The gentlemen commemorated in the names of the new barracks were all distinguished Anglo-Irish soldiers. Major Gen. Sir Frederick Cavendish Ponsonby (1783–1837), a veteran of the battle of Waterloo, was from county Kilkenny. All of the others had served in the Peninsular War. General Charles William Stewart (1778–1854), 3rd Marquess Londonderry was a historian of that war. Marshal Viscount William Carr Beresford (1768–1854), was from county Waterford. Field Marshal Hugh Gough (1779–1869) was from county Tipperary; General Lord Keane (1781–1844) was from county Waterford.[47]

While the buildings of the hutment camp of the 1850s had largely begun to disappear by the end of the nineteenth century, a few of the original stone structures had survived, most notably the clock tower, racquet court and pumping station. Of the wooden buildings the east and west churches were the principal ones remaining.

Following the original plan of Lieutenant Colonel Lugard, the new camp had the same road system and open aspect and into which the Victorian red-brick barracks were placed.[48] The spacious and elegant officers' messes, imposing barrack squares surrounded by billet blocks, drill sheds, offices and guard rooms, the extensive grassed exercise areas rimmed with broad-leaved trees, and the terraces of married quarters, some with verandahs, gave the Curragh camp a character unique in Ireland. Building and improvements continued in the camp until 1922, and before their evacuation of the place the British army gave instructions that any projects in hands were to be completed.[49]

In fact the camp was a self-contained town with churches, hospitals, libraries, recreation centres, canteens, sports facilities, including a swimming pool, shops, bakery and abattoir and with a permanent population of about 4,000 men, and women and children. With its extensive ranges of two-storey red brick buildings, large wooden churches, clock tower and massive water-tower the camp rose prominently from the flat green sward of the plain as a permanent obtrusion. The planting of belts of conifers in the 1940s may have hidden much of the red brick, but the trees themselves were an imposition which failed to mask the fact that the settling of the military on the Long Hill in 1855 had been a major invasion of the age-old openness of St Brigid's pastures; her cloak had forever been rent.

However, if the imposition of the now permanent camp on the plain had serious environmental consequences, its existence had created a whole and thriving new society there and in the neighbourhood. A consequence of the creation of that new society was referred to during the collection of folklore in County Kildare in January 1939 when Maud Delaney from Brownstown Girls' School wrote to the collector saying 'I regret to say that the people living in this district have no stories of the old days to tell, no customs to keep up, few superstitions and only one worthwhile "cure" for a disease they call *mion-féarach* . . . the reason is that the parents of the children, occasionally the grandparents, came here to work in stables, in officers' houses, in the military camp or settled here near the camp after service in the British army . . .'[50]

If the folk life of the district may have been damaged by the advent of newcomers, there was no doubt but that the financial status of the county had been enhanced by the military residents, and even if the soldiers had brought some negative social consequences to the area, they were far outweighed by the advantages gained.

Chapter 12

BOER WAR, 1899–1902

Early in the month of January 1899 when the reliefs for the year were announced, it was ordered that the 6th Inniskilling Dragoons would move from Dundalk to the Curragh, and that the 8th (King's Royal Irish) Hussars would go from there to Ballincollig. From the latter station the 17th (the duke of Cambridge's Own) Lancers were to transfer to the Curragh. From there the 2nd Battalion Royal Fusiliers (City of London Regiment) was to go to Athlone, and the 17th Field Company Royal Engineers was to arrive at the Curragh from Aldershot, with the 7th Field Company R.E. going to that station from the Curragh Camp. At Newbridge barracks M Battery R.H.A. was to arrive from Woolwich, and Q Battery R.H.A. was to leave Newbridge for return to Woolwich.[1]

For the artillery units 1899 was to be a memorable one as the ranges in the Glen of Imaal in west Wicklow were to come into service.[2] Improvements in the range and accuracy of the guns required that longer practice ranges should be available; gone were the days when firing had taken place on the plain of the Curragh. Up to 1896 all guns in Ireland had been practice-fired at sea, but floating targets were not found to be good training for war. In 1887 artillery officers, at their own expense, had started traversing Ireland in search of suitable ranges, and eventually the glen of Aghavannagh was investigated. Simultaneously the Glen of Imaal was rediscovered (there was a small barracks there at Leitrim, built after 1798) and it was thought to be much more suitable. It had a variety of firing positions to which batteries could be moved, and the mountain of Lugnaquilla made an ideal butts. But some of the tenant farmers in the valley objected to the ranges, and the search for a range had to be resumed.

Eventually an area at Glenbeigh in county Kerry was arranged, and artillery practice commenced there, and, even though it did not have much variety, it was better than firing the guns at sea.[3] However, a serious disadvantage was that the local people 'concealed themselves over the range to gather bits of brass and copper from exploding shells before range parties could do so'.[4]

Some five years later another attempt was made to procure the Glen of Imaal. Major Guinness surveyed the lands of the earl of Wicklow, and tenant farmers held a meeting at Knockanarrigan to protest at the compulsory purchase of their lands for a range.[5] However when the earl of Wicklow, who was the major landowner there, was threatened with compulsory purchase, the Glen of Imaal was acquired. Tenants' interest in 14 townlands was purchased for £13,000, and they were readmitted as War Department tenants with a reduction in their rents.[6] It was even rumoured that, despite the original objections to the range, farmers then came offering their land for inclusion in the firing area to avail of the generosity of the War Office.[7]

The new ranges in the Glen of Imaal were designated as the location for the artillery practice in that season of 1899, and a detail of the periods when the brigade divisions would fire there was issued. Each division was to have 2½ weeks in the glen, commencing on 20 May. As the batteries moved to the ranges the town of Baltinglass experienced the passage of the artillery from the south of Ireland, and Dunlavin was also soon to enjoy the excitement of gun trains.

At 10.20 a.m. on 23 May 1899 the first shot was fired in the glen, the inhabitants of which had lost their 'happy valley'.[8] The glen village of Donard, and the towns of Baltinglass and Dunlavin, were to become accustomed to the military presence, and inevitably they benefited financially from it. One old woman who lived near Dunlavin recalled the guns and waggons passing through the town, and the enjoyment of the entertainments given in the camp in the glen. Then drays were sent down to convey the people from the town to the band concerts and parties, occasions they anticipated annually.[9] When the regiments were *en route* to the glen they halted outside the town to feed the horses from their nose-bags. One local family used to collect the scattered oats when the men had left, and in those frugal times a free bucket of oats was a welcome bonus. In local lore the military were seen as generous, and when the Kildare Hunt met in Dunlavin the young men of the area sought to hold the horses, or open gates, as they could earn up to half a sovereign, which was then almost a week's pay. The visitor's book of the Tynte Arms records the names of many of the gentlemen who patronised the hotel.[10]

But there could also be annoyance, and when the authorities decided to close approaches to the glen during firing practices, for reasons of safety, there were complaints about loss of rights from some of the people.[11]

A visitor to a shoot in the glen a couple of years later was Mrs Hotham, the wife of Major John Hotham R.H.A., and a grand-daughter of the La Touches of Harristown. In June 1901 she described the shooting, the galloping guns over rocks and mountains, with shells bursting in the air, watched by 'all those ladies, royal [the duke of Connaught and his entourage were present] and otherwise seem to have rushed and flown about, climbing kopjes, until the duchess fell and sprained her ankle. Then the long drive out of the glen by Dunlavin'.[12]

The proximity of the Curragh Camp and Newbridge, about 18 miles, to the artillery ranges in the Glen of Imaal meant that both of those stations were to be intrinsically involved with the glen as it evolved into an important training ground for other arms in addition to its prime role as an artillery training ground. It gave an additional status to the barracks in county Kildare, and it was to be administered from the Curragh District.[13]

The usual summer militia training commenced in the Curragh in the first week of June 1899 with Col. Patterson, O.C. 102nd Regimental District, Naas, in charge of the 1,500 strong brigade which consisted of the 3rd, 4th and 5th Battalions R.D.F.; the officer in charge of the 3rd Battalion (the old Kildare Militia Corps) was Col. Tynte, an extensive landholder from near Dunlavin.[14] There were also several line battalions, as well as the militia, under canvas on the Curragh in weather conditions which were so good that the camp was likened to 'a foreign station'. From Aldershot had come the 1st Battalion R.D.F. with a strength of 18 officers, 2 warrant officers and 673 N.C.O.s and men, causing one observer to remark that 'seldom had regular and militia battalions of fusiliers encamped together in such strength . . . it had happened only once before about 12 years ago when the 1st Bn. R.D.F. was stationed in Ireland'. The trophies of that unit were admired, many dating from the time when it was the Madras Fusiliers; there were numerous pieces of plate, prizes for shooting, and two addresses, one from the Madras people in 1858, and another presented after the Indian Mutiny, in which the battalion had played a distinguished part.[15]

Later in June 1899 the Military Tournament at Ballsbridge again engaged men from the Curragh in various capacities. During that week General Gosset was heard to regret the loss of Lord Frederick Fitzgerald from service; he had been a major in the King's Royal Rifles, in which regiment his brother Lord Walter also served. They were sons of the 4th duke of Leinster, and as with the men of most of the other county families, they served in the army as part of their career structure. On their return home to Kildare they would have maintained their social relationships with the officer corps, especially on the hunting field and in other sports. A brother-officer of Lord Walter Fitzgerald once remarked that 'General Bob's and Lord Walter's brogue was one time the most popular thing in the army'.[16] During the month of June 'Vonolel', Lord Robert's grey Arab charger of 23 years died at the Curragh. 'Vonolel' had been decorated by the queen with the Afghan Medal with four clasps, the Kandahar Star and the Jubilee Medal; he was buried at the Royal Hospital Kilmainham.[17]

Lord Roberts was to supervise the manoeuvres in which over 9,000 troops were to participate in August, and for which the troops were to assemble at the Curragh before moving to the exercise area on the borders of Tipperary, Kilkenny and Queen's counties, the same as that which had been successfully fought over three years previously. Accompanied by Lady Roberts he had moved to the commander in chief's quarters on the Curragh for the drill season, and one of their first engagements there was to receive the Duke of Connaught on Monday 3 August when he arrived by special train to inspect the Inniskilling Dragoons, of which corps he was colonel. The royal party was to stay for some days in the camp, as guests of Lord Roberts.[18]

For the inspection the duke was on horseback, and after the royal salute and the national anthem had been played, the dragoons cantered past, ranked past (during which the duke could scrutinise every man and horse as they walked by), galloped past in squadrons, performed lance and sword exercises and finally charged past. During the exercises a band played martial airs, accompanied by a drummer on a piebald horse. After the duke had complimented the colonel, all of the soldiers in the garrison gave a display of field manoeuvres to conclude the grand parade. The duchess and Lady Roberts watched the proceedings from their carriage in the review enclosure, and huge crowds of lesser folk enjoyed the spectacle from outside the circle.

In the afternoon the lord lieutenant, accompanied by his *aide de camp*, the Earl of Granard (and whose brother was Resident Magistrate on the Curragh), arrived at the headquarters to pay his respects to the duke, and then Lord and Lady Roberts entertained the royal and viceregal guests to lunch. The next day the duke and duchess had a quiet day; he watched a divisional exercise for a time in the morning. It was the first brigade field day of the season and the manoeuvres of the horse artillery, cavalry and infantry stretched from Mile Mill near Kilcullen to the hills of Grange and Dunmurry. The capture of the Gibbet Rath was the aim of the exercise [whether the royal party or the troops were conscious of the historic significance of the rath to the people of Kildare is not known], and after that had been accomplished there was a march past taken by the Duchess of Connaught on her pony; she was accompanied by Lord Roberts and the Marquis of Ormonde. The only unpleasant part of the extensive exercise was when an unfortunate officer from the Leicesters was unseated just as he rode past Lord Roberts. That afternoon the royal couple were taken for a drive around the plain to see Donnelly's Hollow, and to have a view of Old Kilcullen and 'the fortress of Knockaulin'.[19]

Another treat for the duke and duchess was their first ride on a jaunting car. Mr E. Harrigan of Newbridge had the privilege of driving them in his 'splendid pneumatic outsider, drawn by the prize winner 'Highland Lass'. Harrigan was considered to be the 'first carman in county Kildare', and as the duchess sat in the car Lord Downes assured her that she was 'on one of the best pneumatic-tyred cars in Ireland, with all the latest improvements and with Harrigan she would be safe and comfortable'.[20]

Scarcely had the royal party left the Curragh when the 11th All Army Military Athletic Meeting[21] and the All Ireland Army Rifle Meeting were held; both were successful, with 'a fashionable throng' attending the prize giving by Lady Roberts after the shooting competitions.[22] A week later the manoeuvres began, but Lord Kitchener of Khartoum, who was staying as a guest with the Roberts, was unable to stay for them; however, a Field Day was given in his honour.[23] To be near the manoeuvres Lord Roberts was invited to be the guest of the Viscount De Vesci at Abbeyleix. The general staff were to stay at Napton, which was then unoccupied, and lent to the military for the period by the viscount.[24]

At the beginning of August, as the rumblings of war were heard, the winter stations of the cavalry were announced; both the 6th Inniskilling Dragoons and the 8th Hussars were to be in the Curragh, and the 14th Hussars at Newbridge. Another announcement that month was that Maj. General Sir H. Chermside K.C.B. was to take up command of the Curragh District in September, and his *aide de camp* was to be Lord Fincastle V. C., a lieutenant in the 16th Lancers.[25]

On 23 September 1899 the *Kildare Observer* carried an editorial asking Peace or War in the Transvaal?, and elsewhere in the paper announced that officers had already left the camp for the war, and that the 9th Company A.S.C. had departed for Natal. Within a couple of weeks the Inniskilling Dragoons were ready to depart for the 'seat of action', and two more companies of the A.S.C. had left for Dublin *en route* to Africa.[26] The newspaper was soon condemning expressions of disloyalty such as those expressed at the District Council sitting at Naas where praise for Kruger had been expressed. Irish sympathy for the Boers, the *Kildare Observer* held, was misplaced, and the sense of the Naas Town Commissioners who showed sympathy for the Irishmen at the front preparing to fight for England was commended. It was known that a large number of Kildare men, especially natives of Naas, were at the front with Irish regiments, principally the 2nd Battalion Royal Dublin Fusiliers.[27] A proof of the loyalty of the people of Newbridge was the large number of them who assembled along the route as the last of the A.S.C., accompanied by military bands, marched to the railway station. The men had been a long time in the Curragh, and were very popular.[28]

In the Curragh camp the troops were being daily exercised in field movements,[29] ready for a moment's call to duty, and the first-class army reserve was called out.[30] Already the football fields were being deserted as little football was being played. Some of the best men had gone to the Cape;[31] in Newbridge there was the same decimation.[32] The departure of 500 men of the R.D.F. from Naas coincided with the call-out of the militia, and another outbreak of disloyalty.[33]

A handbill was being circulated through the country which said that 'Enlisting in the British Army is treason to Ireland'.[34] Again the *Kildare Observer* condemned 'certain so called Irish speakers and journalists' who were critical of the war or supported the Boers.[35] The editor's belief was that 'yet there are Irishmen, and what is more, the greater part of them are most likely nationalists, and we feel sure better Irishmen, though wearing the red coat withal, than most of those men who are responsible for the blatant pronouncements that have recently been made against the British'.[36]

Welfare societies for the families of the men at the front were being activated. At Naas Lady Mayo and Gertrude de Robeck sought subscriptions from the county families for the Soldiers' and Sailors' Families Association,[37] and in Monasterevan the Countess of Drogheda initiated a fund for the families of soldiers.[38] In the Curragh the wife of a man serving in South Africa was charged with being drunk and with cruelty to her five children. The children were put into care and she was given a month in Mountjoy.[39]

The training of regiments prior to departure for the war was the main activity on the Curragh. Lord Roberts came from Kilmainham to inspect the 1st Battalion R.D.F. and the 14th Hussars before they departed.[40] The R.D.F. battalion was 850 strong when it took train at the Stand House siding for Queenstown. About 200 other members of the battalion, including recruits, remained behind in the Curragh where the 21st Lancers had also arrived, from Egypt. A 280 strong company of the Royal Engineers left the camp, and A and B Squadrons and H.Q. 14th Hussars, 348 all ranks, were played off by the bands of the 8th Royal Irish Hussars and the 21st Lancers from Newbridge railway station. Their departure aroused 'wild scenes of enthusiasm. Crowds of civilian and military friends lined the barrack square, the street and every point of vantage to give the gallant 14th a good send off, and the barrack's walls were lined by friends of the R.H.A. who cheered them. At the station the scene is not easily forgotten, wives, children, friends, it was affecting in the extreme . . . they sang *Auld Lang Syne*'. But there must have also been some joy as 8 officers, 424 N.C.O.s and men, 32 women and 59 children of the 21st Lancers were back in the Curragh camp. They had returned from the India, where they had gone from the Sudan.[41]

With the opening of the hunting season in November the absence of the officers was obvious as the men gone to the war or preparing to go were missed. It was accepted that the Kildare Hunt, as many others, would suffer due to their absence 'besides which the families of the Irish gentry give an extraordinarily large number of officers to the army'. The family of Sir Anthony Weldon of Kilmorony, Athy, was given as an example as, at the time, he had three brothers, four sons and five nephews at the front.[42] That number was soon to be reduced when Capt. George Anthony Weldon R.D.F. was killed while trying to assist his servant who had been shot at Glencoe. At home his mother was said to have felt anxious for several days, and she thought she heard a voice calling her on the day before she got notification of his death.[43] When Lieut. Betram de Weltden Weldon, 1st Leicestershire Regiment, another son of Sir Anthony's, was invalided home in May 1900 he left two brothers and seven cousins at the front. Each week the county newspapers printed lists of Kildare men who were casualties at the front (Figs. 31 and 32).[44]

The second Kildare-born recipient of the Victoria Cross earned his honour at the siege of Mafeking, culminating in his valour on 26 December 1899 when, in a hand-to-hand encounter, he killed four of the enemy with his sword. He was

Fig. 31. *'F' or Capt. A.A. Gorringe's Squadron Irish Horse in the Curragh Camp, April 1902. Two such mounted infantry Yeomanry Units were formed in County Kildare in 1900, for service in South Africa.* Photo: *Charleton & Son: Newbridge, Curragh and Dublin. Courtesy B. Siggins and T. Kinsella.*

Fig. 32. *Capt. George Anthony Weldon R.D.F., who was killed at Glencoe in South Africa in 1899. His mother at home in Kilmorony, Athy, had a premonition of his death. Twelve members of his family fought in the Boer War. Courtesy Ger McCarthy.*

Captain (later brigadier general) Charles Fitzclarence, Royal Fusiliers, who was born in 1865 at Bishopscourt, near Kill, the home of his mother Lady Maria Henriette Scott. She was the eldest daughter of the 3rd Earl of Clonmell and his wife Hon. Anne de Burgh, daughter of General Sir Ulysses de Burgh, of Bert, Athy. Lady Marie Henriette had married Capt. George Fitzclarence R.N., Knight of Medjidie in 1864, whose grandfather was the eldest illegitimate son of King William IV. All four sons of the Fitzclarences joined the services; the youngest died in the Crimea, another was killed in action with the Egyptian army, and his twin brother rose to the rank of brigadier-general. The hero of the family, Charles, went from the militia into the Royal Fusiliers, and he accompanied Kitchener on the Khartoum campaign. When the regiment of the Irish Guards was formed in 1900 he transferred to it. Later he was battalion, and then a regimental, commander of the guards, and he served for a period on the Curragh. He was killed while leading the 1st Guards Brigade with the Expeditionary Force in France on 11 November 1914.[45]

There was little Christmas cheer in the Curragh in 1899. Lady Downe, the wife of the Cavalry commander, departed for Windsor to take her turn in waiting at court, and her husband and daughters also went to England. Major Cyril Torrance

Pallin formed a pack of beagles to hunt during the Christmas holidays. In Newbridge the 21st Empress of India's Lancers gave a smoking concert and a big send-off party to the 8th Royal Hussars who were leaving for Aldershot before sailing for South Africa.[46]

The new year of 1900 saw the duke of Connaught assuming the appointment of commander in chief of the army in Ireland, in succession to Field Marshal Roberts who had been sent in December to South Africa to take command of the British Forces in the Boer War. Official duties brought the commander in chief on frequent visits to the Curragh camp, and when the status of the duke occasioned even more preparation and anticipation for all ranks there. During his stay in the camp in the autumn of the previous year the duke had familiarised himself with the area, and he had expressed satisfaction with 'the famous camping ground of Ireland'.[47]

During the war years from 1900 to 1902 there was a constant movement of troops into and out of the Curragh, Newbridge and Naas, with a resultant frequent change of command.[48] Sometimes local officers, such as Major St Leger Moore of Killashee, or Lieut. Col. T. J. de Burgh, Col. de Robeck or Commander de Robeck, all from Naas and related through marriage, were mentioned.[49] De Burgh was raising a troop of yeomanry from the county, and seeking volunteers who were to offer themselves to the government as mounted troops who would live together as troopers under ordinary service conditions. The requirements for the honour were that the men should be good riders and rifle shots, provide their own horses, pay £100 towards the expenses of the troop, and not receive any pay.[50] Col. de Burgh, who had already served in the 57th Regiment and the 5th Dragoon Guards, was later to command the 17th Battalion Imperial Yeomanry in South Africa. His younger brother, Col. Ulick de Burgh, had commanded the 7th Dragoon Guards, served in the Egyptian Campaign, and commanded the 3rd Dragoon Guards before becoming an Assistant Director in the War Office. He was recalled for service in World War I.[51] An Irish Hunt Company of the Imperial Yeomanry was formed in Newbridge, with many hunting men from Kildare and Carlow enlisting. However, when a local well-known sportsman failed to join-up he was said to have received several letters with a white feather enclosed.[52]

In January 1900 the nationalist *Leinster Leader* reported that a batch of 60 volunteers who were 'hurrying to England's aid under the banner of what is known as the Imperial Yeomanry, attended at Newbridge barracks to go through some evolutions in riding and shooting, so that they might be able to storm the Kopjes around Ladysmith, drive the Boers back and rush forward to embrace, or more properly to feed, Sir George White and his beleaguered soldiers, when they arrive at the scene of hostilities. The large majority of the volunteers proved themselves equal to the task put to them by Col. Wyndham, Capt. Mann and Lieut. Cecil. The firing was conducted at two ranges, one at 200 yards and the other at 500 yards. Of the 60 candidates some 35 passed'.[53]

The fact that the military were not universally popular in the rural area was evident from a row which developed over the Newbridge water supply in January 1900. The water supply in the town was in a critical state, and it was proposed that a new supply should be laid, which would include the rural areas of Newbridge and Morristown-Biller. Objecting to have to subscribe to a supply which included the town the residents of Morristown-Biller held that 'the residents of Newbridge were composed of people from all parts of the world, and that they lived on British gold, and that they could well afford to pay for their water supply without appealing to outside sources'.[54] From that statement it could be deducted that some of the old residents of the locality, and who did not benefit from the presence of the military, remembered that the town of Newbridge was founded for and flourished on the army. The townspeople would have been loyal in their views, either for patriotic sentiments or mercenary motives.

Within the Curragh camp there could also be friction between the military and the civilian workers, and for reasons which had also caused problems almost half a century earlier when the original encampment was being built. Eight or nine years before, when construction work had commenced in the camp in the early 1890s, a large number of civilians, tradesmen and labourers had come to the area. Due to the normal friction between the soldiers and the workmen the military authorities required that the civilian canteen, which was run by Dickeson and Co., should not cater for soldiers. Messrs McArdle of Dundalk, who were extensive military contractors, took out the licence for the canteen. McArdles were prosecuted in January 1901 for having the canteen opened on a Sunday, and their solicitor said that there seemed to be a misunderstanding as to whether the canteen was under the control of the civilian or military authorities. Henceforth, he added, the canteen would be closed on Sundays, and that there would be a new manager appointed to it.[55]

The building work in the Curragh had by the month of November 1901 completely changed the appearance of the camp. The new buildings which had replaced the wooden huts then extended from the Kildare end of the camp to the Kilcullen end, and across to the Newbridge side, but it was reported that 'the work was not yet nearly complete. When finished the Curragh will be one of the finest military centres in the UK'.[56]

That year of 1900 saw the origin of a new tradition in the army when the queen ordered that on St Patrick's Day all ranks in H.M. Irish regiments should wear a sprig of shamrock in their head-dress to commemorate the gallantry of her Irish soldiers in the recent battles in South Africa.[57] At the annual quadrille to celebrate the national feast at Naas barracks 80 people sat down to supper, a smaller number than usual due to the absence of those in the wars.[58] Perhaps during the evening they discussed 'the ungenerous references' of Anna Parnell to the queen's authorisation of the use of the national emblem, but her brother John took another view, and thought that a visit to Ireland 'by the queen and all foreigners might help Home

Rule'.[59] The distribution of the shamrock became established in the army, and was to be adopted by the Irish army in due course.

Within a month of St Patrick's Day the royal visit to Ireland took place, but that time the queen did not visit county Kildare. Instead a contingent from Naas, led by clergymen and Lady Albreda Burke, travelled to Dublin to see her,[60] and the daughter of the sergeant-major of the 1st Bn R.D.F., recently in the Curragh, presented the queen with flowers at the Viceregal Lodge.[61] From Newbridge a detachment of 11 officers and 228 men of the 21st Lancers, together with the Imperial Yeomanry from the Curragh, were amongst the troops who lined Dame Street and the quays for the royal procession.[62]

In April 1900 some 550 men joined the 9th Battalion Royal Irish Regiment being formed in the Curragh, and in the same month[63] the 3rd Company Irish Imperial Yeomanry left for the war; a couple of weeks later the 3rd Bn R.D.F., 450 strong, were cheered off from Naas station as they too went to battle.[64] The Jesuit Clongowes Wood College, Clane, could then also claim to be represented on the field of war by 42 officers and volunteers.[65] When the Irish Hunt Contingent of Yeomanry was reported captured by the Boers in June 1900 several men from county Kildare were amongst them, including a de Robeck of Naas, Mackey of Athy, Greene of Millbrook, Athy, Odlum of Kildangan and Forbes of Athgarvan Lodge, on the Curragh.[66] When they eventually returned home they were presented with their War Medals at a parade in the Curragh Camp in September 1901.[67]

While the serious business of training and preparation for overseas continued in the barracks, the recreational and entertainment needs of the troops were not neglected. Though two of the military races at Punchestown in 1900 had to be cancelled, due to lack of potential entries,[68] and the reduction in the attendance of the military there was regretted, the Military Steeplechase was held. The attendance of royalty once again brought glamour to the Meeting with the presence of the Princess Henry of Battenberg, Princess Christian of Schleswig-Holstein, and the Duke and Duchess of Connaught, as well as the Lord Lieutenant and Countess Cadogan.[69] The annual Curragh District [Race] Meeting,[70] and the All Army Point-to-Point were held as usual.[71] During the otter hunting season of 1900–1901 the early morning meetings of the King's Regiment Otter Hounds were led by Capt. Hastings, in the absence of the Master Capt. Sheppard, who was in Africa. The river Liffey at Blessington, Kilcullen and Newbridge received most attention of the sportsmen.[72]

Nor were the polo and cricket grounds deserted,[73] despite the absence of many players, and when men returned from the front to the football field they were noticed by their supporters: 'we are glad to see old familiar faces in the ranks of the Inniskillings . . . missing from Curragh football for some time,' wrote a sports correspondent in a local paper.[74] A new arrival at the Inniskillings Regimental Headquarters in Ponsonby barracks in the month of June was Lieut. Lawrence

Oates who had been gazetted into the 6th (Inniskilling) Dragoons at the end of May. Then he wrote to his mother 'I am very pleased about it as it is the best Heavy Cavalry Regiment in the British Army which is saying a good deal when it has to compete with the Greys and the Royals'. In the large garrison at the Curragh he would have found, to quote another officer, that 'off parade the various units kept very much to themselves. Precedence was not only rigidly observed, but it was also carried to ridiculous lengths. Thus, the artillery and engineers affected to look down on the infantry, and the infantry looked down on the ordnance, medical and commissariat branches, while the cavalry from their superior heights looked down on everybody'.[75] Oates was to spend less than six months on the Curragh before he was drafted to join his regiment in South Africa.[76]

The participation of a large number of the Royal Dublin Fusiliers in the Corpus Christi Procession around the convent grounds at Naas in June 1900 further enhanced that regiment's image in the parish,[77] while in Kildare town the construction of a large artillery barracks was welcomed as creating better times for St Brigid's cathedral as the influx of gunners and their dependants was expected to increase the small attendance at the restored cathedral.[78]

The proposal to build an artillery barracks on the site of the former Lock hospital at Kildare was confirmed, and it was known that, due to the war, the military establishment was being enlarged and soon the plans for Kildare barracks were revealed.[79] It was to be completed in five months and to cost £90,000. The plan was for iron-roofed wooden huts which would accommodate from 6–8,000 men.[80] When the construction work had commenced the 'quiet and plod-along town was given a fillip'; not alone were the town water supply and sewerage system being extended to the barracks, but electricity was to be provided to both the civilian and military areas.[81] By January 1901 the barrack square had been laid, and the progress on the huts was such that it was expected that they would be occupied by the month of May.[82] In fact, the work had reached a stage by the middle of February that an entertainment was held in the new barracks and a large number of guests had their first opportunity to see the place.[83]

In Newbridge at the United Services Masonic Lodge the absence of many of its members was considered 'an enormous drain' to the otherwise prosperous craft. But when the lodge met on 10 January 1901 there was 'a right muster of brethren from almost all quarters of the globe, who as visitors, got a great welcome' as well as a few members who were returned from South Africa.[84] No matter where the military masons were posted there was always a lodge where they would be welcomed, and probably meet old comrades. The order was an important social outlet for its members, and a comfort to them in foreign lands.

The strong patriotic feelings and keen interest of the gentry in the war and in the yeomanry was shown in the work of Lady Annette La Touche of Newberry, Kilcullen, who solicited subscriptions for a Kildare Hunt Bed to be provided in the

Imperial Yeomanry Hospital.[85] The casualty lists continued to appear in the county press, including the names of members of the Kildare Hunt.[86] In the summer of 1900 Thomas Conolly M.P. of Castletown, Celbridge, lost one of his two sons serving in Africa, the eldest, Lieutenant Tom of the Royal Scots Greys.[87] The son of Naas Town Sergeant Gallagher, serving with the 1st Bn R.D.F., was also killed.[88]

In the meantime the usual round of barrack concerts continued, sometimes in aid of charities such as the Lady Roberts War Fund[89] or the Soldiers and Sailors Families Association,[90] or just for fun, like the Smoking Concert given by the Royal Reserve Regiment of Dragoons in June 1900, and their series of concerts given that autumn.[91] The band of the Royal Dublin Fusiliers played in the Market Square, Naas to an 'enormous attendance of inhabitants of the town and district'.[92]

The 'famous camping ground', as referred to by the duke of Connaught, was beside the permanent buildings on the Curragh. It was mainly occupied during the summer training months when the important All-Ireland competitions were held. For the rifle meeting in 1900 a new trophy was approved by the commander in chief; it was a massive silver cup to be called the Lord Roberts Cup.[93] It was the intention of the duke to be in residence in the camp for the drill season, but that the duchess and princesses were to stay at Castleblaney in county Monaghan.[94] At the end of July the duke presented the prizes for the shooting contests on a Saturday morning, and for which ceremony all officers in the camp were required to be present. As Saturday was normally a half-day for the military there was some resentment on the part of the officers 'who had to cancel their arrangements' [for the afternoon]. It was believed that if the duke was aware of the order that he might not have approved.[95]

At the end of the year Major General Chermside took over the Curragh command. Both he and his *aide de camp* were returning from South Africa.[96] So was Capt. Meade Dennis R.H.A. of Fort Granite, near Baltinglass; he was greeted by his tenantry there with great rejoicing.[97] When Lieut. Col. de Burgh, of the 61st (2nd Dublin) Company Imperial Yeomanry, came back to Oldtown, early in June 1901, the flags were flying over the house, but there was disappointment in the town when he detrained at Sallins railway station, and not at Naas where a large crowd had waited to greet him.[98]

By St Patrick's Day 1901 Queen Victoria was dead. In county Kildare her passing was mourned, and there were memorial services in the Curragh and at St David's church in Naas, which were attended by the troops and local residents, and the annual dinner of the Kildare Hunt and the annual general meeting of the County Kildare Archaeological Society were cancelled. The accession of Edward VII was marked with a 21 gun salute at the Curragh on 24 January 1901, and his period of training there 40 years before was recalled.[99]

As training for the battlefields continued in the barracks, so did the recreational occasions. In August 1901 the sergeants of the Inniskillings, with the permission of John La Touche, held their annual picnic at Harristown. A cavalcade of waggons, ambulances and cars of local construction, preceded by two trumpeters, brought the sergeants, their ladies and families to the riverside estate where dancing, games and a ladies' cricket match were enjoyed. On return to barracks an *impromptu conversazione* was held in the sergeants' mess.[100] Before Christmas there were more dances and concerts, including a performance by the Minstrel Troop [sic] of the King's Liverpool Regiment; that regiment justified the starting of their troupe as being due to 'the dearth of camp amusements'![101]

A parliamentary paper published in July 1901 gave the role of the army. It was to give effective civil power in the United Kingdom, and find garrisons for fortress and coaling stations at home. It was to mobilise rapidly for home defence two army corps of regular troops, and one of regulars and militia, and subject to that, be able to send abroad two complete army corps with a cavalry division. Men for India were also to be provided.[102] Part of the plan included that, after a large number of militia units had returned from Africa, the Irish militia infantry was to be organised into brigades which would train every year at the Curragh, Kilworth camp, and Ballyshannon.[103]

Her Royal Highness the duchess of Connaught presented the prizes after the Shooting Competitions in August 1901,[104] and in September the duke was promoted a general and to command the Third or Irish Army Corps, on the Curragh.[105] It was to be a Cavalry brigade of three regiments; one battery R.H.A., one battalion mounted infantry, three infantry divisions totalling 24 battalions. The divisional troops were to consist of one battalion of infantry, one regiment of cavalry, nine batteries of Royal Field Artillery. Corps troops were one battalion of infantry, one regiment of cavalry, eight batteries of artillery, and sections of all other services. Provisional regiments of Dragoons, Lancers and Hussars were to form the Cavalry Brigade in new quarters which were being prepared, and the bulk of the 25 battalions were to be quartered at the Curragh.[106]

The decline in population in county Kildare since the census of 1891 was shown in the returns of April 1901. At Newbridge there had been a fall from 3,207 to 2,903, including 353 soldiers. At Ballysax (including the Curragh Camp), there had been a fall from 6,237 to 5,273, of whom 2,900 were soldiers in the camp.[107] However, in Kildare the figure had risen from 1,172 to 1,576, probably because of the workers employed in the army building programmes, and at Naas from 3,735 to 3,836, including 193 military.[108] Concern at the decline in churchmen was expressed at the Dublin Diocesan Synod in October 1901 when it was observed that in that year there were much fewer soldiers at the Curragh than when the previous census had been taken: 'The absence of two or three English regiments would account for the falling off in the numbers of churchmen and Methodists, while a single Scottish battalion would probably cause most of the reduction in the numbers of Presbyterians.'[109] In the Curragh Camp the low strength was having an adverse effect on the various institutes which were described as 'neglected'.[110] Dr Patrick Foley, Bishop of Kildare and Leighlin, was also to express concern at the 'depopulation of the country . . . which had reached a vast national evil'. He was especially worried about the decline in rural areas, where it was 'as if they were stricken with famine or pestilence'. In his own dioceses, which extended into no less than seven counties, the bishop saw that 8% of his flock had been lost, while non-Catholics had lost 20% of theirs. He observed that 'the disappearance of soldiers from the Curragh camp owing to the war no doubt accounts largely for this abnormal decrease'.[111]

However, early in December 1901, the arrival of a new chaplain 'brought new life to the Church of England's Soldiers' Institute having been lying dormant for almost two years'. The Rev. Blackbourne organised a tea, under the auspices of the Independent Order of Good Templars which was attended by 160 members and friends. Then the wife of Col. Tyler C.R.E., Curragh District, arranged a bazaar in aid of mission work in China, and 'Tommy Atkins' got his turn in a smoking concert given by the Church of England Choir, and which included a large number of members from the band of the 14th King's Hussars. The intention of reviving the Army Temperance Association in the camp was announced during the smoker.[112] Lady Chermside gave a Christmas treat to the children of the District Staff,[113] and at Naas Col. Patterson invited the children of the workhouse to the barrack pantomime.[114] All of these activities boosted the morale of the troops, and encouraged a positive relationship between the military and civilian populations.

Even if garrison life had suffered from the war, the local civilian population continued to benefit from the military. Building work continued on the new barracks in the Curragh,[115] where the roads had deteriorated due to the construction traffic,[116] and in Kildare, though in the latter station plans for the erection of permanent buildings were never to come to fruition.[117] The Soldiers' Home in the Curragh had been enlarged and improved at a cost of £1,000, not including furnishings, which cost a further £100, and at Newbridge a site had been secured to build a Soldiers' Home which was to cost £1,300.[118] Indeed, the town fathers in Newbridge were censured by the *Kildare Observer* for not catering properly for the increase in population owing to the work on the Curragh and in Kildare. There was not enough accommodation for the men in the town, and enough working class houses had not been built. But private enterprise was gaining from the presence of the workmen. Mr J. J. Moran of the Globe Supply Stores had enlarged his business with a bakery and bottling enterprise, and he had vans making deliveries in the camp and in Kildare.[119]

The regular publication of advertisements seeking tenders for supplies to the various military stations[120] did not always mean that the contracts went locally, though a certain amount of new business was generated by the war. Civilian contractors were invited to undertake various works, including painting and repairs,[121] and to supply materials, in the barracks at Newbridge, Kildare, in the Glen of Imaal, and at the Curragh, including the ranges.[122] Remounts were purchased from the Kildare Hunt Kennels,[123] and tenders were sought for the washing of horse blankets.[124] Oil, hops, potatoes, meat, live cattle and sheep, were all required,[125] and even such routine chores as clothes washing, bedding repairs, the removal of ashes and the deodorizing and emptying of privies were out for tender, and quotations for the purchase and removal of manure were invited.[126] At the camp abattoir and bakery civilian staffs were employed to replace the Army Service Corps men in South Africa.[127]

While undoubtedly a considerable amount of business was generated locally through the army purchase system, it also happened that contracts which could have been supplied from the district went elsewhere. This anomaly was created by the war and the necessity to increase purchases for the troops being assembled at home and despatched to Africa. It was sometimes more practical to fill requirements in Ireland from contracts placed in England and elsewhere as the number of troops fluctuated and their locations changed.

When it became known in the county Kildare early in 1901 that the contract for the supply of beer, porter and stout had been declared in favour of an English firm there was a public outcry. A change in the contracting system was blamed for the deprivation of trade from the Irish suppliers. It had been the custom that each regiment contracted for its own supplies, but under the new system there was a combined purchase for the regiments. Since that system was introduced the contract had been awarded to an English firm, reputedly at a higher price than that of the Irish firms. One journalist commenting on the business accepted that 'there might be some excuse for getting English beer for the troops, but surely in the case of porter it is like bringing coal to Newcastle . . . it is grossly unfair to Irish contractors, the way they are being treated by English army officials'. The loss of business to the farmers who grew the barley, and to the breweries, was considerable.[128] The fact that in his evidence to the Royal Commission on Contracts the duke of Connaught had expressed a wish that Ireland should get her share of army contracts, and that as an army corps was soon to be based in Ireland, some hope was given to the business community.[129]

Causing an equal amount of dismay in farming circles was the purchase by the army of meat imported from Australia and Canada. Questions were asked in the House of Commons, and the Naas Board of Guardians passed a resolution protesting at the practice. Mr Talbot from Kildare spoke of a man who had served with the Imperial Yeomanry in South Africa, and who could not now sell his sheep in his own country. He deplored the importation of frozen carcases to an agricultural country, and he knew that the troops did not like the meat. The government, he held, should be encouraging Irish agricultural trade. A copy of the Naas motion was to be sent to the members of parliament.[130]

But if the army purchasing system was causing dismay in trade and farming circles in county Kildare, the crowds still turned out for the grand occasions on the plain. Such an occasion was the inspection there prior to departure for South Africa of the 29th Battalion Imperial Yeomanry (Irish Horse) by Field Marshal Earl Roberts, commander in chief of the forces, accompanied by the duke of Connaught, commanding the forces in Ireland, on Tuesday 19 March 1902. A large party of staff officers accompanied the commanders by train from Kingsbridge to Newbridge where they arrived at 11.30 a.m. and were met by Col. Yorke, commanding the Curragh District. They then rode to the Curragh where, as they entered the plain, over 500 Irish Horse (some of whose mounts were borrowed from the Dragoons and the Hussars), under Colonel Lord Longford, saluted Lord Roberts and the duke as they rode by. Then the battalion dispersed over the plain to carry out skirmishing operations which were designed to test their horsemanship. Roberts and the duke, with their attendant officers, rode hither and thither over the sward observing the operations. The yeomanry were dressed in the new khaki uniform, which had been introduced as a result of experience gained in South Africa; an observer of the Curragh exercises remarked that the 'advantage of khaki as uniform against even the background of green Irish hills was obvious'. Lord Roberts next proceeded to the camp where he inspected the 2nd Battalion of the Liverpool Regiment, 750 strong, including its cycle corps. Then the Yeomanry were inspected, after which they formed a hollow square and were addressed by Roberts.[131]

Early in the month of June 1902 the bell in the Protestant church at Athy was rung to announce that peace had been proclaimed in South Africa, and that the Boer War was over.[132] Gradually the troops began to return home, including Lieut. Oates of the Inniskillings who came back to Ponsonby Barracks at the Curragh, but later transferred to Marlborough Barracks in Dublin. The years from 1890 to 1910 have been described as 'the doldrums of Irish political affairs, a kind of Indian summer for the old Order'.[133] Enjoying the peace, Oates purchased several horses on which he played polo, hunted and raced. Riding his amusingly named mount 'Titus' he was placed third in the Irish Grand Military Steeplechase at Punchestown in 1904.[134] In the following year, as an owner, he ran his 'Angel Gabriel' in the same race, and won. He afterwards admitted that his eyes got so full of tears 'I could not see the horses and had to keep asking the man next to me how my horse was going'.[135]

In 1910 Capt. Oates was seconded from his regiment to accompany Capt. Scott to the South Pole. There he died at the age of 32 on 15 March 1912; frost-bitten and ill he believed he was a burden to his comrades. Before going out into the blizzard he said 'I am just going outside and I may be some time'; and he was never seen again.[136]

In Newbridge there was a decrease in court fines as the soldiers back from war rejoined their spouses,[137] but the return of the militia to civil life caused some disruptions locally.[138] Normal training continued in the Curragh Camp, and social life was resumed. The Elster Grime Opera Company was playing in the gymnasium in August 1902 to packed houses. Staged in 'first class style, the scenery and dresses being particularly good . . . the principal characters in the hands of most accomplished artists', sang their way through Balfe's *Satanella*, Verdi's *Il Trovatore* and Mascagni's *Cavalleria Rusticana.* Not surprisingly, the company was retained for a second week.[139]

By Christmas 1902 it could be reported that while 'the Curragh, of course, is not as bright as it used to be, there is still a sufficient crowd in the Curragh to make the pursuit of pleasure a possible occupation, but here, as elsewhere, Christmas is only referred to in the past tense'. Nevertheless, 'extensive preparations were being made for the festive season. All the messes are brightly decorated, and theatricals were scheduled for the gym.[140] There is always a keen rivalry between the different messes at the Curragh for Christmas time, and even amongst officers this friendly endeavour to excel prevails. The result is that the balls and parties given by the rank and file are liberally subsidised, and considering its much depleted population, the Curragh will be as jolly a place this Christmas as the gay sparks amongst its population know how to make it, many of whom have served in South Africa have not known a "home" Christmas for three years'. The decoration of the shop windows in the camp was admired, and though there 'are very few shops on the Curragh, in matters of this kind they compete keenly. It is, however, in the messes that one looks for really authentic decoration. Each regiment possesses its own artist, whether non-commissioned officer or private. One would wonder how all the holly and ivy which goes Curraghwards at Christmas is disposed of, but an inspection of the messes arouses wonder.'[141]

It was thought that the only people who looked forward to the festive season with anything akin to a shudder were the military police,[142] and perhaps the people of Athy. In 1909, for the second year in succession soldiers from the Curragh came to the town and created a nuisance, causing the correspondent for the *Leinster Leader* to write: 'What a contrast was the conduct of those pampered individuals when compared with the general good behaviour of the poor of the town, amongst many of whom there is indeed not a little want and suffering, which is rendered all the more distressing at the season of peace and joy.'[143] The favourite Saturday night saying amongst the military that 'many men make merriment'[144] was not appreciated by the Athy folk.

Despite the apparent Christmas good will expressed in the newspapers there was persistent local annoyance with the distribution of army contracts. Vegetables for the Curragh, Newbridge and Kildare barracks were being imported from England, or from Dublin. The fault, however, may have been a local one as the town of Newbridge was described by a journalist as 'the worst vegetable market in Ireland, at least it would be if Kildare were not in the running'.[145] Yet, in the opinion of the *Leader* scribe, if the local people made the effort they could supply to 'the thousands of men who have not even got a front garden patch for flowers'. And the question of hawkers offering their wares for sale in the camp and barracks was again being considered. Only licensed hawkers could enter the military areas, an obligation necessary on account of the 'unsuitability of the vendors, rather than of the goods vended'. The chairman of the local Technical Committee had been in communication with the War Department to seek the more general granting of licences, and the reply he received was that facilities would be given for the erection of greengrocers stalls on the Curragh.[146]

By the summer of 1903 normal military routine had returned to the Curragh area, and in the town of Newbridge the town's people were again becoming 'accustomed to Saturday night orgies'. When a large pool of blood was noticed on the pavement of the main street one Sunday morning in May, and the shutters and wall of Mr Kearns' house were bespattered with blood, it was believed that 'a battle royal had been fought out there the night before'. The following day it was rumoured that a soldier had died at the Drogheda Hospital as a result of injuries received in Newbridge, where, it was known that 'little importance was attached to the incident because midnight brawls were not exactly a rare occurrence [in the town]'.[147] In Kildare, where the social impact of the new barracks had not yet become defined, it was reported in March of that year that the price of houses had risen due to the enhanced status of the town.[148]

The importance of the plain to the military in time of war had become apparent, and in July 1901 the War Office again expressed an interest in acquiring all of the grazing rights there, despite the apprehension of the Turf Club whose tenancy there pre-dated that of the military. In 1894 a similar proposition from the War Office had not been proceeded with. Now in 1902, though it was realised in London that the Irish members of parliament would oppose any legislation which would be required to end the rights of the Turf Club on the plain, when the Baron de Robeck, who had been the Curragh ranger since 1892, died in 1904 he was not replaced. Despite further efforts by the War Office to gain control of the crown's rights on the Curragh the influence of the racing gentlemen and the Turf Club succeeded in scuttling the proposal, but the office of ranger was not to be filled until 1910.[149]

Chapter 13

Between the Wars, 1903–1914

By 1904 the accommodation for all military ranks in county Kildare was 6,179. It was distributed between Naas, which had accommodation for 260, Newbridge for 777, Kildare for 892 and the Curragh camp for 4,250. The accommodation in the Curragh was for 181 officers, 24 warrant officers and 4,045 non-commissioned officers and men (of whom 403 were married). Stabling for 162 officers horses, 1,466 troop horses and 58 infirmary horses was also provided. There was stabling for 614 horses in Kildare, for 524 in Newbridge, and for 6 at Naas.[1]

At the Curragh that year the 19th Hussars rotated with the 6th Dragoons in the month of February, and remained there until October 1907. The 11th Hussars were also there, since 1903, and remained until 1906. Each of these regiments had a notional strength of 707. The 4th Battalion Lancashire Fusiliers were there from September 1902, and the 1st Battalion South Staffordshires from January 1904 to November 1906, each with notional strengths of 820. In addition there were elements of other units there, the Army Service Corps, Medical Corps and Royal Engineers, all of which made a resident population of about 4,000 men in 1904. As it is probable that the married quarters for the 403 other ranks were fully occupied the actual population of the camp, including women and children, could be estimated at about 5,000, excluding civilian employees.[2] Such a figure might be accepted for the years until the advent of the Great War in 1914. The census returns for April 1911 show the military population of the Curragh Camp as 4,488; of Kildare barracks as 808; Newbridge barracks 747, and Naas barracks 149.[3]

Another major royal occasion for the military and the loyal citizens was the visit of King Edward VII and Queen Alexandra to Punchestown on 26 and 27 April 1904. The royal party travelled from Kingsbridge to Naas by special train, and the town was *en fête* for the passage of the entourage through the main streets. Photographs show the streets bedecked with Union Jacks, bunting and garlands of flowers, some arranged by Switzers of Dublin in regal style with crown-topped pyramids or the lion and unicorn. The Protestant and Catholic clergy combined with the local gentry and the town worthies in the preparations which were summarised by the chairman of the reception committee, Mr William Staples, who said: 'Naas of the Kings should be making ready to welcome its king'. Mr La Touche of Harristown asked the Urban District Council to have the streets sprayed with water to keep down the dust, and over 300 members of the R.I.C. were drafted in to keep the peace.[4]

An indication of the importance of military etiquette at such major social occasions as Punchestown was given in the dress regulations issued to officers for attendance at the races in April 1907. On the first day silk hats were to be worn, but on the second day the ordinary felt hat would do. No doubt, but the dress requirements for attendance at the Irish Command Military Tournament, which was held in the camp a couple of weeks after the races, were equally detailed.[5]

The social life of the officers in the county was comparable to that in Dublin, and which was described during those quiet years by the G.O.C. Ireland General Sir Neville Lyttelton who was in Dublin from 1908 to 1912. In retirement he recalled 'we did all we could to keep up the tradition of hospitality at the Royal Hospital, many balls, dinners, and parties for Punchestown races, and the yearly Horse Shows. We, in turn, were most hospitably entertained at many country houses, and in Dublin itself'.[6]

That some of the problems which had arisen with the building of the Curragh encampment in 1855 had not changed half a century later was a matter for discussion at a meeting of Kildare County Council in March 1906. A letter from Col. G.K. Scott Moncreiff Commanding R.E. Dublin District,was read. It was asking about the maintenance of three roads; that from Herbert Lodge to Kildare, the road from Brownstown to Crockaun, near Ballyshannon, and the road from Ballymanny, through the camp, to Brownstown. He made the point that the Treasury, in lieu of local rates, contributed an annual grant. From the year 1900–01 that sum had been increased from £1,734 to over £3,000 in 1903–04, outside of the special grants also paid each year by the War Department to Kildare County Council on account of the extra War Department traffic between Kildare, the Connelmore brickfields and the Curragh Camp, and for other roads in the neighbourhood.

The County Surveyor, Mr Edward Glover, who was responsible in that area, explained that the roads to which the colonel was referring were (1) Moorfield to Lumville, (2) Moorfield to Ballymanny, (3) Ballymanny, through the camp, to Brownstown, (4) from three roads near Herbert Lodge on the Curragh to Kildare and (5) from Kildare railway station; in all 6 miles, 1 furlong and 12 perches of roadway. To strengthen those roads to take the heavy traffic would cost £5,596.18s.7d, of which he proposed the county council pay half, and the War Department the other half. 'It was remarkable,' Glover held, 'that ordinary rural roads were able to support such traction engine traffic'. The three mile long road to Crockaun was subject to heavy cavalry traffic, with a regiment of soldiers passing over it every morning. Consequently it cost more to maintain than it had previously. [At Crockaun, near Ballyshannon, each year in the months of June and July, regimental competitions, including tent pegging and equestrian feats, attracted crowds of onlookers to the tented encampment there.]

The roads near Newbridge were subject to extra military traffic, with locomotives of the heaviest type continually passing, even more so than on the Curragh or at Kildare. Glover believed that assistance from national resources should be given when military camps were being formed and built, but that no such aid was given to Kildare except £128 a year for repairs for the Connelmore brick traffic, and about £280 for one single year to improve a road to prevent accidents. The road from Connelmore brick works was in a shocking condition from the 41 carts drawing material from the yard there.

The County Surveyor recommended that all roads around the Curragh Camp subject to military traffic should be strengthened by steam rolling. The cost each year of the upkeep of the roads was £1,700, and he thought that the War Department should contribute more. The Chairman of the County Council took an even stronger hand on the question; he considered the letter from Col. Moncreiff to be an insult to the council, 'but he was not referring personally to the colonel, for very likely he did not know the way the council had been treated by the War Department'.[7]

In the meantime further building was to commence at Kildare barracks with the erection of 30 permanent married quarters 'along the same lines as the new barracks at the Curragh'.[8] In 1911 it was estimated that work costing a total of £48,500 was to be undertaken at the Curragh, and in that financial year £28,000 was to be spent; it included £20,000 to complete the lighting of the camp, £3,000 for quarters for general officers, £1,300 for the heating and hot water system in the military hospital, and £1,000 for quarters and a mess for the R.A.M.C. The sum allotted that year to the Curragh was the highest for barrack works in Ireland; Fermoy came next with £31,000, while £6,250 was to be spent on the Dublin barracks.[9]

While the Curragh could be described as 'bleak', which no doubt it was in bad weather, it was generally regarded as being a healthy station. The hospitals there were new and well equipped, and in the summer of 1906 they had to cater for an outbreak of scarlatina. A few days after a party of the Royal Fusiliers had returned to the camp from having encamped at the Moat of Ardscull, near Athy, two men went down with scarlet fever. Rumour was that they had contracted it in Athy, but the local sanitary sub-officer said if so, that it was the military who introduced it there.[10] There may have been truth in the sanitary officers claim as a month before there had been an outbreak of 'the scarlatina scourge' amongst the artillery in the new barracks at Kildare. Twenty men were ill in the barracks, and the fever had spread outside the barrack wall where ten cases were confirmed. The barracks was medically inspected, and it was decided to make wards available in the Curragh hospital for civilian patients if necessary.[11]

At the end of 1906 there was a scarcity of water at the Curragh Camp and in Newbridge barracks. For the camp it was proposed to pump water with a steam engine from the old well at the Lumville police barracks, and in Newbridge it was intended that the town water supply and sewerage scheme would be extended to the military, despite the objections of some of the local people.[12] Even in matters of health, and essential commodities such as water and sewerage supplies, there was interaction between the civilian and military populations.

Another aspect of the use of the highways by the military was also causing annoyance in the Curragh area. The exercising of horses, three abreast, on the public roads in the vicinity of Newbridge was seen as 'a gross scandal'. The people of the town were warned to beware as 'the military at the Curragh are at perfect liberty to ride over them, and if they are injured, why, then it is their own look out, because though the heaven's should fall, His Majesty's horses must be exercised . . .' Despite a resolution at the District Council, and remonstrances to the military authorities, the reply given was there were not enough men to exercise the horses separately.[13]

The confidence of people living around the Curragh must have been further shaken in the autumn of 1909 when Reverend Mr Wilkinson, rector of Ballysax, was assaulted by two soldiers while he was cycling on the footpath on the road from Kilcullen to the Curragh. The soldiers, one of whom was in uniform, were on horses and they refused to leave the footpath as the parson approached. When the clergyman did not get out of their way the uniformed man dismounted and 'administered a personal chastisement to the reverend gentleman'. Mr Wilkinson went to take shelter in the gate lodge at Castlemartin, but the soldier followed and assaulted him, and then both soldiers rode off.[14]

Badly-lit military vehicles could also be dangerous to the public, as an inquest found into the death of an officer of the Irish Land Commission in December 1913. The man had been killed by a military ambulance waggon, drawn by two horses, and with only one light which was carried by a man of the 4th Hussars.[15] At the Curragh Petty Sessions there were frequent cases involving the absence of lights on bicycles or motor-cycles, and the usual defence was that the light had gone out, a common failure of the carbide lamps then in use.[16]

That there was some change in the road offences before the Curragh Petty Sessions in 1914 was proven by the appearance of Lieut. Basil B. Falconer 4th Hussars on a charge of dangerous motoring.[17] In the army the mechanisation of transport had begun and, in 1910, tractors, trucks and lorries had been used during the summer manoeuvres.[18] In 1911 the Secretary of War told the House of Commons that he was contemplating the replacement of horses by motor transport on a large scale in the army.[19]

A cause of considerable annoyance to some local people was the continued use of the military vehicles to convey soldiers and their wives between the camp and the railway stations, and a meeting was called in Newbridge to discuss 'the plying for hire by troops at the railway stations at Kildare and Newbridge'. Mr PJ. Doyle J.P., in the chair, said that 'the carmen were a very important body of men in the district, and that they were the means of circulating a lot of money in the town of Newbridge. They benefited themselves from the traffic, and the shopkeepers as well as the farmer benefited . . . they would not be able to pay for fodder or hay if their business was lost [to the soldier drivers]'.[20] The meeting led to a question being raised in the House of Commons. It was asked if there was a charge for the use of the military cars? The point was made that they were in competition with the hackney-car owners who had to pay a weekly amount to the G.O.C. for their stand at the Curragh Camp. A report from the army in Dublin was to be requested.[21]

Another constant local aggravation, particularly to the traders, was the question of the disposal of army contracts. When John O'Connor M.P. asked in parliament 'through which department supplies for the Curragh of Kildare were procured, and was it the custom to purchase supplies, where possible, from the farmers of the district?', the reply he was given was that 'supplies for rations of provisions and forage were obtained under contracts, and that it was not customary in such a case to specially purchase supplies from local farmers. However, during the manoeuvre season, purchases were frequently made from places not covered by existing contracts. All supplies were inspected on delivery by qualified officers of the A.S.C.' This response brought little new information to light, and did nothing to promote the placing of tenders in the county.[22]

An impression of the quality of business generated in the county by the presence of the military can be gleaned from the 1910 edition of *Porter's Directory.* Advertisements for firms in Newbridge, and mainly directed at the military, included those of J.M. Waters, a saddler and harness maker to H.R.H. duke of Connaught; T. Cooke and Co., late shoemaker to His Majesty's Forces in Ireland; J. Scanlan, confectioner, supplied officers' messes on special terms, while Stynes' Bakery, J.C. Price, Ironmonger, and the Dublin Laundry Co. Ltd., delivered on the Curragh. Moy-Mell Domestic Employment and Servants' Registry, catered for the Curragh, Newbridge and Kildare. Thomas Berney, Kilcullen, made military saddles, harness, etc., while Dardis and Son, with a shop in both Kilcullen and the Curragh camp, were meat contractors to H.M. Government, and William Fanning, Master Plasterer and Slater, was also a contractor to the government.

There were several hotels, including the Prince of Wales in Newbridge and the King's Arms in Kilcullen, and that of John Mallick at Athgarvan and the Stand House Hotel of Joseph Whelan. In the Curragh camp there were branches of the Dublin shops of Eason's, Stationers and Newsagents; Todd, Burns and Co. Ltd., Drapers, and McCabes, Fishmongers, as well as those of Llewellyn, Coal Merchants, Richard O'Mahony, Builder, H.W. Church, Cycle Agent, Photographer, Jeweller and Motor Engineer, and the Army and Navy (Junior) Co-operative Stores. At Brownstown there was a coach builder, and D.E. Williams, Vintner, and there were other public houses in the neighbourhood of the camp at Ballysax, Blackrath and Suncroft, as well as 25 licensed premises in the town of Newbridge, 14 in Kildare and 10 in Kilcullen.

Of the towns just mentioned Newbridge merited 17 pages of entries in the guide, Kildare had 9 and Kilcullen 6. The county town, Naas, had 15, Maynooth 19 and Athy, the centre of a prosperous farming community, 23. Monasterevan had 6, and Celbridge and Leixlip had, respectively, 4 and 3. The dominance of Newbridge as the shopping centre for the military might be expected on account of the barracks there, and its proximity to the camp. The variety of businesses reflected the varied requirements of the garrison, with several high-class grocers, confectioners, drapers, laundries, pawn-brokers, and a forage contractor (the only other one in the county was at Naas). There was a Commercial Club 'for the recreation of the inhabitants, non-political and non-sectarian', a Masonic Hall, market hall, court-house, and Sheridan's Hall, which accommodated 500.[23] The surviving records of Cummins' Public House and Grocery, which is situated directly across the street from the site of the barracks, indicates that the family did a steady trade with the military, held contracts for the town barracks, and delivered goods to the Curragh camp in the years from 1910 to 1913.[24]

The account book for the years 1907–1909 of Thomas Berney, the Kilcullen saddler, gives ample evidence of the importance of the military to his trade. Amongst his customers were officers from the various regiments which served in the county, and sometimes their addresses in England were pencilled in above the local one. Payments and non-payment were shown. Examples of the entries are: Major Cubbet, Brigade Major, 5 South Road, Curragh, 19 March 1908: Stable kit complete £1.2s; Two tins saddle soap 1/8; Two chamois 7/–; Two martingales 13/–; Two curry combs 1/8. Major Loyde [sic] of the Black Watch, in January 1908, paid £1.15s. for two under-blankets, and £2.14s. for two sheets bordered blue. In the following month he paid £6 for a hunting saddle to order, with patent spring bars. A set of brown gig harness with a breast collar was the purchase at £8.5s. of Major Crookshank R.E. on 4 September 1908. The account of Mr Burke R.H.A., Newbridge, from January to October 1910 amounted to £5.3s.6d, and General Fanshawe, Newbridge, was billed for the sums of £3.9s.9d, 10/– and £1.18s. in the autumn of the previous year. The requirements of such a large number of hunting men, in addition to those of the resident gentry, made the skill of the saddler a busy and lucrative one.[25]

An annual boon to merchants, farmers, hunting people, hotel owners, jarvies, and trades people were the dispersal sales of horses held in the camp each summer. That in June 1911, by order of Col. the Marquis of Waterford, commanding the South Irish Horse, offered 150 horses, about 100 of which were being sold without reserve. Described as originating in Clare, Mayo and Galway, and being strong and hardy, and the best ever draft offered for sale, they had been examined by a Board of Officers and Veterinary Surgeons.[26]

Of Newbridge it was said that while the town prospered on its contracts with the garrison, 'few of the townspeople were directly employed [in the barracks] . . . only carters delivering supplies, or farmers taking out loads of stable manure, or washerwomen working in the laundry were regularly admitted to the barracks'. But local people could also be employed there to clean the chimneys, light lamps or wash the bed clothes. Repairs in the barracks were done by a builder from the town named John McElwain (and after whom a terrace in the town was subsequently named). He had the contract for repairing barrack damages, caused by and paid for by the soldiers, or for damage done by the horses.[27]

Barrack security required that contractors entering the barracks had to be signed in and out, and while inside they were kept under supervision. There was an instance in October 1910 when a local farmer who called in to discuss the letting of his fields at Ballymanny to the army for trenching practice was arrested, on suspicion of espionage. When identified he was released.[28] One Newbridge firm whose business with the military was widely recognised were the photographers Messrs P. Charleton and Son. In the face of opposition from Dublin and London they won the contract for photography at the military tournament held in Ballsbridge, Dublin.[29] Charletons also supplied most of the photographs of the camp for Messrs Eason's book *Curragh camp and district, illustrated and described*, which was published while General Lord Grenfell was commander in chief of the forces in Ireland (1904–1908), and whose portrait is a frontispiece to the book. As well as giving photographs of the principal old and new buildings in the camp, the book includes splendid impressions of the reviews, the artillery exercises and the encampments there.[30]

The cases appearing before the Petty Sessions and other courts showed little change over the years. Early in 1906 a soldier from the 11th Hussars was charged with the rape of an old woman, he pleaded guilty and was given four years penal servitude.[31] A couple of months later Lance Corporal Arthur Wilson of the Mounted Police, a veteran of the South African war, was discharged and given nine years for the indecent assault of a nine year old boy named Patridge.[32] Private Purvis of the Warkwickshire Regiment was found guilty of stealing a watch from Violet Harrold, a member of a theatrical company playing in the camp,[33] and a man of the 3rd Dragoon Guards got four months for stealing a diamond pin from the home of Lieut. Col. Herbert Mercer in the camp.[34] A more serious case was that of Sapper Taafe 59th Coy R.E. who shot dead Sgt. Major H. E. Theobald after he had reprimanded him for not having the mess room clean. Taafe believed that he was being persecuted by one of the officers, and when it was established that he had 'two relatives lunatics' Taafe himself was found to be insane.[35] However, when Pte. Leonard Booth 19th Hussars shot at Sgt. Maj. Barnett on the barrack square at Ponsonby Barracks it was suggested that the culprit had been reading too many 'Penny Dreadfuls', and he was remanded for the Assizes.[36]

An internal scandal at Christmas-time 1906 was the desertion of Lieut. Stephen Mansfield A.O.C., who lived at McElwain Terrace in Newbridge. Married and aged 47, he could soon have retired with a pension, but he disappeared with over £200 from the monies due to the men of his unit. His photograph was published in *Hue and Cry*.[37] Desertion of other ranks was a common military offence, but to find a sailor from the Royal Navy being so charged before Mr Ronaldson at the Curragh Petty Sessions in November 1907 was most unusual. John Chambers East Lancashire Regiment had deserted from H.M. Pembroke in the belief that 'he could serve H.M. in a way that was far removed from the dangers of sea-sickness, so he joined the land forces. But a few months on the bleak plains of the Curragh, with a six o'clock 'jump-up' in the morning to learn to goose-step, sickened him. Even the idea that he was, and could still be, one of those who are popularly supposed (amongst our British friends) to Rule the Waves, was somewhat better, and he surrendered'. Sending Chambers back to his ship the magistrate admonished him: 'It is rarely that young men get such an opportunity of getting tired of both services in such a short time, though there are many in either both tired and sick'. If that report showed an appreciation by the R.M. of life in the services for some members, it more clearly revealed the attitude of the journalist to both the crown and those who were loyal to it.[38]

Sometimes the circumstances of desertion could be amusing, as was the case of Driver Frank Cartledge R.H.A., which was reported in the local press. Awaiting court-martial in the barracks at Newbridge he managed to escape no less than four times, and on each occasion he took refuge in the bog at Roseberry. The local people sympathised with him, and food, or warning messages if the military police were around, were sent to the fugitive tied to the collar of a dog.[39] When finally captured by the R.I.C. papers were found in his possession describing his escapades and referring to himself as 'the hero of the bog'. He was sentenced to 168 days imprisonment by a court-martial in Newbridge.[40]

In Kildare barracks in 1907 a strange story had circulated in the month of May, while most of the artillery men were firing the annual gun practices in the Glen of Imaal. It was believed that the body of a deserter was in the barrack water tank, and the alarmed personnel commenced drawing their supply by bucket from the town. Finally it was decided to drain the tank, and no body was found.[41]

A crime which was all too common in the barracks was that of suicide, and it led to a reference in the *Leinster Leader* of 18 September 1909 to 'the recent epidemic of suicides at the Curragh' and the ease with which men could get ammunition.[42] One of the cases referred to was that of Private John Vivian Crowther 18th Hussars who had recently shot himself. At the inquest he was described as 'a cultured and educated Oxford graduate who had inherited a large property'. He described himself as an actor, and was married to an actress. Off duty when he died in the wash-house, his rifle, sword and kit-bag were left lying on his bed. He was found to have been of unsound mind, but the question was asked as to how he could have got the ammunition?[43] The same query was posed following the death of Private Walter Thorne 5th Dragoon Guards who shot himself in his bedroom at Piercestown House, Newbridge, in 1911. His officer replied that the bullet might have been picked up on the range.[44]

In those doldrum years between the wars the improvements to the camp continued, including the recreational facilities. An old resident of the county, and a member of the Kildangan Cricket Club, recalled that in 1903 Mr Tinsley (who was to be Green Keeper at the Curragh Golf Links for 54 years) was employed in the laying-out of the new Curragh Cricket Pitch.[45] In June 1906 the first match on the new cricket ground at Kildare barracks was played between the 32nd Brigade (Curragh) and the 33rd Brigade R.F.A.[46]

The Curragh golf club which had opened its doors to civilian members in the late 1890s 'and this was restricted to the local gentry', had also allowed ladies to participate in the game there since at least 1890 when the wife of the camp chaplain, Mrs Norman Lee, won a handicap competition. In 1904 the club was graced with a royal visitor, and an amusing story is remembered about the occasion. The duke of Connaught, then commanding the forces in Ireland, was playing golf with an officer of the South Irish Horse and when they came to the first green, the latter gentleman was six inches from the hole, and the duke 100 yards. Observing the lie of the ball the duke remarked 'I always count two on greens and thus avoid putting . . .' Perhaps influenced by Connaught's enjoyment of the Curragh course, in 1910 the title royal was conferred on the club, confirming the high status of the links.[47]

In 1910 the other ranks rejoiced at the opening of Sandes Soldier's Home in Newbridge. It was described as 'extremely large, and built through the generosity of Miss Sandes, on a site on Edward Street at the western end of the barrack wall obtained from the government by Gen. Fanshawe. It contained a concert hall and recreation room, dining rooms, mess rooms, a kitchen and bathrooms, all fitted up in the most modern manner and hit by electricity. Although there were some 700 men of the Royal Horse Artillery then in the barracks there was plenty of room for those who from time to time wished to spend a few hours pleasantly at the Home.'[48] Elsie Sandes' intention to remove the men from public houses and bad company was again being realised as another of her final total of 22 homes throughout the country was opened.[49] It was in October 1911 at the Curragh camp. Then Miss Sandes wrote to *The Irish Times* asking that illustrated papers and magazines might be sent there, or to the Newbridge home. That year too she was left a legacy of £500 by a widow from Bray, county Wicklow.[50]

During the summer training period of 1910 the field exercises on the Curragh again culminated with an attack on the Gibbet Rath in which a battery of the R.F.A. was sited. Both the 5th Dragoon Guards and the 18th and 20th Hussars had the objective of capturing the guns, from different routes. Due to what may have been a bad battle order the Hussars and the Dragoons attacked without knowledge of the other's timings. A party of Dragoons, failing to stop in time, rode close to the guns which were firing, and an officer and two sergeants were wounded; simultaneously the Hussars charged into the Dragoons causing mayhem, and injuring several men and horses, four of which had to be put down there and then.[51] While there would have been serious enquiry by the military into the cause of the accident, the intensive field training and exercises which were practiced helped to reduce such mishaps to a minimum.

That year of 1910 was seen by one commentator as the end of the 'Indian summer for the old order', with Irish political affairs in the doldrums,[52] or as General Sir Neville Lyttelton G.O.C. Ireland from 1908–1912, described his term 'a haven of rest'.[53] The Irish party had the balance of power in the Liberal government after the two general elections, held at the beginning and the end of the year 1910. For some years before new political forces were emerging in Dublin and Belfast; *Sinn Féin* was gaining strength, *Fianna Éireann* had been formed, and the trade union movement was established. Home Rule was again a vital issue with the third Home Rule bill being passed in the House of Commons in January 1913, only to be defeated in the House of Lords, a procedure to be repeated again in July of that year. However, new restrictions on the veto power of the House of Lords meant that the Home Rule Bill should come into effect before the end of 1914.

However in 1911, even if the sun did not shine as brightly for the military as hitherto, they had a memorable year. At the end of March instructions were received for the numbers of troops from the Irish Command which would be required for the coronation ceremonies in London in June. The cavalry regiments at the Curragh, the 5th Dragoon Guards and the 20th Hussars, were each to furnish a detachment of 100 non-commissioned officers and men, with a proportion of officers. In addition the Curragh regiments were each to send one officer and 17 other ranks for the royal procession.[54]

The spring and summer months on the Curragh were, as always, very busy with the 32nd Irish Command Bronze Medal Military Tournament being held in the camp for three days in March,[55] the celebration of the king's birthday in May which included the firing of the guns at midday,[56] and the visit to Dublin and Maynooth by King George V and Queen Mary, accompanied by the Prince of Wales and Princess Mary, in the first week of July.[57] If the review in the Phoenix Park, and in which troops from the Curragh participated, was 'the most popular and spectacular incident of the royal visit',[58] for the people of Kildare, the reception by the hierarchy of the king and queen at the Royal College of Maynooth was the most memorable. Cardinal Logue led the archbishops and bishops in their homage. Special trains brought loyal subjects from Dublin and elsewhere, and the town of Maynooth was extravagantly decorated for the great day. There was no military presence.[59] It was to be the last such royal visit to southern Ireland.

The training season continued in the Curragh area with camps for a detachment of the City of Dublin Cadets being held at Donard, county Wicklow and the Curragh,[60] and a battalion training camp for units of the 13th Infantry Brigade at Rathmore, near Naas.[61] The general commanding in Ireland, Sir Neville Lyttelton, came to the Curragh to visit the 5th Dragoon Guards and the 20th Hussars during their training period,[62] and the Dublin University Contingent Officers' Training Corps held its annual camp there.[63] The 3rd Cavalry Brigade held field operations in the vicinity of the plain,[64] and in mid-August the All Ireland Rifle Meeting had 'a heavy but exceedingly interesting week' on the ranges.[65] The Gibbet Rath was again to be the assembly place for the R.F.A. Brigade of the 5th Division to practice ground movements,[66] and soon after the completion of that exercise Lyttelton was back again to observe the training of the 14th Brigade.[67] By the month of October the programme for artillery practice in the Glen of Imaal in 1912 had been issued,[68] and in the Curragh the indoor war games and lectures had commenced.[69] So ended another year for the soldiers in county Kildare, and if it was remembered by them for their part in the royal pageants, it was also recalled by the general who commanded the cavalry brigade there as a pleasant one.

The first decade of the new century had been a time of consolidation for the military in county Kildare with the new barracks being opened in Kildare town, and the continuation of the major rebuilding programme in the Curragh camp. It imposed ranges of barracks on the skyline of the Long Hill which were to become a permanent feature of the landscape. Military traffic on the roads through the plain disturbed the pastoral nature of the place, and in the Newbridge area it had become an annoyance to the local people. The maintenance of the roads was a vexed subject, with the costs being disputed between the county council and the military.

But, despite complaints about the placing of tenders by the army, business in the district was booming. The construction work in the camp gave additional employment to a large work force, and if the soldiers sometimes caused trouble in the towns or villages, the swelling of the ranks during the training period brought increased trade to the shops and hostelries. The military reviews on the plain, and the annual sports and entertainments in the camp and other barracks, were firmly established as social occasions for the many civilians who appreciated the presence of the military in their midst.

In January 1911 Brigadier General Sir Hubert Gough was appointed to command the 3rd Cavalry Brigade at the Curragh. That year his brigade flew its banner over their reserved quarters at the Punchestown April meeting, as did the Royal Horse Artillery, 1st Battalion Army Service Corps, and 3rd Rifle Brigade, 2nd Bn Wiltshire Regiment,[70] 2nd Bn Connaught Rangers, 2nd Bn Essex Regiment, Royal Sussex Regiment and the Royal Wiltshire Regiment. In later years General Gough was to look back on 'the first three years of my command of the 3rd Cavalry Brigade, life was pleasant but uneventful. As usual in Ireland hunting, races and horse shows were our chief amusement when we were not soldiering, as they were of the people among whom we lived. None of us cavalry soldiers were much interested in Irish politics, as far as we were aware, the people apart from the professional politicians, were not much more interested either'.[71] If the isolation of the officer class from mainstream Irish life was reflected in the general's musings, so was the pre-occupation of his peers, both lay and professional, with the horse. As the dominance of that animal in army life was declining in 1911, so was that of the major sporting class in national affairs.

The establishment of the Ulster Volunteer Force at the end of January 1913, and the formation of a provisional government in the province of Ulster in September, preceded the raising of the Irish Citizen Army and the Irish Volunteers in November. While the loyalty of a substantial part of the population was still to the crown there was a major renewal of nationalist feeling amongst a cross-section of the people. Units of the Irish Volunteers were organised in 'nearly every parish in the county of Kildare' during the early months of 1914.[72] A member of the volunteers was Michael O'Kelly, who had succeeded his brother Seamus as editor of the *Leinster Leader* in 1912. He remained with the Irish Volunteers when the movement split later that year. His strong nationalist views were to be reflected in his newspaper, especially in reports concerning the British army.[73]

However, the traditional entertainments organised by the military were still much sought after by many of the local people. A large audience appreciated a concert given by the garrison at Newbridge in March 1913. The patron of what was regarded as a very successful evening was Brig. Gen. E.A. Fanshawe R.A.[74] A decade later his only daughter was to marry the future General Sir Eric de Burgh of Naas (Fig. 33). By then General Fanshawe was living at Rathmore, near Naas. [75]

The writer Sean O'Faoláin, who about that time as a boy spent holidays with an uncle living 'in the military town of Newbridge . . . always a garrison town, dyed pink and khaki', in later years recalled his time between 'the high walls of the barracks . . . [and] the shops and houses which had once been the original sutlers' booths'. The youth was impressed by the elegance, accents and clothes of the married officers who rented the top floor of his uncle's house, and by 'their dressing for dinner every evening, their silverware, their general proposal of distance, wealth and romance, especially when the war began in 1914'. O'Faoláin remembered being awakened in the mornings by bugle calls or by the jingling and rattling of horse-drawn gun waggons ambling down the street into the countryside on some military exercise, their drivers astride the leaders, chin straps fixed, wearing the steel shin guards that kept their legs from being crushed between the frightened horses in the road of battle'. 'At night there were always soldiers in the street, patrolling before the lighted windows, up and down, up and down; poor bored, bloody English swaddies, as I now see them, fed up with this country town where there was nothing to do but get noisy in a pub or try to lure a girl under a hedge for a

Fig. 33. *General Sir Eric de Burgh K.C.B., D.S.O., O.B.E., Oldtown, Naas (1881–1973). In 1966 at the 75th Anniversary Dinner of the Co. Kildare Archaeological Society (of which he was president) in Devoy Barracks he recalled that he 'did his recruit training as a subaltern in the 3rd Royal Dublin Fusiliers (Militia) [in the barracks] and was chased up and down the square by two excellent Sgt. Majors French and Brumby, in 1901, before going out to Africa'. The singer Chris de Burgh is his grandson.*

bit of mugging. There was no cinema, but every known denomination had its tin chapel'.[76]

By the end of 1913 the ultimate demise of the concert and other home-made entertainment was sealed. The Curragh Picture house, with new technology installed, was open, and it included a studio in which film taken locally could be processed to be shown the next day. War was to come to the Curragh, twice nightly, with the promised screening of *The Battle of Waterloo*, but in the meantime the troops and their friends were enjoying *From Circus to Racecourse*.[77] The Grand Stand at the Curragh Racecourse was a limit of the course over which the All Ireland Military Cross-Country Championship which was run in April 1913. It was won by a team from the Connaught Rangers.[78] The sporting calendar for the year included all of the usual events, including hockey. It was during a hockey game in July 1913 that Pte. James Duffy, 4th Hussars was accidentally killed.[79]

That summer the military became involved in an unpleasant incident which had its origin in a dispute concerning the renting of grazing lands. In the giving of evidence before the King's County Assizes in July 1913 details were given of one of several 'cattle drives' in the Ballynowlart area, it was mentioned that 40 horses belonging to the military at the Curragh had been driven off land which had been leased for six months. They were found the following morning wandering over a radius of three miles. But there would not appear to have been any special ill-will directed towards the military, the aggravation was with a middle-man who was acting without the knowledge of the landlord.[80] More serious was the stampede of 75 horses of the 4th Hussars which had been put out for two months grazing on the lands of a farmer at Kildoon, near Suncroft, in November 1913. Soon after they were let loose in the field the horses stampeded, broke down the gate and galloped along the road back towards the Curragh. An unfortunate farmer's wife who was bringing home a load of turf from the bog was knocked down and trampled by the horses. She died the next day. The inquest found that there was no blame on any person for the accident, and it was recommended that the victim's aged husband should be given compensation.[81] (However, in the Suncroft area a persistent rumour is that the horses were released intentionally, for political reasons.)

But for the imaginative young Sean O'Faoláin the Curragh was a special place: 'above all, I loved the great open plain of the Curragh, two miles from the town [of Newbridge]. This immense rolling plain was empty except for its grass, a few grazing sheep and tufts of yellow gorse. In the distance one saw an occasional car slowly crawling across it like a beetle over a golf course. Ireland's most famous racecourse has written an oval in one corner of it . . .' He admired watching the horses being worked out in the early morning, and he observed on the eastern edge of the plain 'a faint red line as if drawn by a pencil marked yet another military settlement still known as the Camp, years, indeed three full centuries, after its canvas had turned to red brick buildings, tarred wooden huts and tin sheds. Here too one heard distant bugle calls wavering on the wind, the dull smack of rifles from the pits, the faint smooth brass of a military band. On a sunny day the heliograph winked across the bottoms'.[82]

There was an ominous start to the new year of 1914 for the military in the Curragh Camp when, on the night of Monday 26 January, the sentry on Magazine Hill was attacked and disarmed. He was found bound and gagged at his post, but with his rifle taken. He had been without ammunition, as it was then not usual to supply it to duties there. The authorities took a very serious view of the fact that such an incident could occur at an important post, and ordered that in future the sentries were to be doubled there.[83] Undoubtedly the disarming was achieved by men from one of the volunteer units newly active in the district, but no further mention of the incident has come to light.

While the daily routine of the regiments in the camp proceeded as normal, the senior officers there were, in the month of March 1914, faced with a very serious situation. Troops then stationed in the Curragh were from the 14th Infantry Brigade, commanded by Brig. Gen. Rolt, and one of the three brigades in the 5th Infantry Division, commanded by Maj. Gen. Sir Charles Fergusson, and from the the 3rd Cavalry Brigade. It was commanded by Brig. Gen. Hubert Gough, and as it was not part of the 5th Division, it came under the direct jurisdiction of the General Officer commanding Ireland, Lieut. Gen. Sir Arthur Paget. Two regiments of the 3rd Cavalry Brigade were stationed in the Curragh, the 4th Hussars, a large portion of which was Southern Irish, in Stewart Barracks, and the 16th Lancers in Ponsonby Barracks. Three battalions of the Infantry Brigade were in the camp, 1st Battalion Duke of Cornwall's Light Infantry in Beresford Barracks, 2nd Battalion Suffolk Regiment in Gough Barracks and 2nd Battalion Manchester Regiment in Keane Barracks. Also in the camp were 4th Field Troop R.E., 3rd Signal Troop R.E., while the 3rd Brigade R.H.A. was in Newbridge. There also, under the command of Brig. Gen. J.E.W. Headlam, was the 27th Brigade R.F.A. In the barracks at Kildare were the 7th (Howitzer) Brigade R.F.A. and 15th Brigade R.F.A. The Divisional Engineers at the Curragh were the 7th Field Company R.E. and 59th Field Company R.E.[84]

The early months of 1914 were rife with rumour; that the Ulster Volunteers were going to attempt to illegally import arms, or that the government was about to take a heavy hand with the volunteers.[85] Opposition to the implementation of the Home Rule was considerable, and a speech delivered at Bradford on 14 March by Winston Churchill, an ardent supporter of Home Rule, further inflamed the Unionist.[86] A couple of days later when orders were received by the G.O.C. Ireland to increase security at depots in the north of the country further unease permeated Ulster.

When the order to issue live ammunition to sentries, followed by an order to so equip all men in barracks, was received in the Curragh camp Major General Allenby was on a visit to Gen. Gough's brigade, and which he was commending for the high standard of training reached by the squadrons. That evening of 19 March the Goughs had arranged a dinner party for their visitor at Brownstown House, but two of their other guests had to be excused. Alice Fergusson, the wife of General Fergusson, sent a note saying that they would be unable to dine as he was 'occupied with movements of units'. That evening Gough, Fergusson and Rolt received messages asking them to attend Army Headquarters at Parkgate street, Dublin, on the following morning.[87] At the meeting, General Paget briefed his commanders, first giving details of the movement of troops to Newry, Dundalk (which were to include the 1st Battalion Duke of Cornwall's Light Infantry from the Curragh) and elsewhere in the north. Then he told them that the Secretary of State for War had made it clear that officers ordered to act in support of the civil power could resign their commissions if they did not choose to obey orders; if they refused they would be dismissed from the army.[88]

The general officers were shocked at the latter condition, which was to become known as 'The Ultimatum'; Fergusson immediately sent by hand a written detail of the orders to his units in the Curragh, while Gough walked to nearby Marlborough Barracks to give the officers of the 5th Royal Irish Lancers the message. Seventeen of the 20 officers present decided that dismissal would be preferable to active service in Ulster.[89] General Rolt drove directly back to the Curragh Camp to brief his battalion commanders before lunch, and Gough arrived back in the camp in mid-afternoon. The gentlemen of the 16th Lancers, 4th Hussars and 3rd Brigade R.H.A. were gathered in the Officers' Mess at Ponsonby Barracks. All 16 members of the 16th Lancers present decided to accept dismissal, as did 17 out of the 19 officers of the 4th Hussars. Six of the 13 officers of the 3rd Brigade R.H.A. doing duty, and the officer on duty in both the 4th Field Troop Royal Engineers and 3rd Signal Troop Royal Engineers, as well as two officers of the brigade staff, and 17 officers of the 5th Lancers in Dublin, also accepted dismissal, bringing to in all, a total of 61 gentlemen, including the brigade commander, who 'respectfully, and under protest, preferred to be dismissed' rather than to be involved in the initiation of active military operations in Ulster'.[90]

General Fergusson did not have an opportunity to meet his officers until the morning of Saturday 21st; after he had spoken to the officers of the Suffolk and Manchester Regiments in the camp gymnasium he drove to Kildare, Newbridge and Dublin to address the units there. His message to the officers was that military discipline, and loyalty to the king and government was their duty.[91]

In the meantime the War Office had learnt of Gough's response to 'The Ultimatum', and a message was sent requiring him and his cavalry officers to report to the War Office,[92] but the G.O.C. Ireland decided to go to the Curragh to meet the disaffected cavalry officers himself. His words to them were unclear, but the message was that he would not go to war against Ulster, and that if fighting did occur that he would order his troops not to return fire. He asked that the officers should make their decisions known to General Fergusson; when they did the only change of mind was on the part of the officers of the Royal Horse Artillery and the 4th Hussars.[93]

When General Gough presented himself at the War Office in London on March 23rd it was soon apparent that during the week-end there had been a flurry of nervous activity there. The realisation by the civil authorities that the military could not be coerced raised spectres of the days of the Stuarts,[94] and a realisation of the truth of a warning given 20 years before by a commander in chief Ireland, Lord Wolseley, that 'if ever our troops are brought into collision with the loyalists of Ulster and blood is shed, it will shake the whole foundations upon which our army rests to such an extent that I feel our army will never be the same again'.[95] At the War Office Gough was told that there had been 'a misunderstanding', and he sought and got a letter, approved by the Cabinet, acknowledging that fact, but restating the obligation of the military to support the civil power in the maintenance of law and order. When he saw further problems with that undertaking, he got a further written statement to the effect that 'troops under our command will not be called upon to enforce the present Home Rule Bill on Ulster'. That codicil to the Cabinet paper was later disputed, as it had not received Cabinet approval.[96]

The consequences of the 'Curragh Incident' were that an army order was issued on 28 March stating that in future no officer or soldier would be questioned about the attitude which he might adopt if 'in the event of his being required to obey orders dependant on future or hypothetical contingencies',[97] and a deepening of the normal suspicions which existed between the civil and military administrations. [98]

Some observers saw the stance taken by the officers as a reflection of the 'socialized and isolated' class to which they belonged, or the influence of the Anglo-Irish officers.[99] General Gough was greeted as a hero by *The Irish Times*, '[his] fearless and honourable conduct has added lustre to the laurels of a great Irish family',[100] and a later historian of the 5th Lancers believed that the attitude adopted by the 5th Lancers and the 3rd Cavalry Brigade 'ought to be blazoned in letters of gold, and thus transmitted to history'. However, when in March 1918 General Gough was removed from his command of the 5th Army (which had suffered heavy casualties), it was thought that 'certain elements of the government' who had never forgiven him for his part in the 'Curragh Mutiny' were responsible for the decision.[101] General Fergusson, for his abidance of traditional army loyalty, was denounced as 'a cur'.[102] General Sir Nevil Macready [G.O.C. Ireland 1920–23] believed that 'the whole affair was regrettable and unnecessary'.[103]

In Ireland the failure of the British Government to control the army was seen as a denunciation of parliamentary procedure, and a confirmation of the traditional belief that nothing had changed in London. For many nationalists the words of Daniel O'Connell at the Rath of Mullaghmast in 1843 could be seen as prophetic. Referring to the massacre at the Gibbet Rath in 1798 he said 'the Curragh of

Kildare afforded an instance of the fate which Irishmen were to expect who confided in their Saxon enemies'.[104]

How some of the other ranks, who were confined to barracks, saw the predicament of the officers was afterwards to be revealed when letters of the period were published. Private C. Smith 16th Lancers, writing to his brother in England, said 'last Thursday all the guards on the Curragh were doubled and served out with ball ammunition, as if for active service. Also extra guards were put on at different points, also the roads guarded by armed men. On Friday [March 20th] my regiment was ordered to the north of Ireland, destination unknown. Officers of my regiment, many distinguished men among them, the Colonel, Lord Holmpatrick, and all others held a conference and decided not to go to Ulster to fight such loyal men, as they are, and were in Africa'.[105]

The wife of a non-commissioned officer serving in the camp gave her impression of the emergency to her parents, also in England: 'Things have been very serious here this last week, and we are still under guard with fixed bayonets all over the camp and at our quarters, every married man has ammunition and rifle in quarters, and no one allowed out of camp. Of course there are many miles of camp on the Curragh, we can get plenty of walks, but no one is allowed to enter camp. What do you think of all our cavalry officers resigning . . . it is one of the finest things that they could have done. They new [sic] the men could not do it, so they done it themselves, although the men's hearts were the same way if they dare, but all our men say they would have been shott [sic] first than take arms against Ulster . . .'.[106]

Some of the officers' ladies also left their accounts of the crisis as it was seen in the Curragh camp. Mrs Rosalind Howell, the wife of Major Philip Howell 4th Hussars, thus described her impressions of the crisis: 'Sir Arthur Paget came down to the Curragh, and all the interest of Europe was focused on our quiet little camp. A breathless atmosphere of suspense seemed to surround the H.Q. hut that afternoon, where all the officers were closeted with the generals. Outside journalists were driving up in side-cars or wandering about like pelicans in the wilderness, presumably looking for General Gough, and seeking vainly for any information, the ordinary Tommy not having the least idea what it was all about and apparently taking little interest. Wives of officers in the brigade were in tears in the main Curragh street, and I, foolishly, trying to comfort them with a brief sketch of the history of Ireland.'[107]

Nora Gough, the wife of Gen. Gough, also left a memory of the local reaction to her husband's stance. She remained in their home, Brownstown House, south of the Curragh, when he went to the War Office on 23 March. A letter which she wrote to her husband told of the messages of congratulations which had come for him, and the promises of support for his actions from brother officers. At church on Sunday Mrs Gough had 'prayed so hard [for her husband], I felt so thankful that I stayed to fight your cause, I sang at the tip top of my voice, other women were on the verge of tears. So was I, if truth be known'.[108]

When General Gough returned to the Curragh camp on Tuesday 24 March he was given a magnificent welcome by his regiment and other supporters. He had travelled by the morning train from Dublin and arrived at Kildare. Private C. Smith's squadron of the 16th Lancers was sent to the station to escort him back to barracks; we 'were in review order, which looked lovely. The regimental coach was drawn by four horses of the Maxim Gun Section, drivers also in Review order. All men in barracks waited to cheer him in and my word it was a reception. Then the colonel [Holmpatrick] and Gen. Gough stepped out of the coach and gave a speech thanking men of all ranks (the mounted men with erect lances formed a ring around the carriage)[109] for backing them up in their fight with the War Office, which was a bloodless fight, and saved many lives'.[110] As the cavalcade of 50 lancers, and the coach with two postillions drove across the plain towards the camp it passed a section of recruits who were drilling. They cheered so loudly that Mrs Brett, the wife of Lieut. Col. Brett, 2nd Suffolks, heard them in her married quarters.[111]

An anonymous non-commissioned officer's wife also saw the welcome: 'All the cavalry turned out in review order and escourted [sic] him home in his carriage as if he were a king. I cannot describe what it has been like here this last week, everything and everybody up to war strength and ready to move off in five minutes notice, but still we are smiling at it all'.[112]

The boredom and uncertainty of army life, and the constant waiting for something to happen, can be detected in a letter which Private Charles Smith 16th Lancers wrote to his brother Harry who was art editor of the *Illustrated London News.* Writing just after the 'Curragh Incident' of March 1914 he expressed thanks for the gift of shamrock and other presents, and he said that if 'we had received orders from our officers to go [to Ulster] and fight I would not trouble in the least. It would be a change to do something exciting, in fact I would love to go on active service. No doubt you think I am a very bad writer by this letter, well I am doing it in drill time 1–1.2 time, as I am rather busy with parades. Last Tuesday we had a funeral, a young soldier was on night guard over the horses and whilst walking through the stables a horse kicked out and caught him on the right temple, am sorry to say, and died a quarter of an hour later. I was bearer, one of six. His name was Private Mercer. On Wednesday I was field officer's orderly in review order. Plenty of work getting ready'. He also mentioned that if his brother wanted photographs of any of the regiments in the camp that he should send to Mr Charleton, [c/o] postmaster, Hare Park post office, Curragh. The photographer referred to was Peter Charleton, Newbridge.[113]

The possibility of trouble in Ireland was always of concern to the military commanders. Both the duke of Cambridge and Lord Wolseley had doubted the loyalty of Irish soldiers, and Wolseley in 1895 had advised the War Office that even in peacetime troops of the Irish regiments should not be in their own country. Lord Roberts had a definite belief that it was necessary to keep an army corps in Ireland, in case any disaffection arose. Then the militia units raised in Ireland should be transferred to Britain. When the militia was superseded by the Irish Special Reserve Battalions in 1905 the stationing of those units in Ireland was again being debated.[114] However, the exigencies of the service did not always permit ideal postings to be made, and during the Great War Reserve Battalions of the Royal Munster and Dublin Fusiliers were in the Curragh.[115]

Chapter 14

Onset of the Great War

With the immediacy of a crisis in Ulster postponed for a time, and the ominous clouds of war gathering over Europe, the generals in the Curragh continued with their regimental affairs. Time was found to attend the usual two-day meeting at Punchestown in the last week of April 1914, and the Maiden Military Hunter's Race and the Irish Grand Military Race again provided the main interest for the army spectators. In the following month, for the third time, the Home Rule Bill was passed by the House of Commons. But the formerly peaceful atmosphere and leisurely style of military life was changing. In May also units of the Irish Volunteers [a nationalist organisation founded in 1913] were formed at Athy and Monasterevan.[1]

In the first week of June 'mimic warfare' was held in the neighbourhood of Naas, to the alarm of the townspeople. Cavalrymen and cyclists, of opposing forces, forced their way into the town and came into contact. When they opened fire on each other the startled residents rushed to their doors and windows, and frightened passers-by sought shelter. Mounted soldiers pursued each other through the main street, and 'enemy captives' were taken.[2] It was all very exciting for the onlookers, and probably good training for the military. That same week a party of Irish Volunteers were drilling near the town, and at Athgarvan, about a mile from the Curragh Camp, another unit was in training. Frequently the drill instructors were ex-British army, and very often the training was observed by officers and men from the barracks.[3]

On Monday 22 June the king's birthday was celebrated in the Curragh with a Grand Review. Major General Sir Charles Fergusson took the review in which troops from the Curragh Camp, and the Newbridge and Kildare garrisons took part. The weather was threatening, but luckily it stayed fine until the celebrations were over.[4] In July the 22nd Irish Command Athletic Meeting was held at the camp, and it was described as successful and enjoyable, and with a big attendance.[5]

But there were less joyous events also that month; Gunner Williams D Battery R.H.A., a well-known footballer, and goal-keeper for his unit, died. His funeral was attended by the entire battery in uniform and on horseback, as well as a large number of civilian friends.[6] Within a week there was another funeral, that of Private Albert Mace R.F.A., Newbridge, who was drowned while swimming in the Liffey.[7] Another death about then was that of Major Eustace Loder J.P., of Eyrefield Lodge on the Curragh. He had come from Sussex, and served as adjutant of the 12th Lancers. In 1896 he had purchased Eyrefield, and, having resigned his commission he became a horse trainer, and employed a considerable number of men there.[8]

In Kildare town there was trouble in a public house where soldiers were drinking. A 'Red Cap' [military policeman] who tried to have the place vacated was given abuse by a soldier's wife, and when the affray came before the R.M. he said that 'respectable people must be protected, and the town was not a bear garden'.[9] In Newbridge a military engine of the A.S.C. in the Curragh swerved when crossing the bridge over the Liffey and crashed into a Mission Stand in front of the parish church where the Redemptorist fathers were conducting a Mission. The army agreed to compensate for the damage[10].

By July units of the Irish Volunteers had been established in Kilcullen, Carbury and Ballymore Eustace, and a section of *Cumann na mBan* [nationalist women's organisation founded at the same time as the Irish Volunteers] in Athy.[11] In mid-July on the Curragh over 7,000 people gathered at the Gibbet Rath at a nationalist rally, and the Maddenstown, Kildare and Newbridge Volunteers paraded nearby at the Stone barracks. The troops in the cavalry barracks, on the west side of the camp, and the Dublin Fusiliers who were under canvas, were reluctant recipients of the loudly delivered patriotic messages.[12] A county Kildare committee for the Irish Volunteers was founded in August, and the effective strength of the members then in the county was given as 6,000. That month also a *Cumann na mBan* unit was founded in Newbridge, and a branch of *Fianna Éireann* at Athy.[13] Such nationalist activity was known to the military authorities, and when the 5th Battalion of the Leinster Regiment Reserve was being stood down after annual training the men were warned that if they joined the Irish Volunteers they would be dealt with by the military authorities.[14]

Early in August the signs of war were more clear, and on Sunday 2nd mobilisation was ordered; at midnight word of the mobilisation came to the troops training at Coolmoney camp in the Glen of Imaal, and to the ear of Miss Sandes who had opened a Home there for the summer. She immediately arranged to return to the Curragh where she knew that the demand on her services would be great.[15] War was declared on 4 August and immediately the whole military establishment was directed towards preparation for combat.

In the Curragh camp the proposed arrival of the King's Own Scottish Borderers at number 8 camping ground was not being welcomed by the other troops, whose relationships with the people of the neighbourhood was good. The Borderers had taken part in the shooting which had followed the Howth gun-running on 26 July, and in which three people were killed and many injured. It was anticipated that the advent of the Scots would inflame the men of the Manchester Regiment and those of the 4th Hussars, both of which had been recruited from Irish nationalist centres, and had already expressed their strong disapproval of the actions of the Borderers. It was also known that the Suffolk Regiment, also in the camp, had an old cause of quarrel with the Borders, and consequently the diversion of the regiment elsewhere was welcomed by everyone.[16]

A visitor to the camp soon after the declaration of war found that all of the units were very busy; stores were being augmented and the cavalry were sharpening their swords. At Newbridge railway station waggon loads of arms and ammunition were being loaded for the Dublin garrisons, with only a small guard on duty 'unlike during the "mutiny" some months ago when mounted soldiers followed the waggons, fully armed, and others marched, as ammunition was being sent to different barracks'.[17] As several trains left the Curragh siding with men for the battlefield there were rumours circulating that the Curragh was going to be used as an internment camp for prisoners of war,[18] or that Canadian regiments would be based there.[19] Some 2,000 men, including the 27th Brigade R.F.A. and the 2nd and 3rd Batteries R.H.A. from Newbridge, were reviewed by General Fergusson, and General Gough left the Curragh for Aldershot where he was to assume command of a cavalry corps.[20]

As the biographer of Miss Sandes wrote: 'In the next four years [of war] her tasks included reorganisation and enlargement of almost every Home as reservists were called up and crowded the camps. There were repeated farewells as these men were sent to the front only to be replaced by thousands of recruits.'[21] Such, indeed was the situation at the Curragh for the duration of the war; at one period alone there was estimated to be 30,000 men billeted and under canvas there.[22]

At Naas there were pathetic scenes at the barrack gate which was surrounded by relations and friends of the reserve who were reporting to the Depot, and when 300–400 men of the Special Reserve marched to the railway station they were led by the local Volunteer band, carrying a banner with the words *Nás na Riogh* and *Nás Abu.* Big crowds of people followed the parade singing *A Nation Once Again* and *Who Fears to Speak of '98.* The Volunteers had been given permission to drill on the barrack square at Naas.[23]

For two days in Kildare the Royal Field Artillery, with waggons drawn by six horses, were unloading arms and taking them to the barracks of the 46th and 47th regiments in the town. The men in the barracks were said to be 'ready for the front at any moment', and soldiers on leave were called back to their barracks. Members of the R.I.C. were on duty at the bridges over the river Liffey and on railway bridges near Newbridge and the Curragh.[24] Military Intelligence had taken over control of Kildare railway station where thousands of reservists and recruits were arriving daily, and fully equipped units were departing. Touching scenes were reported from Athy railway station when a couple of hundred reservists left to join their regiments.

An early volunteer for service was 45 year old Hervey de Montmorency, from an Anglo-Irish Kilkenny family, who had served in the Royal Regiment of Artillery in the Boer War. As he was too old for the reserve of officers he decided to join the new Kitchener Army, the first unit of which to be formed in Ireland was the 10th Division.[25] One of the brigadier-generals in the division was Charles Fitz Clarence V.C., an old friend of de Montmorency. Fitzclarence was to die at Gheluvelt on 11 November 1914.

Given a commission as a captain in the 7th Dublins de Montmorency was back in the army for the third time in his life. His company was recruited from the Irish rugby football clubs, and he was to enjoy their games at the Curragh. There he wrote, 'I found myself once more in the old routine; recruits' drill on the square,[26] barrack inspections and meals in the officers' mess; it seemed very strange after five and twenty years' absence. For four months we all trained zealously on the Curragh, eagerly looking forward to the day when we might be sent to France to fight for our great cause. In February 1915 we moved to the Royal barracks in Dublin, being relieved by a battalion of the Irish Rifles who were in need of the better facilities for training afforded by the Curragh of Kildare'.[27]

Another gentleman who was mobilized early in August 1914 was Lieut. Col. Charles O. Head R.H.A.[28] Then living in King's county, he reported to headquarters in Dublin from where he was posted to a quartermastering appointment in the Expeditionary Force. He soon applied for a transfer to regimental duty and was posted to the new artillery unit forming in Newbridge. When he arrived there he found 'the barracks packed with men in civilian dress, the raw material for the 10th (Irish) Divisional Artillery of Kitchener's first army; but only a small percentage of them were Irishmen. With my allotment of men, representing one brigade of artillery, I was soon pushed over to the empty barracks at Kildare. The brigade consisted at first of three batteries, one of which I commanded, as well as the brigade. A captain on leave from India got the second battery, and an old subaltern of mine, who had retired some years previously, got the third . . . He arrived under the idea that he alone represented the battery, but I gave him 250 men to help him in his responsibilities'.[29]

At Kildare the brigade found a good store of uniforms, left behind by a brigade which had gone to France. It was issued to 'the unkempt, variegated crowd which had trudged from Newbridge to Kildare', making them unrecognizable in their new attire. After about a week in Kildare the brigade was moved to a camp on the Curragh 'on a cheerless spot known as Donnelly's Hollow. It was near the end of September, the weather was bad and many of our youngsters suffered severely from bronchial catarrh'. The shortage of training materials made training difficult and boring, but when a few horses and guns were procured things picked up. In October, when conditions under canvas were deteriorating, the brigade moved to Dundalk.[30]

Colonel Head survived the war and returned to care for his estate at Borris-O'Kane only to find that in the Troubles of 1921 he was, as a resident magistrate, a marked man. Forced to take shelter in the military barracks at Birr, he soon learnt of the burning of his house, and he and his family moved to England.[31]

In the Curragh Camp the staff of the Ordnance Depot had to work overtime to cope with the mobilisation, and a Cyclists Corps was being formed from the Irish Command; it was to train at the camp before going to the front. In Newbridge more civilians had been taken on to work in the barracks, and Sheridan's Concert Hall in the town and the Stand House at the Curragh racecourse were requisitioned for the accommodation of mobilized troops.[32]

The requisitioning of part of the Curragh racecourse led to an allegation, in a publication called *Truth*, that the Turf Club had refused permission to the military to tie horses up at the rails, or for the use of the grand stands. *Truth* called the Stewards of the Turf Club 'a notorious, bigoted, conservative body, the very worst of the Kildare street set whose excuse was that the horses might pull down the railings, and that they would interfere very much with the condition of the gallops. If the nationalists in Ireland refused anything which the government asked for in the present crisis, I wonder what would be said?' A minute in the file of the Office of Woods, from whom the Turf Club held a lease, carried a note that 'although the statement emanates from *Truth* it does not necessarily acquire that virtue'. In a letter to that office the Keeper of the Match Book in Ireland said that there was no truth in the assertion, and that the stewards had immediately placed the Curragh race stands at the disposal of the military authorities when asked to do so, 'and have since received the thanks of the army Remount Department in whose charge the stands are at present'.[33] That department required the Stand House as quarters for the 14 officers and 100 men who would manage the 5,000 horses about to be mobilised, and the magnitude of the exercise caused the cancellation of the September Meeting, but by October there was more organisation, and that month's Meeting went ahead.[34]

The acquisition of mounts for the army led to all of the stables in Naas being visited by military officers and horses being commandeered for service; even the jarveys' horses, and other steeds met on the street, were held for examination.[35] Numerous horses were being purchased around the county for the army Remount Department, and good prices were being given. Up to £70 for a horse was being paid, and one man boasted that a broken-down animal, which he had paid £1 for shortly before, had fetched £45.[36]

An early alarm of war was raised in the second week of August when it was reported that a foreign aeroplane was seen over the artillery barracks in Kildare; the army response was to increase security in all stations. Later two men allegedly found making plans of the barracks were arrested.[37] On the actual battlefield at Mons the mettle of the regiments which had become familiar to the world during the

'Incident' on the Curragh five months before was tested. The 5th Infantry Division and the 3rd Cavalry Brigade had been among the first units to land in France in August, and on the 22nd of the month the first British shots of the war were fired by E Bty., R.H.A. During the retreat from Mons the units provided cover, and they 'fought with great courage and endurance'.[38]

By the month of September the Naas U.D.C. had offered the use of the Town Hall to the military,[39] and men of the 15th Battalion R.D.F. had arrived back from the front. Recruits from all over the country, and from England and Scotland, were daily arriving in the town where they 'could be seen walking aimlessly up and down the main street, and on the roads. Some Scotch laddies lying on a grassy bank said they could get nothing to eat in the town, and were going to a farmer's house for milk. When they were asked as to why they came to Naas one replied "Because we wanted to join the Dublin Fusiliers, but if we had known this was to be our treatment we would have joined a Scotch regiment".' Recruits were accommodated in the barracks until they were dressed, when they were sent to the Curragh for training (Fig. 34).[40] Some weeks later another unexpected recruit presented himself at the barracks. A Frenchman from Paris, he also said he was attracted by the 'fame of the Dubs', and wanted to join them![41]

In Naas barracks the man from the *Leinster Leader*, with a jaundiced eye, saw:

> squatting about in all directions, the future defenders of the British Empire, waiting to go through the ordeal of fitting on the khaki uniform. The congestion is so great that many things appertaining to the comfort of the recruits have to remain unattended to, but the poor fellows grin and bear it. You would pity one poor Irish lad as he lay stretched out on the grass humming to himself a few lines of Kickham's ballad *The Irish Youth*:
>
> 'Dear countryman, take heed of what I say, If you ever join the British ranks, you'll surely rue the day.'

To ease the overcrowding in the barracks it was intended to set up an auxiliary training camp for recruits in a field opposite the barracks, and huts were to be erected to accommodate 500 men. Up to 700 would be accommodated in the barracks itself. These proposals were seen by the townspeople as 'a great boon to the town, the traders of which are sure to benefit by having such a large number of men located in their midst'.[42] The prices of commodities in the shops in Newbridge and in the Curragh camp had risen in price, 'due to the war'.[43]

But the rumours of bad conditions in the Naas barracks persisted; apart from the barracks being overcrowded, with the sleeping accommodation impossible, it was believed by the journalist that a health inspection should be made of the place. So inadequate were the rations there that a house-to-house collection was made in the town of Naas on behalf of the men during the first week of September. The serving

Fig. 34. *View looking west from Donnelly's Hollow, with the Cavalry Depot encamped beside stables surrounded by an earthen bank,* c. *1914*. Photo: *National Library of Ireland, Eason Collection. Inset is a sergeant major of the Royal Dublin Fusiliers outside his tent on the Curragh; his cricket bat is on the ground, with his bath and bucket behind,* c. *1918*. Photo: *National Library of Ireland.*

of notice to quit married quarters on the wives and families of men at the front caused further consternation.[44] Another rumour circulating in the town was that the old jail had been cleaned out and was to be taken over by the military, either for recruits or to contain German prisoners.[45]

In an attempt to alleviate one of the problems in the barracks a request was made to Naas Union that cases of infectious diseases should be taken into the Union hospital, as the fever hospital in the Curragh had been converted for the use of the wounded. It was decided that 15 men could be received at Naas, with room for five or six small-pox cases. The average cost of maintenance of three patients per day was £1.[46] Before Christmas that year a letter was received by the board of the Union from the clerk of the Gosport Board of Guardians concerning a widow from Morristown Biller, near Newbridge. She had married her husband while he was in Newbridge, and now her son, Sergeant Bedding, was gone to war and she had no means of support. She was told that she should be maintained from Gosport.[47] The wives of men at war received a marriage allowance, and if they misbehaved it could be stopped. Such was the case of Mary Reilly, the wife of a Dublin Fusilier, who was before Naas Petty Sessions in November 1914, charged with assaulting Annie Downey. She got a month in jail and was told her allowance would be stopped.[48] Domestic problems increased in the absence of the men, and the court appearances of the women folk became more common than in peace time.

In Newbridge the bustle of preparation for overseas was continuing to fascinate the townspeople. One evening, after dark, when 800 cavalry men from the Curragh marched through the town at a very fast pace to the railway station, and then marched back again, they were cheered by a big crowd. With a brigade of the R.H.A. and R.F.A. already gone to war the civilian population had grown accustomed to the constant passing into the barracks or to the Curragh of parties of troops. The good conduct of the men was admired by the local reporter for the *Leinster Leader* who wrote: 'Passing along Newbridge at night you find a few thousand men quiet, orderly, but cheerful. There has not been the slightest necessity, in connection with the large body of men who are constantly about the town when the work of the day is over, for the interference of the constabulary or military police. There has not been a complaint of an individual soldier out of the thousands which have arrived or passed through for the past week to the Curragh Camp. In Newbridge barracks alone there are some 2,000 men, in addition to the necessarily reduced staff of the artillery'.[49] The newspaper man might not then have found access to the Curragh Camp easy, and a journalist who was motoring through the Curragh was arrested and placed in the guardroom on the suspicion that he was a spy. A soldier there told him that he was lucky that he had not been shot by the sentry, but the Irish sergeant-major accepted the man's credentials and released him.[50] It was estimated that in October there were some 14,000 men 'lying at the Curragh Camp', in addition to those in Kildare and Newbridge.[51].

Recalling his holidays as a youth in Newbridge the writer Sean O'Faoláin wrote of the romance of the military which came 'gloved in the mud of Flanders trenches, breathing the cordite of its guns, smelling of the brown smell of India, the snows of the Himalayas, Egypt's sand (with Kitchener in the Soudan), all embodied in the tiny Union Jack on the tip of the far-off [Curragh] tower high over the empty grass, the few cropping sheep, the rattle from the pits, the elegant powerful horses . . .' and 'the prattlings of his cousins Lil and May Boyhan about which regiment was coming and which was going, the gay doings of the "boys" in the camp (between 1914 and 1918 all soldiers were "boys") . . . my cousin Tom was the first of us to answer the call'.[52]

While the journalist from the *Leinster Leader* might have been impressed by the incident-free nature of daily life in Newbridge, he would have known that elsewhere in the area such was not the case. Two men of the duke of Wellington's Regiment who were fed up with the service hoped to achieve their discharge by breaking windows in Sallins,[53] and in a Naas pub Sergeant J. Malone R.D.F was attacked by John Fennessey, an ex-soldier of the same regiment, who claimed it was all a mistake as he wanted to re-enlist in the service.[54] On the Curragh plains Private Matthew Hannan R.M.F. attacked Stella Harpur in the furze bushes, but the screams of her girl friend led to his capture.[55] By October there was even trouble in Newbridge when a soldier of the Manchester Regiment broke windows, again in the hope that he would be discharged, but instead, as an example, he got two years hard labour.[56] A more daring crime in the locality was that of Private Patrick Clare, 6th Battalion R.D.F., who was aided by his wife and his mother, in stealing jewellery from the home of Sir Joseph Tichbourne Bart at Ballymanny Lodge on the opening night of the Curragh races.[57] Crimes committed by civilians against the military were equally common.[58]

Even though the economic benefits to the locality of the vast army of men was accepted by the community, it was but to be expected that such numbers of troops, and the war situation, would have some adverse repercussions. Both in the Curragh and Newbridge indignation was expressed in early November at the number of civilians arrested by the military. William Murphy of Ballymanny, who had gone to the artillery barracks in Kildare about a milk contract, had been detained until five o'clock in the evening. Thomas Mirrelson, a Newbridge shopkeeper, had been held for a longer period. As a boy he had come as a refugee from Russia, but when someone in Military Intelligence decided that he was an Austrian they had him lodged for three days in the guardroom. Hundreds of people turned out for a popular demonstration in Newbridge before he was released.[59] Another man in trouble with the military was James O'Neill, a carman, from Carna, Suncroft. He appeared before the Curragh Petty Sessions charged with being drunk and with obstruction, when he was halted by the sentry at the gate of Ballyfair House, the residence of the General Officer Commanding. O'Neill informed the sentry that he was a Fenian, and would pass that way [to his home], even should he be shot in his attempt to do so! He was bound to the peace and fined 5/–.[60]

As the numbers of recruits increased daily the training areas became inadequate, and in the autumn of 1914 the making of new shooting ranges became necessary. The Secretary of State for War had the powers to enable a range to be opened at the Little Curragh, towards Friarstown and Rathbride. As it required the closing of a road when firing was in progress persons who might find themselves injuriously effected could make redress to the War Department.[61] Another range was being constructed at Castledillon, near Straffan, and 80–90 men were working on the project.[62] When completed the range at Castledillon was used by troops from Dublin, and in May 1916 the 4,000 men of the North Staffordshire Regiment who were encamped for the summer training at nearby Lodge Park were expected to create 'a boon to local shopkeepers'. There was a noticeable stimulus to business being created in Celbridge on the day of the arrival of the advance party, consisting of Army Service Corps, medical and veterinary personnel.[63]

Oswald Mosley was amongst the subalterns who arrived in the Curragh that autumn. He was to join the depot of his regiment the 16th Lancers.[64] Proud to be part of such a brilliant regiment he was soon manoeuvring on horse back on the plains, and every morning he participated in 'stables', that was the grooming, watering and feeding of the horses.[65] In his autobiography Mosley wrote 'those days at the Curragh in the autumn of 1914 confirmed the impression of the regular army which I had originally derived from Sgt. Maj. Adam and Sgt. Ryan at school [Winchester], and I became deeply attached to that way of life. Some years later, in the light duties of convalescence, I was to know very happy and relatively relaxed days at the Curragh. But in those early days of the war all was serious and arduous training'.[66] Another young cavalry officer who trained there that year was Alex Craigie from Harristown in county Dublin. Taking part in a dress parade he had the misfortune, when marching forward to greet his colonel, to fall flat on the square when his sword caught between his legs. Subsequently confined to barracks, himself and a friend decided to go to Newbridge. Taking a Model T Ford from the camp they drove off across the Curragh only to grind to a halt when the car overheated. In ignorance they opened the radiator cap, and Craigie was scalded and admitted to Naas Hospital where he spent ten days. Route marches were a vivid memory of his training on the Curragh to Henry Crookshank in later years; joining the Irish Fusiliers there from Trinity College, Dublin, his fitness from the sportsfield enabled him to enjoy the marching, but he pitied the unfit 'townies' who found the road work heavy.[67]

While the town of Kilcullen did not have a permanent military presence, it was a popular place of recreation for the men off the Curragh. By the month of October 1914 its attraction to the troops appears to have got out of hand as the clerk of the Kilcullen Petty Sessions District made an order closing all licensed premises in the district at 8 p.m. on Saturday nights, and at 9 p.m. on other nights, 'to suppress drunkness'.[68] A month later it was in Newbridge and Kildare that the public houses were causing trouble. No doubt, but the presence of up to 50 soldiers in a licensed premises at Maddenstown during the day, and for which the proprietress Mrs Walsh was fined £2, had annoyed the military authorities.[69] An order was made which stipulated that the pubs could not open until 10 a.m., and were to close at 8 p.m.; after the intercession of the publicans with the G.O.C. it was agreed that the premises might open at 8.30 a.m. and close at 9 p.m., and at 8.30 p.m. on Saturdays. The complaint from the publicans was that the hours were causing much inconvenience to farmers as they could not do their shopping until after their day's work was completed, and most of the public houses in the county stocked groceries.[70]

The Chairman of the Newbridge Town Commissioners had another angle on the business when he expressed his opinion on the budget increases, and their effect on the town's licensed premises. Newbridge, he said, 'was connected with one of the largest military districts in Ireland. There was a considerable amount of employment at present in the neighbourhood of Newbridge and the Curragh Camp. The workmen could not afford a pint or two, one of which was usually their mid-day meal . . . [but] the military as a rule got what they required inside the barracks, to a very great extent'.[71] One way or the other, it would seem, the licensed trade was expected to suffer, although with the thousands of men in the neighbourhood that seemed an unlikely prospect.

On 18 September 1914 the royal assent was given to the Third Home Rule Bill, but its operation was suspended until the war was over. John Redmond M.P., at Woodenbridge in county Wicklow, called on the Irish Volunteers to come to the assistance of Britain by joining the army, a proposal rejected by militant nationalists. The volunteer movement split, with the majority following Redmond and becoming known as National Volunteers, a regiment of which was raised in Kildare. The minority, dominated by the Irish Republican Brotherhood, kept the title Irish Volunteers. The supreme council of the I.R.B. had decided that there should be an insurrection before the war ended.

Sporting life in the county was again being damaged by the departure of men for the war. Captain Talbot Ponsonby, Master of the Kildare Hunt, had rejoined the R.H.A. in August,[72] and at the general meeting of the Hunt Club in November the loss of many members, families and supporters due to the war, was regretted. However, 'as in the [famine] years 1846 and 1847, the hunt would keep going to maintain employment'. Financial support was requested.[73] Within a month a letter from J.W. Dane, the secretary of the hunt, appeared in a local paper. Explaining that for some years past the expense of running the hunt was about £4,000 a year, he was afraid that 1914 would 'bid fair to spell absolute ruin. A large proportion of funds have been provided by the military [in 1913 regimental subscriptions alone had amounted to over £477]. This source of income is now cut off as, owing to the exigencies of war, the officers in training are not allowed to hunt government horses, and should officers possess their own horses they are obliged to sell them to the government. I am very sorry to say that we have already lost many of our members, killed in action. A large majority of the members of the hunt are either at the front or training with their regiments, or otherwise working in connection with the war. Our M.F.H. personally has lost a considerable sum through so many of his horses being taken for the war'[74].

In the month of November both the County Council and the Naas Urban Council came to the assistance of the hunt. The Urban Council gave a favourable lease of premises in the Town Hall to the club, and the County Council, appreciating that the hunt contributed to the economy of the county by its encouragement of the breeding of fine hunters, agreed to solicit the public to support the fund for the Kildare Hunt Club 'many of whose members had died while serving their country'.[75] Consequently the Kildares managed to survive the war, and the subsequent years of turmoil, and remain today one of the leading hunts in Ireland.

Activity on the polo field and the cricket pitch was also adversely effected by the absence of the military, and the 'Royal Golf Club was denuded',[76] but they all regained some of their status after the war ended.

As winter approached the ladies of the district showed their concern for the comfort of the troops in a practical way. In what a local paper described as 'a very pretty incident' the colonel's lady, Mrs Downing, presented the regimental police, cooks and other N.C.O.s and men constituting the regimental staff of the 7th Special Service Battalion R.D.F., with warm clothing, woollen shirts, socks and jerseys, mittens and muffles.[77] The men on the battlefield were never very far away from the minds of those at home, nor was the fear that the war might come to military installations in this country. In October when the tanks of the Anglo-American Oil Company at Ballymanny, Newbridge went on fire, and the brigades from the town and the Curragh camp were summoned, the immediate alarm of the local people was that the Germans had attacked the camp.[78]

One soldier who trained on the Curragh, arriving there on the last day of December 1914, left an account of his experience. His first impression was one of desolation when he saw the wide expanses covered with snow. If the billet, with its plank and biscuit bed was uninviting, the fellowship was good. The only comforts available in the camp were to be found in the cinema and in the various soldiers' homes, and these the men fully availed of.[79]

It was during the second year of the war that the Curragh lost one of its best known features when Sgt. Gunner McEvoy, whose family lived in Newbridge, had the honour of firing for the last time the great gun 'which proclaimed the hour of retiring and was heard in almost every portion of the county Kildare'.[80] The Baron de Robeck, at home in Gowran Grange, near Naas, had always checked his watch by the explosion![81]

If social life was subdued in the camp during the war years, the large numbers of troops and the fervid preparations for the trenches ensured that there was constant activity. Lord Kitchener, who was appointed War Minister on the outbreak of the war, had initiated the creation of a 'New Army', and by May 1915 1,700,000 men had answered the call.[82] At the Curragh tented accommodation was erected between the camp and the western end of the plain, and a strength of up to 30,000 was accommodated between the billets and the tents.[83] From August 1914 to February 1915 some 50,107 recruits were enlisted in Ireland, and before November 1918 a total of 140,460 had joined the ranks. In addition about 30,266 reservists had rejoined their regiments.[84]

At the end of August 1914 the 10th (Irish) Division of the New Army began to assemble. The General Officer Commanding was Lieut. General Sir Bryan T. Mahon from county Galway. One of the three brigades of the Division was in training on the Curragh in February 1915. The 29th Brigade held day and night exercises as well as musketry training there, and Divisional support units, three brigades of field artillery and the heavy battery; three engineer field companies, the signal company and Divisional cyclists, participated in the exercises.[85]

Bryan Cooper of Markree Castle, county Sligo recalled his training on the plain: 'though we criticised them bitterly at the time, these Curragh field-days were among the pleasantest of the Divisions' experiences. By this time the battalions had obtained a corporate existence and it was exhilarating to march out in the morning, one of eight hundred men, and feel that one's work had a definite part in the creation of a disciplined whole. The different units had obtained (at their own expense) drums and fifes, and some of them had pipes as well. As we followed the music down the wet winding roads around Kilcullen or the Chair of Kildare, we gained a recollection of the hedges on each side bursting into leaf, and the grey clouds hanging overhead, that was to linger with us during many hot anxious days'. Cooper survived the war and later served as a T.D. for Dublin County until his death in 1930.[86]

An officer of the East Surreys also had happy memories of his training, especially in the Glen of Imaal; there he 'listened to snipe drumming in the bogs, studied Lepidoptera on the Donard ranges, and watched a golden eagle in majestic flight over Keadeen mountains towards Sugar Loaf'. On leaving Ireland Lieut. H.F. Stoneham acknowledged that he had learned to love the country during his several years service there.[87]

When General Mahon held a divisional parade at the Curragh on 16 April 1915 he inspected the troops, mounted on his favourite hunter: 'his easy, informal manner relaxed them and gave all an added pride. The units then passed in review to the music of their regimental marches. In battalion mass, followed by the artillery in line of close interval, they paraded before their divisional commander. It marked the final welding of units into one fighting formation of all arms and services, ready almost, for war. All were proud to be the first Irish division to take to the field.' One officer of the 6th Royal Irish Rifles wrote: 'it was a tiresome day but being one of such a stalwart body of men filled one with elation. We felt we would be quite invincible'.[88]

In Athy Lady Weldon, assisted by several local ladies, sought to entertain the soldiers living locally, and she sent regular supplies of comforts to the troops at the front (Fig. 35). At a soldiers' sing-song in the town in April 1915 she encouraged 'all young men to go and help their fellow countrymen. Athy, out of a population of 4,000 had sent 300 men, and if other towns did as well the war would soon be over . . . the example of the Leinsters and Dublin Fusiliers should inspire the young men to go and do their part for their King and country'.[89] Lady Weldon had her own Comfort Fund for Soldiers at the Front, as well as being active in the local branch of the Red Cross, H.R.H. Prince of Wales National Relief Fund, and, after the war, in the War Pensions Committee.[90] It is estimated that 567 men from county Kildare laid down their lives 'for king and country' in the war.[91]

January 1916 saw the continuation in county Kildare of various activities associated with the welfare of the troops at the front. Lady Weldon arranged the sending of parcels to the men from Athy, and she in turn received many letters of thanks from the recipients, sometimes giving details of life in the trenches.[92] In Naas flag days were held, and a dance, with music from the band of the 14th Hussars, was being organised for the Town Hall to raise funds for gifts to be sent to prisoners of

Fig. 35. *A shooting party on the steps of Kilmorony House, Athy, the home of the Weldon family. Between 1850 and 1926, 16 men of this family did military service. Courtesy Ger McCarthy.*

war in Germany.[93] Miss Evelyn Moore of Killashee, Naas, rallied volunteers for hospital work in the war zone with the Hospital of St John of Jerusalem.[94] The Royal Dublin Fusiliers gave a concert for the provision of comforts for the troops at Kilmeague,[95] and the Baroness de Robeck sought funds for the sending of hot water bottles and cigarettes to the wounded soldiers in France, and she was prepared to supply wool to anyone willing to knit socks for the men.[96] Her husband was then a rear-admiral and Admiral of Patrols in the Royal Navy.[97] Lieut. Col. R. McCalmont, O.C 1st Battalion Irish Guards arranged for a supply of the provincial paper the *Leinster Leader* to be sent to his men.[98] The Hibernian Novelty Co. in Dublin inserted an advertisement in that newspaper offering everything for the military, stationery embossed in gold, active-service fountain pens (required no ink), writing tablets and soldiers' song books. Regimental crests were also available.[99]

At a fête held at the Naas golf links by the R.D.F. County Kildare Committee £90 was raised with a golf championship, American tea, sport, and games which included 'Bombing the Kaiser'.[100] Lord and Lady Drogheda at Moore Abbey, Monasterevan, raised £100 with their fête for the War Hospital Supply Depot and the District Nurses.[101] There were also fêtes at Carton, the home of the duke of Leinster, which raised £360, and at Coolcarrigan, where £202 was collected by Lady Wright, and at Naas barracks where the sum of £80 was realised, for various charities.[102] Lady Milbanke at Mullaboden set out in October to collect 20 turkeys for the Christmas dinner in the Curragh Military Hospital; she wanted them then to fatten them in time for the feast.[103] Amelia Milbanke was the only daughter of Lieut. Col. the Hon Charles F. Crichton, late Grenadier Guards, and the widow of Capt. Sir John P. Milbanke 10th Hussars. In 1920 she married General Sir Bryan Mahon, and they continued to live at Mullaboden.[104]

The Royal Dublin Fusiliers depot at Naas was the centre for a recruiting area which included counties Dublin, Wicklow, Carlow and Kildare, and by June 1917 a total of 26,611 men had joined up at Naas[105] (by the end of 1916 over 250 men from the town were at the front).[106] Private Frank Laird, who was invalided back there from Gallipoli in December 1915, found the quarters at Naas bare and 'not luxurious by any stretch of the language, and the town itself offered no diversions in off hours except a public house or bacon and eggs in Lawlor's Temperance restaurant'. He looked forward to weekends in Dublin.[107]

A new military presence in county Kildare was that of the Royal Flying Corps at an aerodrome which was commissioned near the Curragh Camp in 1915. It was intended to be a temporary station for one Training Squadron (Day Bombing), and it remained in use until 1919. Canvas hangars were erected, and the station had an establishment of 332 personnel, who were accommodated in the camp. The 19th Training Squadron, which was equipped with 24 aircraft, was based at the Curragh from December 1917, and in October 1918 a new unit, the Irish Flying Instructors School, was established there.[108]

Despite the mechanisation and the introduction of aerial warfare, the horse continued to be a vital part of the war machine. At the Remount Depot at the Curragh Stand House 500 horses and mules, which were imported from Australia, Argentina [those from the former country were known colloquially as 'whalers', and from the latter as 'swimmers'][109] and England, as well as purchases by civilian contractors in Ireland, were in constant training. The animals came by train to the Curragh siding. Stabling for the horses and mules had been made in a double row of tin huts, and in the sick lines there was stabling for 150 horses. An isolation stables for any animals which were found to have infectious diseases were located at Ballyfair House, the residence of the general officer commanding.[110]

Tom Garrett, who worked at the Remount Depot from 1915 to 1920, was one of 200 men employed there. Their task was the breaking-in of the mules and horses for the cavalry and for the wagons and gun carriages. The men, who each had an identification number, checked in at 7 a.m. daily and worked until 6 p.m., for a weekly wage of 25/–, which was considered very good pay (a farm labourer's wage was 6/– a week). The depot was commanded by a retired major, and a retired regimental sergeant-major was in charge of the men. Civilian veterinary surgeons were employed. Hay for the depot was purchased all over the country by local buyers. The location of the depot beside the railway siding and the plains, its proximity to the camp, and the availability of the Drogheda Memorial Hospital on the edge of the Curragh for men injured at work, combined to make it a highly efficient station.[111]

As the personnel in the various barracks in the county departed and returned from the battle front, and the local newspapers were perused for the lists of the dead, the political climate gradually changed. Within the ranks of the Volunteers in county Kildare there was discussion and debate as to the aims of the organisation, and finally, in the summer of 1915, the Athgarvan Company decided to affiliate to the Provincial Committee of the Irish Volunteers. Michael Smyth, Commander of the Athgarvan Company, recorded his remembrances of the period some 50 years afterwards.[112]

By 1915 soldiers invalided in the war were returning home and they were recruited by the volunteers as training officers. One such man was William Jones who, as an army reservist in 1914, had drilled the volunteers before he was called-up for active service. Injured and back from the war, he was given employment in the Curragh Camp until, in September 1915, he was arrested under the Defence of the Realm Act, and given three months imprisonment, which, on account of his army service, was reduced to a fine. But he lost his job in the camp, and his army pension. He set about the training of the volunteers with enthusiasm, instructing them in drill, rifle and revolver practices, and field exercises which were carried out on the Curragh, only a few hundred yards from the camp, 'in trenches which were made by Kitchener's army'.[113]

As the Athgarvan Company of the Irish Volunteers was the nearest one to the Curragh Camp it was given the task of procuring arms and ammunition from the soldiers there, with cash provided by volunteer headquarters. In this it was successful, acquiring a good supply of 1915 pattern Lee Enfield rifles, but it was more difficult to get .303 ammunition for them. Senator Smyth recalled that all of the soldiers from whom he got rifles were English, except for one man from Laois. That soldier revealed the fact that in the previous year, while encamped on the Heath of Maryborough, a case of ammunition had been buried there, and that he knew where. Wearing his khaki, the soldier accompanied the volunteers to the Heath, but they failed to locate the ammunition. But the Athgarvan unit managed to get a good supply of ammunition from the civilian workers on the Curragh ranges, as most of those men were sympathisers of the cause. A year later when Smyth was in Wormwood Scrubbs he was charged with purchasing rifles and ammunition 'with £300 of German gold', and he was given a two year prison sentence.[114]

In fact details of the volunteers were known to British Intelligence. For the year 1914 a confidential report on the state of the country found county Kildare to have been 'in a satisfactory and peaceable state . . . free from agrarian trouble. The National Volunteers made no progress until May, when some branches were formed and drilling started. Thereafter its progress was rapid, and in July the membership was placed at 3,000'.[115] Records of the *Buro Staire Mileata* give twice that figure as the effective strength of the volunteers in the county on 18 August 1914.[116]

When it became clear that the Home Rule Bill was not to become law there was disappointment among the people, but the outbreak of the war, and the recall of the instructors [reservists who were called-up], as well as harvest operations 'served to damp the enthusiasm of the volunteers, and the action of some of the branches in discouraging enlistment caused some of their unionist and non-political supporters to drop out of it. At the end of the year it was reported that very few drills were being held, and that interest in the movement had practically died out'.[117]

In September 1914 the strength of the volunteers in the county Kildare was estimated by British Intelligence as 4,492 in 32 branches,[118] but after the split the number of Irish Volunteers in the county was given as only 344, armed with but 24 rifles.[119] During 1915 the county was again found to be 'peaceable, orderly, and free from agrarian trouble'. It was believed that but 13 members remained in the Irish Volunteers, while the National Volunteers numbered 2,778 at the end of the year, of whom 290 were *Sinn Féiners*. Recruits from the county who had joined the army during 1915 numbered 689.[120]

However, the discovery of tins of gunpowder, detonators, safety fuse, cartridges, milk cans, and scrap iron, some of which was made into bombs, at St Catherine's, Leixlip, early in February 1915 gave cause for concern, and the inspecting Ordnance Officer believed that the bombs were 'undoubtedly intended for the destruction of human life'.[121]

The finding of explosives at Leixlip demonstrated the opposite side of Irish life to that displayed by the men who flocked to the colours, and the women at home who supported them. In both respects county Kildare was not any different to the rest of the country. The resurgence of nationalism as expressed in the Sinn Féin movement was to gain momentum in the coming years, and eventually to change completely the face of Irish society.

Amongst the regiments Christmasing in Kildare in 1915 was the 3rd Cheshire Yeomanry, and some of the men of that unit wrote a letter to their local paper at home. In the *North Cheshire Herald*, under the caption 'Hyde boys at the Curragh', appeared: 'we are doing our duty for king and country. We spent our Christmas at a little village called Kildare, where the houses are just like stables. You talk about being amongst the pigs, we are amongst them . . .'. Clearly they did not appreciate the antiquity of that town, or the streets of thatched mud cottages, but they did express thanks to the St Stephen's Sunday School at home which had sent them parcels of comforts.[122] Nor did the war-time men appreciate the winter training on the Curragh as their time there was limited and the opportunities to socialise with the local people would not have been as available as in peace time.

While the comforts of the fighting men were a priority with many well-intentioned people, the welfare of the men at home was not always satisfactory. In mid-February 1916 a soldier patient in the Curragh hospital wrote to the Board of Mountmellick Union complaining that they were trying to extract payment from his mother who was in hospital. He informed them that she had four sons in the army, two of whom had been killed. He was told that she, and another such woman patient, had not and would not pay.[123]

It was not just the dependants of soldiers who had cause for annoyance. At Naas, according to a report in a local paper, there was a feeling of 'class and sectional differences in the town. While all the gushing activity for the comfort and recreation of the military is taking place . . .' the police had interfered with a collection for the Sallins Pipe Band. In Naas itself the fife and drum band had a room in a lane. Described as a harmless and inoffensive body, including several men who had served in the army, they had twice been refused permission to play at Baltinglass.[124] Clearly the security forces saw the assemblies, and possibly the tunes played, as being potentially disloyal.

Apart from the countless soldiers constantly moving through the Curragh camp, there were also changes amongst the permanent staff. When Barrack Warden J. Dunne, Chief Clerk of the Barrack Department, was transferred to London he was given a presentation at which he spoke of his happy days at the camp, and he thanked all of the civilians who had worked very hard since the start of the war, putting in much longer hours than in civilian life.[125] Another indication of the status of the civilian employees of the army was the funeral of Mr L.G. Puttock, manager of the Church of Ireland Institute in the camp. It brought together many Catholics and Protestants, as well as some of the members of St Conleth's Club in Newbridge.[126]

Early in the month of March it was announced that Col. Hall's stud, Tully, Kildare, had been bought for £60,000 by the government. Described as a first class stud for the production of horses for army purposes, it was intended that during the war the stud would be looked after by the Board of Agriculture, and that after the war it would become a War Office establishment,[127] but the latter intention was not to be realised due to the evolving political situation.

St Patrick's Day 1916 was not neglected in Newbridge. In the barracks there was a parade, and at Sandes Home the celebrations included a concert, a lecture on St Patrick and the showing of slides of Irish views.[128] At the Curragh Mr W. L. Dobell had just finished a ten day season at the gymnasium where two performances nightly were given of dramas and other plays.[129] Elsewhere on the day, the Athgarvan cadre of the Irish Volunteers had a parade and rifle practice. That month the unit also sent a number of the rifles acquired from the soldiers to the G.H.Q. of the volunteers.[130] A resident of Ballysax recalls stories of his father's involvement with the volunteers and the purchasing of rifles from the soldiers until such time as locked rifle-racks were placed in the billets, and the trade halted.[131]

The months before Easter 1916 in the county were, in the eyes of British intelligence forces, 'very peaceable and orderly, the supply of meat and potatoes was plentiful, and those people engaged in the cattle trade were thought to be doing extremely well'. Recruiting for the army was only very fair, if in some districts 'decidedly creditable'.[132] It was not expected that attendance at Punchestown in April would be very large. But it proved to be satisfactory though 'needless to remark there was an absence of that gaiety and brightness which we were accustomed to in the pre-war times, and the hunt stand bore a particularly empty look. Khaki was strongly in evidence, and how thoroughly keen are the soldier sportsmen in assisting at the great chasing carnival was manifested by the large number of officers who spent portion of their brief respite from trench work at the meeting. They included several capable horsemen.'[133]

Just about a fortnight after Punchestown, on Easter Monday, 24 April and the day of another great race meeting, the Irish Grand National at Fairyhouse, the rebellion started. The postponement of the rebellion from Sunday gave the volunteers an advantage as a great number of the military were at the races. At the Curragh the first indication of the emergency was a telegram received at midday by the general officer commanding. It ordered the mobilization of the column which had been arranged to meet any emergency, and to dispatch it dismounted to Dublin by trains which were being sent from Kingsbridge.[134] At 4.45 p.m. the first train from the Curragh arrived in Dublin, and by 5.30 p.m. the whole cavalry column from the 3rd Reserve Cavalry Brigade, 1,600 strong under the command of Brig. Gen. Portal, was there.[135] They were given the task of forming a line of posts from Kingsbridge station to Trinity College via the Castle. That line was intended to drive a wedge along the Liffey between rebel positions.[136]

In the weeks before Easter, Michael Smyth, Company Commander of the Athgarvan Company of the Irish Volunteers, had tried to raise additional members in Newbridge, Kildare, Kilcullen, Naas and other places, but he found it impossible to form a company 'in those places as they were all garrison centres, and as the majority of the people in those areas were pro-British, it was very difficult to recruit members . . . large numbers from those areas had joined the British army'. Instructions had been received by the volunteer leaders that the Kildare units were to be 'used as outpost groups between the Curragh Camp and Dublin when the outbreak took place'. It was expected that north Kildare would provide between 100 and 150 men to demolish rail lines, roads and other communications between the city and the Curragh, and the R.I.C. barracks at Sallins and Kill were to be attacked. But, as the district had only five small companies of volunteers, it was not a feasible objective.[137]

On the night of Good Friday the Athgarvan Company, was mobilised. The 16 men who attended were told to stand by for a call-out, and that in the meantime they should go to confession and holy communion. On Saturday the knowledge of a split in the Executive Committee of the Volunteers became known to Smyth, but he decided to obey the mobilisation orders which he had received.[138] On Sunday morning 16 members of the Athgarvan Coy paraded, but when the conflicting orders from Dublin were received the company stood down; they again paraded on Monday, but as no further orders were received, they again stood down.[139] The volunteers from Maynooth had commenced their march to Dublin that morning.

On the afternoon of Tuesday 25 April Brig. Gen. W.H.M. Lowe, who commanded the Reserve Cavalry Brigade at the Curragh, arrived at Kingsbridge with the leading troops of the 25th Irish Reserve Infantry Brigade. Lowe assumed command of the forces in the Dublin area which were then estimated at 2,427 men of the Dublin garrison, 1,600 dismounted cavalry men of the Curragh mobile column, and 1,000 men of the 25th Irish Reserve Infantry Brigade, 1,600 from the Belfast, Athlone and Templemore garrisons,[140] totalling in all some 6,627 men. When the Chief Secretary, Mr Augustine Birrell, was afterwards commenting on the events of Easter 1916 he said 'the Rebellion failed from the beginning, because the soldiers were there before the end of the day in quite sufficient force, from the Curragh and Belfast'.[141] On the afternoon of 29 April Patrick Pearse surrendered in Moore street to General Lowe, who was accompanied by Capt. Henry Eliardo De Courcy Wheeler A.S.C., of Robertstown, and who was also serving on the Curragh.[142] He came from a county family which gave many sons to the colours, and he was a member of the County Kildare Archaeological Society.[143]

An official summary of the result of the rebellion in county Kildare found that it put a stop to recruiting for the army, and 'subsequently a feeling of sympathy with the rebels was entertained by many who had previously condemned the movement. The R.C. clergy were not in sympathy with the rebellion and thought it politically stupid. In some cases they openly condemned it. The Reverend Fr O'Brien C.C. at Kill, was an exception and went as far as he could in favour of it. The community of Maynooth College adopted the same attitude as the local clergy generally, but a good many of the students were strongly in sympathy with the rebels'. The only physical evidence of the rebellion in the county had been the holding up of a police patrol near Maynooth, and the cutting of a telegraph post on the railway line between Kildare and Athy.[144]

Though not mentioned in the formal report there was one more tragic and lasting effect of the rebellion in the county. Alfred Warmington, manager of the Munster and Leinster Bank Naas, lost his only son, Edward, in the fighting. A major in the Royal Irish Regiment, he was killed in the attack on the South Dublin Union.[145] He had survived the Boer War and Flanders, as the *Kildare Observer* noted, 'only that he should meet his death in the capital of his native country'.[146] Commenting on the rebellion on the 75th anniversary of the Rising a writer in the *Leinster Leader* found that, as reflected in the pages of the local press, '1916 and the nationalists cause take a second place to the casualty lists, rolls of honour, and letters from the Front, generated by the Kildare participation in the war . . . virtually every household in Naas, Newbridge and Athy having a son or father at the Front'.[147]

A more traumatic effect of that Easter week on some families in county Kildare was the arrest of their menfolk. Some 20 members of the Irish Volunteer force, including Michael Smyth, were brought to the Harepark Prison in the Curragh Camp. From there they were marched under heavy guard to Kildare railway station for transfer to Richmond barracks in Dublin. There they were tried, and in the case of Smyth, who was sentenced to two years imprisonment, dispatched to Wandsworth Jail in London, and later to Frongoch Internment Camp.[148] Back in Kildare the *Leinster Leader* reported the return to the Curragh of the troops who had been rushed to the city on Easter Monday: 'Out of close on 1,000 military who left Newbridge for Dublin during the recent riots all returned quite safely, with the exception of one officer who was slightly wounded'. In the House of Commons Mr Lawrence Ginnell M.P. for county Westmeath asked 'if any officers at the Curragh followed the precedent of their predecessors in 1914 by refusing to obey orders unless required to give guarantees they would not be requested to fire on insurgents?' The reply received was: 'No'. An official denial was issued by Military Headquarters that any disturbance had taken place at the Curragh.[149]

The conservative *Kildare Observer* brought out three editions and a special supplement to report on the 'Dublin outrage',[150] but while the immediate reaction to the rebellion expressed in the *Leinster Leader* was also one of horror, it was soon expressing sympathy for the leaders. In its first editorial on the violence the paper deplored the outbreak of domestic violence, saying 'now that the rebellion has been crushed, we may in common express the hope that we may soon revert to that state of order which only peace, prosperity, and mutual goodwill can give'. It is very unlikely that those sentiments were from the pen of the republican editor Michael O'Kelly, as he, and two other members of the staff were arrested in the first week of May and taken to Naas military barracks. Subsequently they were amongst the 26 prisoners from the county who were transferred to confinement in Dublin by train. A week later the paper was expressing the view that responsibility for the rebellion should be fixed, and blame was laid on the unionists in the north who had spurned constitutional methods. By the beginning of June O'Kelly had been released and was back at his desk. From that time onwards, in the words of Michael Byrne who researched the history of the paper, 'editorial policy during the period from the rebellion to the civil war was generally impartial so far as the Irish factions were concerned, but sometimes very critical of the British administration'.[151]

Evidence of the hostility of one individual to the military resulted in a case being brought before the Newbridge Petty Sessions. A man from Sallins was charged with insulting two men of the R.D.F. by shouting 'up the Germans, you hoppy . . .' to one of the offended, a sergeant who was lame from a wound received in the Dardanelles.[152] The insult given to a medically discharged man in Kilcullen in mid June 1916 was that 'an English soldier would not give one the price of a pint, but a German would'. The remark started a row with what was described as an anti-British crowd. Two soldiers helped to put the offensive person in the R.I.C. barracks, and when he appeared before the Petty Session he was fined 15/–.[153] In Castledermot at the time of the rebellion there was a more passive insult when the Roll of Honour, giving the names of 120 soldiers who had served, or were serving in the army, was tarred.[154]

A story from Newbridge at that time proves that the soldiers had not lost their sense of humour with the Irish situation. There must have been a twinkle in the eye of the Newbridge shop keeper who overheard the following exchange between two customers; they were a senior and a junior English officer, and when the latter saw a blackthorn cudgel for sale asked what it was for. The older officer replied 'oh that is what we Irish use in times of peace'.[155]

Two gentlemen who afterwards recalled their memories of the outbreak of the 1916 rebellion were Percy La Touche of Harristown and Lieut. Col. Charles Marcus Lefevre Clements M.C. of Killadoon, Celbridge. La Touche and a neighbour drove up to Dublin on Tuesday 2 May 'taking revolvers with them just in case'. They found the streets full of troops, armoured motor cars, ambulances and machine guns, causing La Touche only to speculate when racing might be able to start up again![156] Colonel Clements recalled that on Easter Monday he had accompanied his mother on a visit to the wife of Brigadier Portal, Commanding the Cavalry Reserve

Brigade, who lived in a rented house near Newbridge. They were to attend the military sports after lunch, but the lunch was cancelled as the brigadier was very busy. By tea time they knew that there was trouble, and the trains were no longer on schedule. Brigadier Portal hired a car to take his visitors to Naas, as it could not go any further, and a telegraph was sent to the coachman Killadoon to have the sidecar sent to the hotel at Naas to collect them. On the way home the young Clements saw troops, with their equipment and guns, lined up along the wall of the de Burgh demesne of Oldtown on the road to Sallins.[157]

The behaviour of the wives of absent soldiers was causing annoyance in various places in the county. At Naas Mary Reilly had broken the windows in the house of another woman; Reilly was described as 'a hopeless case' whose children had been taken from her, and her allowance from the army stopped.[158] Another woman who was accused of beating her sister-in-law was warned that her allowance would be stopped if she did not behave herself.[159] In Newbridge a soldier's wife was given two months in jail for being vagrant, drunk and disorderly,[160] and in Kildare Anne Murphy, the wife of a soldier at the front, had her charge of drunkenness adjourned.[161] The women who cavorted with the soldiers were continually in the news. Elizabeth Glasgow, in Newbridge who was described as 'unmarried, often drunk, a bad character', and Mary Byrne 'one of the greatest old nuisances in the town', were each fined 5/– on a charge of drunkenness.[162]

The behaviour of soldiers near the barracks at Naas, or in New Row, the street leading to it, was the subject of a letter to the local paper in November 1916. It said that they were interfering with women, 'maybe due to the attitude of a certain other type of woman, and the dim street lights'. It was recommended that the military and the R.I.C. should patrol there, and that the soldiers should go to their club rooms in the town.[163] Earlier in the year the character of some of the denizens of New Row was revealed in a court case concerning a charge of using abusive language; the litigants were the widow and a wife of soldiers. The former, whose husband had been killed at Suvla Bay, denied that she had expressed a wish that the husband of the other 'would be riddled'.[164]

But it was only a minority of women who were unable to cope on their own, and a newspaper reported that 'most soldiers' dependants are making good use of their allowances, their homes and children are models of neatness, but some of depraved habits are causing suffering to the little ones. Stern action should be taken to those entrusted with the welfare of children of soldiers in the firing-line'.[165] From Athy the news was that there the separation allowances were not misused, and a draper from the town was quoted as saying that 'purchases were brisker than ever, and the pubs were closing'. The reporter felt that 'the soldiers' dependants there should be congratulated, not censured, in days of anxiety when so few drowned their sorrows'.[166]

Even if there was an increasing disenchantment with the government amongst many people, and a more subtle change in the public perception of the army as a result of the rebellion and the execution of its leaders, life for the soldiers either stationed in county Kildare, or preparing there for war, went on much as usual. The annual sports meetings were held in the Curragh and Newbridge, with the 10th Reserve Cavalry regiment sports being 'one of the most successful held for years in the large military centre'.[167] Games of cricket and football were played,[168] and thousands of people congregated as was customary on the Curragh plain to watch the military parading. In the middle of August when General French took a parade of a large number of troops, 'who looked in splendid form', the occasion was further highlighted by a spectacular thunder storm. While flash after flash lit up the camp the men returned to their barracks in a deluge, and some thousands of persons who had attended the review were drenched on their way home.[169]

To cater for the constant influx of troops in the garrison towns new facilities were opened. The Y.M.C.A. erected a hut on the green opposite the Crown Hotel at Newbridge, but there was a complaint made locally that the Town Commissioners did not have the authority to sub-let the plot to the War Department. The use of the refreshment hut was to be confined to soldiers and their friends, so that the keepers of such shops in the town would not suffer.[170] At Naas the Urban District Council held a special meeting to consider the provision of a meeting place for the soldiers in Naas, or passing through. It was agreed to give, at a nominal rent, the use of the ballroom and the assembly room in the Town Hall. There were then over 500 soldiers of the R.D.F. and the Sherwood Foresters in Naas, 'a large extra number'.[171] Lady Charlotte Fitzwilliam condescended to be president of the club,[172] and the clerk of the Crown and Peace, James Whiteside Dane, was chairman. In that capacity he rejected some criticisms which had been made. He felt that the objections were 'to an English regiment, and that as many people in England supported the R.D.F. we should return their hospitality'. An anonymous writer in a local paper claimed that his concern was also for the civilian population and natives, and that the [opening of a club for soldiers] was a military and not a civilian obligation. He continued that 'Irish harvesters were offered English hospitality, and the Irish were much wanted for the army, but that there was not much hospitality shown at Reading and Frongoch [where Irish prisoners were incarcerated]'.[173]

In a subsequent issue of the paper the proprietor of the *Leinster Leader* disclaimed those comments, and said that they were not sanctioned to appear.[174] The actions of the Urban District Council in making the rooms available were praised, and the good relations between the military and the townspeople, which had always existed, were a matter of general satisfaction. Since the advent of the troops then stationed at the barracks those good relations had not been disturbed, and 'anything that would disturb these good relations should be discouraged and deplored'. As an additional gesture of good will to the men the board of the Technical School offered facilities to any soldiers who wanted instruction, and the school annual dance, which had been abandoned the previous year, was to be held to raise funds for army welfare.[175]

Undoubtedly but the town fathers and business people of Naas recognised the value of the barracks to trade, as did their contemporaries elsewhere in the county who dealt with the army. One man involved in the horse industry regretted the fact that in the past the best horses had been sold to foreign armies, 'our own army must fill the demand now'.[176] The usual advertisements inviting tenders for army supplies appeared,[177] and a soldier wrote from the front to a friend at Naas telling that when he was feeding hay to the horses in Belgium he found a label reading 'purchased from Mr Henderson, loaded at Athy 22.12.1915'.[178] For the 1916 hay crop the contract price per ton in England and Wales was to be hay £6, oat straw £3.10s, and wheat straw £3, on the conditions that the Forage Department bale or truss, and the vendor deliver to the ports or troop centres.[179] At Athy Lieut. Laurence Heffernan was in charge of forage arrangements, and part of his work was to sell hay or other crops not required by the military at regulated prices.[180] The fluctuations in station strengths, and the difficulties of estimating the requirements of fodder, made the job of the controlling officers more difficult than in peace time.

Farmers who were finding it difficult to get supplies of artificial manure were recommended by a correspondent in a local paper to purchase from the military stables in the vicinity of the Curragh. If distant, the farmers could have it sent by rail. It was the opinion of the writer that 'much more might be done by farmers in the exploitation of this great source of natural manure in the military area'.[181] On another aspect of the supply problem a farmer was giving advice to the military. He suggested that as the army would need the greater part of the year's hay crop it should 'set up centres at weigh-bridges to store the crop, press the hay', and 'it would be about a quarter of the waste of time and expense of the present method. They would also have a better chance of getting good hay, as it has always been the practice of farmers for many years to sell good hay to the Curragh trainers. The last years plan of hay purchase had been unsatisfactory, and the farmers had lost much money.' In a further move to assist the war effort the government issued an order that all 1916 wool in Ireland and Britain should go to War Office contracts.[182]

The County Kildare Agricultural Committee at its meeting in early July notified the attendance that a letter had been received from the military authorities. It stated that a depot for milk, butter, eggs and vegetables had been opened in the Curragh, and the committee was asked to supply the names of farmers in the district who were in a position to supply the required goods.[183] Captain Weigell 59th Division, attended the meeting to explain that 'there had always been a bone of contention as to the price between contractors and consumers, and to avoid this a system had been devised by the Army Canteen Committee whereby they purchased direct. It was successful in Aldershot and other stations'. The views of Sir Horace Plunkett had been sought. They were that regular supplies were vital, and that there was no middleman. It was necessary that farmers should organise to keep up the supply. The captain said that the system was already working in Straffan (where there was a temporary encampment of troops) and in Dublin. There was no commandeering or forcing, and the books of the canteen were open to inspection.[184]

The North Kildare Co-operative decided to call a public meeting. This was held at Naas and the recent commandeering of the hay crop was also discussed. The edition of the *Kildare Observer*, dated 8 July, had carried a notice to the effect that the hay, oat and wheat straw 1916 crop in England and Ireland had been taken over by the Army Council, and that the actual price, exclusive of freight, which the vendor would receive would be £5.10s a ton for hay, £3 a ton for oat straw, and £2.10s for wheat straw. This was less than the set contract price in England. At the Naas meeting it was emphasised that the farmers wanted the first and second crops for their own use, and that the military should take old meadow hay, of which an abundance was available. A decision was made at the meeting that the farmers would co-operate to send a cart to the Curragh on set days, at the cost of the producers, the first of which was to be on 2 August. The government was to be requested to pay English prices for hay and all produce 'as the supplies were as good as theirs'.[185] When hay was purchased in the county there was a delay in payment, causing a loss to the farmers, which led to a question in the House of Commons.[186] Nevertheless, an agricultural correspondent in one of the local papers commented 'facing the third winter of war, we think it will be just as profitable as last season, where the farmer has a supply of home grown goods available'.[187]

At the South Kildare Agricultural Show in Athy in August 1916 the prominence of army horseflesh in the economy of the district was reflected in the competition for chargers, the property of the government, or of any officer quartered in Ireland. There were eight entries, and the winner was Capt. A. Rotheram 4th Hussars, Curragh Camp. Capt. R. O'Kelly R.A.M.C., Curragh had the best 5 year old hunter.[188]

A question was also asked in the House of Commons about the system of purchase of remounts in Kildare. The War Office had re-established an office in the Curragh camp for the acquisition of remounts in Kildare or within 25 miles of the Curragh, and days were fixed on which horses would be seen at the camp. The allegation that a dealer who had bought two animals rejected by Col. Wood, and later sold them to the remounts, prompted the parliamentary question.[189] At that time the Co. Kildare Agricultural Society was considering new schemes to encourage the breeding of horses for the army. In the course of the discussions it was remarked that Mr E. Kennedy of Straffan Station Stud was going to Newmarket selling horses, but that while the French army wanted to purchase Irish chargers to the value of £300, and were offering higher prices than the War Office, the breeders were not allowed to sell to France as all horses were wanted for the British army.[190] Later in the year complaints were being made on the same subject, and the ban on the export of horses to France was giving 'no inducement to farmers to breed horses for army purposes'.[191] One permanent advantage of the large number of horses in service was the availability of manure to the farmers. R.J. Goff and Co. auctioned 'eight large heaps of well-rolled manure' for sale at the first gate of the sewerage farm at Brownstown on 5 December.[192] The yearly agricultural cycle in the county had become interwoven with the inflated needs of the military, a dependence which was to cause financial decline for many people in future years.

The determination of traders and business people to keep contracts in the county, or at least in Ireland, was manifest in a complaint voiced in a local paper that the uniforms for the Officers' Training Corps which had been formed in the Curragh camp in August 1916 was being outfitted from London. It had been understood when the corps was formed that the contract was to go to a tailor in Dublin, where there were many suitable firms.[193]

Another annoyance was caused by the military printing office in the Curragh undertaking work for civilian traders and others in Newbridge. In July 1916 there were posters for plays and picture houses, as well as advertising for business firms, displayed with 'printed at the H.Q. Printing Office, Curragh Camp' on them. A writer in the *Leinster Leader* asked that 'the military authorities may very well consider whether it is reasonable at this, or indeed any, time to allow such very unfair competition with regular printing firms which are, in so many ways, much handicapped, especially since the opening of the war. We are told from time to time that everyone in the army at present is engaged one way or other in war work, surely general printing for civilians is not war work?'[194]

Sometimes there was friction between the local authorities and the military when matters of mutual concern arose. Captain de Courcy Wheeler A.S.C. complained to the Naas No.1. District Council in October 1916 about the waste of water in Kildare barracks. Despite the fact that the water tank in the barracks was overflowing had been reported to the council it had not been rectified, and now the captain sought a reduction in the quarterly water charge.[195] The officer commanding in the Curragh camp was also seeking a reduction in water charges for Newbridge barracks on the basis that large sums had been paid for water in the previous two years. His request was refused.[196]

The firing on the ranges on the Curragh was much intensified during the war as the training of men continued, and sometimes the old complaints of danger to the public were repeated. Two ladies in a pony and trap who were passing the firing station on the plain were thrown from the vehicle when the pony was frightened by the sudden floating of warning flags. The ladies were rescued by a party on its way to a wedding.[197] A funeral was the cause of another man's frightening experience. He was a reporter from a provincial paper who had attended the burial of a soldier in the military cemetery on the plain. After the funeral when he took a short-cut by the back of the butts to Brownstown Constabulary barracks he found bullets whizzing around him. He afterwards alleged that he did not see any warning signs, and that his short-cut might have been the long way home![198]

Behind the bustle of trade the political activity continued as the outrage generated by the aftermath of the rebellion spread through the population. In the month of September 1916 the R.I.C. took down posters which had been displayed by the National Aid Society to solicit help for the families of those affected by the rising. In Brownstown other sentiments were displayed when a retiring policeman was honoured. A presentation was made to Sgt. J. Clinton,[199] who had taken up the appointment of Petty Sessions clerk in Newbridge, following a long service in the Brownstown. It was described as a difficult place 'where the civilian and military population came so much in contact, and the excellent relations that existed between them was in no small way due to Sgt. Clinton'. The over-lay of military life and activity on the daily lives of the civilian population was as complex as it had ever been, or possibly even more so as the movement towards national determination accelerated.

That Christmas of 1916 was a white one on the Curragh; in the snow a Hussar fell on the frozen ground and broke his leg, but his comrades celebrated the season by tobogganing on the hill at Lumville or skating by lantern light at night near Donnelly's Hollow.[200]

Chapter 15

The War of Independence

During Christmas 1916 soldiers from the Curragh camp gave a concert and entertainment, organised by Reverend W. Pearson C.F., to a full house at the school-house in Carnalway in aid of the Red Cross.[1] Such public performances brought the military and civilian populations together, as well as enhancing the uniform in the eyes of the people.

Oswald Mosley, serving in the camp in 1917, found that 'the regimental depot was not quite so peaceful as usual because the Easter Rising had occurred the year before. The genial company of the hunting field by day was divided by the sniper's bullet at night. The Irish revolutionaries were much blamed by the army at that time for their method of fighting, but guerilla war was clearly the only possible means to carry on their struggle against an overwhelming military force; it was a method which became familiar all over the world at a later date'. Mosley adds a note to his comment on his time in the camp to the effect that 'as the fighting was then over except for occasional sniping my account is not based on personal experience, but was derived on the spot from recent participants in these events'. [In fact the British Army issued *Notes on Guerilla Warfare in Ireland*, which were 'the result of practical experience in the field and of tactical exercises at Guerilla warfare classes held at the Curragh'.].[2] However, as his duties at the Curragh were light, Mosley was able to travel to Dublin to enjoy the company at the Vice-Regal Lodge and at G.H.Q., Ireland. He appreciated the hospitality of Lord Wimborne, the Viceroy, and General Sir Bryan Mahon, commander in chief, and 'their charming wives'.[3]

Back on the Curragh the soldiers continued with their own amusements, and they enlivened another winter's evening for the public in February 1918 when the Reserve Cavalry gave a concert at Sheridan's Hall in Newbridge, in aid of military charities, to 'a large crowd'.[4] A month later, in the gymnasium at the Curragh, Lady Wimborne, the wife of the Lord Lieutenant, was the patron of an entertainment titled 'In a Persian Garden, a setting of the Rubaiyat', given in aid of the Irish Guards Fund. It was arranged and produced by H.E. Dodgson Lodge and A.H. Lunt of that regiment. For that charity people came from all over the county, as well as 'from the neighbouring borders, in addition to a large military assemblage'.[5] During the winter months Sunday concerts were given in the camp, and dances at the same venue attracted large numbers of guests from the neighbourhood and further parts.[6]

There were over 5,000 spectators at the final game of the Squadron Football League, in which A Squadron 2nd Reserve met D Squadron of the same regiment, at the Curragh camp sport grounds in February 1918.[7] In May a football match was arranged by the Sporting Committee of the Curragh in aid of the Kildare Christian Brothers School, which had been damaged by fire.[8] In Newbridge two picked teams from the Sergeants' Mess of the Reserve Cavalry played 'an enjoyable match' on a Monday evening, and on Whit Monday 1918 the General Parade at the Curragh was the venue for the Garrison Sports. The massed bands of the 51st and 52nd Cheshire Regiments, the Black Watch, and the bands of the 6th Cyclists Brigade and the 9th Lancers kept the crowd in tune during the proceedings.[9] Cricket matches and boxing tournaments brought their own following to the camp when,[10] it might be believed, the traumas of the 'Incident' and the rebellion, as well as the horrors of the trenches, were temporarily forgotten.

While the rebellion of 1916 had halted any furtherance of proposals to introduce conscription in Ireland, and in February 1917 in the House of Commons the Prime Minister, David Lloyd George, had warned that if an Act was passed conscripts would be got only 'at the point of the bayonet and a conscientious objection clause would exempt by far the greater number. As it is, these men are producing food which we badly need'.[11] Inevitably the severe losses inflicted on the British army on the Western Front in the spring of 1918 again forced consideration of the possibility of conscription.

Cabinet discussions, and the advice of Lord Wimborne who believed that conscription should not be introduced before Home Rule was implemented, combined with the opposition of the Irish members of parliament, the hierarchy and the Irish Volunteers halted the prospects. A recruiting campaign in Ireland was undertaken instead. The appointment of Lord French (of a Roscommon family living in Kent) in succession to Lord Wimborne as lord lieutenant in May 1918 heralded a stronger approach to the Irish Volunteers and *Sinn Féin.* Persons suspected of involvement in a plot to secure aid from Germany were imprisoned, but when *Sinn Féin* was victorious in the general election of December 1918 French secured the release of imprisoned nationalists, and the proposals for Home Rule were not acceptable to them.[12]

An attempt to force the men of the Remount Depot in the Curragh into the ranks in March 1918 caused widespread unease locally. The 200 men there, who were members of the Transport Workers' Union, refused when ordered to 'either get into khaki, or be dismissed'. The presumption on the part of the men was that if they had consented that they would be drafted into a Labour Battalion and sent to the front. The military backed down on the threat, and the men were told that 'there is no objection to their joining any worker's union they wish. The men in consequence continued in their employment as usual, a fact which has given much satisfaction in the district'.[13]

Of course everyday life in the barracks continued to be dominated by the war. A visitor to the Curragh camp in June 1918 found that 'those who knew the camp in the early 1900's, and who revisited it without having marked its development meantime, would scarcely recognise it is the same place now. It was thickly populated according to its accommodation 18 years ago, and beyond the few inhabitable brick houses in the immediate vicinity of the flag staff it presented no appearance of permanency. It is now a military city with arms extending towards Kildare and Newbridge, and its population must be close on 80,000. Gymnasiums, theatres and libraries have sprung up as if by magic!'[14] That gentleman's estimate of the strength of the establishment must have included army families as well as civilians, but even then one must allow for some journalistic licence.

Signs of the war in the camp, in addition to those of the continuing organisation and training of the troops for the front, were the making of regulations to save electricity in the barracks and on the streets,[15] and the arrangement of special services for Jewish soldiers in the gymnasium at weekends. They were to be conducted by Rev. A. Gudansky.[16] Also the bringing of motor vehicles commandeered in the Dublin Military Area to the camp for military utilisation, caused 'some little sensation in the district'.[17]

The facilities of the Curragh military hospital were inadequate for the care of repatriated wounded, and an auxiliary hospital was opened at Firmount, near Clane, in June 1917. Twenty-five men were then sent there from the Curragh hospital to recuperate, and soon 40 beds were opened. The average number of patients there was 38, and they were from the Expeditionary Forces or the garrison. In the first year 390 patients, of whom 152 were from the war, were treated there, and there were no deaths in that time. In his first report on the hospital Col. O'Sullivan R.A.M.C. said that 'patients from Firmount always returned to the Curragh full of praise and gratitude for the kindness and good treatment they have received, and it has been difficult to get such patients to again accommodate themselves to the conditions of the ordinary military hospitals such as the Curragh'.[18] There was good local support for the hospital, and gifts of books, cigarettes, vegetables, fruit and eggs were sent there.[19]

The tensions and stresses of the service were sometimes exposed in tragic ways. In June 1916 Lieutenant and Quartermaster William Evans, 5th Leinster Regiment, shot himself with his revolver in his office. He had been worried and overworked, and had sought clerical assistance which was not granted. Corporal George W. Askew, 17th Lancers, shot himself with his rifle after reading a newspaper account of the death of his brother in France.[20] Thirty-five year old Private T. Short R.F.A. was found dead beneath his billet window,[21] and two soldiers were found on successive days on the railway line near Newbridge in February 1918; Private Hickey 60th Battalion, 5th Australian Division was dead, and the man from the Munster Fusiliers was severely injured.[22] Six months later Sergeant Buckley of the Black Watch was found dead on the line near the Curragh Siding.[23]

Nervous tension could also be exposed during training, and on one such occasion the quick action of a non commissioned officer prevented a bad accident. One day during live grenade throwing practice on the plain when a soldier threw a grenade it struck the parapet of the trench and dropped at his feet. 'Seized with fear, he immediately froze. With split second reflexes Sergeant Hutchison made for the grenade, knowing that it had five seconds before it exploded. Three seconds had ticked by before he managed to shove his terrified colleague to one side, snatched the grenade, and toss it over the parapet, where it immediately exploded'. Hutchison, who was in the Highland Light Infantry, was given the Albert Medal in April 1917.[24] Not so fortunate was Lieutenant Edward Shurley, bombing officer of the 52nd Cheshire Regt. In July 1918 when he was giving instruction he was killed when a bomb exploded.[25]

Whether it was war nerves or high living caused the downfall of 2nd Lieut. Colin Coleman R.F.A. in the summer of 1918 is not known. He was responsible for Land Claims, that was dealing with proprietors whose land had been injured by the military during training. He was found not only to have altered the claims book and misappropriated £70, but also to have defrauded his colleagues. His own major claimed that the sum of £120 which he had entrusted to Coleman to invest had not been invested, and other borrowed monies had not been repaid. While his fellow officers had covered for him several times, he had eventually been caught out. Arraigned before the Rt. Hon. Justice Ross at the County Assizes he was found not guilty of defrauding the major. Justice Ross thus expressed himself on the case: 'In view of the very large army drawn from all parts and classes it was to be expected that offences of the kind would be committed, and it was a matter for wonder and congratulation that the offences were so few.'[26]

Offences which were causing more concern to the establishment were those involving the subversion of soldiers by civilians. At the end of May 1918 William Shaw, a sculptor from Stradbally, appeared before a court-martial in the cavalry barracks, charged with soliciting arms from Private J. Lennox, 4th Hussars. He refused to recognise the court, and was found guilty and given 18 months hard labour.[27] A couple of weeks later a combined patrol of military and police stopped a farmer from Stradbally and searched him for despatches.[28] A young volunteer from Tullamore, Seamus Clarke, was court-martialled in the camp for having been found wearing a volunteer uniform in Edenderry a few weeks before. His sentence was three months.[29] At Newbridge Petty Session an ex-sergeant of the Connaught Rangers was given two months on a charge of obstructing the police when he was found taking eight boys for illegal drilling.[30] Clearly the status and role of the military throughout the country was changing, and as the necessity for greater security became obvious. In May 1918 the commanding officers of units in the Curragh, Newbridge and Kildare were asked to submit rolls of their men 'who had an aptitude for police work,' as the transfer [of suitable men] to a corps of military foot police in the Irish Command is at present a matter of urgency.'[31]

The routine work of the police proceeded as usual, and the old problem of the unfortunate women remained. Even if they no longer squatted in the furze, they still frequented the plain. But now they could be charged with trespass or vagrancy as was Anne Doran before a special court on the Curragh in May 1918. She was given a month for the latter offence, while her two female companions were fined, one for vagrancy, the other for being drunk.[32] In Naas when Mary Hartigan, a soldier's wife, was brought before the Petty Session by Private Thomas Kehoe R.D.F. for abuse and filthy language towards him and his wife in March 1918, she was bound to the peace.[33] Amongst the magistrates in the county in 1910 were four former officers, Col. T. J. de Burgh D.L., Col. St Leger Moore, Col. F. Wogan and Major M. Thackeray.[34]

In the summer of 1918 Sean O'Faoláin holidayed with his uncle who lived in Number 33, Row 6, Married Quarters, Block B in the Curragh Camp. He afterwards remembered wandering about the plain, or playing handball in the ball alley below the quarters. 'I discovered,' he wrote, 'during those vacant days that this camp which had always stirred my imagination as a place full of every sort of excitement and gaiety was as silent and seemingly empty as a ghost camp. I never saw more than two or three soldiers at a time. All I saw was a sentry or two, a nursemaid or two wheeling perambulators, an occasional truck, two or three men lugging big dixies or big garbage cans across empty barrack squares. Even at night there was only the mildest stir around the cinema, its blue carbon lights sizzling in the open doorway of the projection box. Where do soldiers in big settlements like this go during the daytime? If they were on the plain, then the plain swallowed them. Do they go on long route marches by night and sleep by day? It might have been a monastery where all the monks rose before dawn, were busy indoors or in distant fields all day long and never seen after that. Or was it that I, wandering like a sheep on the plain, was lost in my own dreamy mind . . . ?'[35]

O'Faoláin's perception of the camp was an accurate one. The routine of military life kept the other ranks occupied all day, either on the ranges or the training and sports grounds, or away on route marches. There was no time for loitering in the lines until after duty hours, and then men would have rested in their billets or travelled to a neighbouring town. The families in the married quarters had no recreational areas, and as there was no attractive shopping or social area for them they also stayed in their homes, except perhaps for a weekly foray into Newbridge or Kildare. The officers' quarters were separated from those of the other ranks, and their social activities rarely involved communication with the men or their families.

By the month of September 1918, when the war situation was changing with the imminent collapse of the German front, the sale of troop horses commenced at Kildare,[36] and soon remounts were on offer by R.J. Goff in Newbridge.[37] The acceptance of forage by the military was also being simplified with the A.S.C. at the Curragh giving permission for the weighing of forage at public ouncils. The procedure previously had been that when farmers brought produce to the barracks for sale it had to be weighed there. The acceptance by the military of the documentation of the public councils was not alone seen as an improvement in the arrangements for the farmer, but also a matter of convenience to the military themselves.[38]

Another indication of the changing times was a strike in September 1918 for higher wages by the 100 men employed by Dowling and Co., Sheridan Bros., and Laverty and Sons, the building contractors working at the camp, and in Newbridge and Kildare. The men were also members of the Transport Union. The dispute was settled, and work proceeded.[39] The rising power of the trade unions was apparent since the calling of the General Strike in April, and it had been a decisive factor in the anti-conscription campaign.

Armistice Day, 11 November 1918, was a day of celebration throughout the Empire, and especially in military centres. But in county Kildare the altered political climate caused the occasion to be muted. Apart from leading articles in both the *Leinster Leader* and the *Kildare Observer* commenting generally on the signing of the armistice, there was little local notice. Both papers still had their war-time imposed reduced number of pages, and the *Leinster Leader* found only a space to note that Celbridge had been brilliantly illuminated on Monday night, and that there was dancing on the street into the small hours.[40]

Jubilation in Naas, Newbridge and the Curragh was reported in the *Kildare Observer*. In both of the latter places the military had celebrated 'with flag waving, cheering and other demonstrations of joy', and the Newbridge Town Commission had passed a resolution of congratulations to the Allies, and of satisfaction of the splendid triumph which had been gained. When news of the signing of the armistice had reached Naas the Union Jack had been hoisted over the Court House. The local reporter saw that 'there was a general feeling of relief, if there was little outward manifestation of jubilation. Towards evening there was some flag waving by the military, who later indulged in pranks and demonstrations to show their joy at the termination of the war'. Some of the houses in the town and its vicinity also displayed flags.[41]

What was a time of celebration and joy for many people was a time of sorrow for the families of the 49,000 Irishmen who had died in the war, out of the 150,000 who had served in the British forces.[42] In county Kildare it was estimated that by the end of 1916 over 250 men from the Naas area were in the army, and that a total of about 1,600 men from the district around Athy[43] fought at the front, out of which as many as 200 died.[44] The county newspapers carried weekly lists of the casualties and the honours.[45] Amongst the latter was a Victoria Cross to Lieutenant John Vincent Holland, the eldest son of a veterinary surgeon who lived at the Model farm, Athy. Serving with the Leinster Regiment, he was decorated for 'fearlessly leading his bombers through our own artillery barrage and clearing a great part of the village in front', at Guillemont in France on 3 September 1916. Given a heroes welcome when he returned to Athy, he died in Australia in 1975.[46] Private William Francis Scully A.O.C., was one of the three sons of a Warrant Officer of the Durham Light Infantry and Mrs Scully, of the Town Hall, Naas, who were on active service. William Francis was awarded the Gold Medal by the King of Serbia for distinguished service during the Egyptian Expeditionary Campaign in 1916.[47] Mr Arthur Dease, Celbridge, who had been working with the motor ambulance section of the French army since the beginning of 1915 was awarded a Croix de Guerre.[48]

The war caused desolation to many of the landed families with the deaths of sons and heirs. In county Kildare the 5th Duke of Leinster lost his son Lord Desmond, a major in the Irish Guards, on the Western front in March 1916, and a nephew Captain Gerald Fitzgerald of the 4th Dragoon Guards, in 1914.[49] Lieutenant Colonel Thomas de Burgh of Oldtown, Naas, had his son Thomas, a lieutenant in the 31st Lancers (Indian Army) killed in July 1917,[50] and a cousin John Maurice de Burgh R.N. was killed in action in 1915.[51] Brigadier Gen. Charles Fitzclarence V.C., Irish Guards, who had been born at Kill, was leading the 1st Guards Brigade in November 1914 when he was killed.[52] Second Lieutenant George Hubert Medlicott South African Infantry, of Dunmurry, Kildare, was killed in 1916.[53] So was Capt. R. J. Smith, of Jigginstown House, Naas, an engineer, and who merited mention in despatches,[54] and Lieut. J. Tynan of Monasterevan.[55] Captain Harry Greer, of the National Stud, Tully, Kildare, and who had retired from the Highland Light Infantry in 1889 to concentrate on the bloodstock industry, lost two sons; the elder, Col. Eric, aged 25 was commanding a battalion of the Irish Guards.[56] Captain Greer, a keen golfer, had himself paid the wages of the greenkeeper at the Curragh club during both the Boer War and the Great War, when the finances of the club were reduced by the absence of officers at the front.[57] By November 1916 Clongowes Wood College was able to list 41 pupils or past pupils killed and 80 wounded, and with 500 students or past students in the army and navy. Lieutenant Holland had been educated at Clongowes, and eight other old boys had received the D.S.O., 14 were awarded Military Crosses, and 51 were mentioned in despatches.[58]

In the euphoria of the war's end there was sometimes sorrow as maimed, injured or shell-shocked men returned to their families. Or there might be estrangement due to absence, such as that of a family in Newbridge. One veteran, walking from the railway station into the town, encountered his wife and her sister in Charlotte street. Inflamed with jealousy, owing to rumours of his wife's infidelity, he shot her dead on the street; later he was said to have killed himself.[59]

The war-weary troops returning from the horrors of the trenches to the barracks in Ireland were not to find again the generally welcoming and peaceful atmosphere which had formerly characterised Irish garrison life. The political and national scene had changed irrevocably. In the 1918 December elections republican candidates had taken 73 out of the 104 seats, and as a golf historian remarked, that after the armistice was signed, 'the military members [of the Royal Curragh Golf Club on the Curragh] found themselves at war with Ireland'.[60] The War of Independence which started on 21 January 1919 was to last until 11 July 1921. At the Sandes Home in the Curragh camp 'Miss Sandes and her helpers quietly carried on (during "The Troubles") the work among the soldiers. Not for the first time her calm expression hid a heart that was torn and distressed'.[61]

The social round of the New Year of 1919 started in Naas with a Victory Golf Dance, organised by the County Kildare Club, at Naas Town Hall, a venue which before the war 'was the most popular in Leinster'. Colonel the Hon. Eyre Massey brought a party of three, and in addition to several captains and lieutenants there was a strong attendance of local families, including Odlums, Osbornes and O'Connors.[62] About the same time there was a public meeting to call for the release of Irish prisoners in English jails.[63]

As the soldiers returned home, and the numbers of civilians employed at the military installations decreased, there was widespread unemployment.[64] The price of horses collapsed, and the knowledge that many dead horses were being sent to the Hunt Club kennels from the barracks confirmed the local belief that bad times were coming. A jarvey in Newbridge purchased a charger for £5, and it was jokingly said that when the horse heard a bugle he had misgivings as to the dignity of his new work. However, an English soldier stationed in the town remarked that when he was in Germany and England the food was short, 'but over here you have what you want, give me Ireland any time'.[65] But some soldiers were returning to England, and Capt. Allen of Ballysax Hills advertised for sale his 15 prize bred goats, 5 Belgian hares, 24 geese, 40 young hens and pullets, rabbit hutches, wire runs, etc. It would seem that his pastoral life in Ireland had come to an end.[66]

In the week that the programme for the 1919 Punchestown races was published, including two military races, a tea and concert for the Yeomanry of Fife and Forfar was given in the Soldiers' Club in the ballroom of the Naas Town Hall. The restoration of the military races to the Punchestown programme after a five year absence was welcomed in a local paper with the comment that 'the stern call for war prevented our soldier and sailor riders from participating in the pleasure of the chase, except perhaps for an occasional day when home on short leave from duty'.[67] It would seem that some normality was returning to daily life.

Such sentiments were soon shattered when it was learnt that the Irish Railway Executive was requesting the curtailment of racing due to the shortage of coal, and it was further discovered that *Sinn Féin* intended to have meetings of the hunt stopped, pending the release of its imprisoned members.[68] Two violent occurrences in the county alarmed the populace in February; the first was the burning of the village hall at Kill, an outrage associated with Lady Mayo's refusal of its use by *Sinn Féin* in August of the previous year.[69] More shocking was the killing of Patrick Gavin from Maddenstown as he drove a cow to Newbridge fair at 5 a.m. one morning. He was challenged by a sentry as he passed the Pumping Station at Brownstown, and when he failed to halt the soldier shot him. At the hearing at which Private Gay was charged with the killing, the Sergeant of the Guard of the Duke of Wellington's Regiment gave evidence that he heard Gay call out the guard, and then a shot. The soldier claimed that Gavin had attacked him with a stick. He was discharged.[70]

The *Sinn Féin* proposal to disrupt hunting was quickly acted on in Kildare and at a Meet of the Kildare Hounds a crowd of people surrounded the members of the hunt; Lieut. Col. St Leger Moore afterwards said that the hounds were beaten, the horses struck, and the riders, including a reverend gentleman, were assaulted and pelted with stones. He added that the windows of the home of the Master of Fox Hounds had been broken, and he warned that if politics invaded sport that not only hunting, but also racing, cricket and football would be affected.[71] Hunting was temporarily suspended. However, it was clear to see that at the Curragh racecourse a return to normality was afoot when, on 6 May 1919 in the House of Commons, the Secretary of State for War, Mr Winston Churchill, was asked by Major Sir K. Fraser, Member for Leicester, 'whether he would consider the question of moving the Remount Depot, at present at the headquarters of the Irish Turf Club, Curragh Grand Stand, county Kildare, to a more suitable place?' Fraser said that the present temporary accommodation of tin stabling at the Curragh was badly situated, as being in close proximity to the racing stables there was a danger of disease; further that there was no grazing on the Curragh and no land sale except at a high price. He suggested that the Curragh Remount Depot should be returned to Lusk, where there was plenty of good grazing and good permanent remount stabling. Churchill's reply was that he had called for a report on the matter.[72]

The reaction of the Hunt Club to the disruption of the hunt, and also to the knowledge that poison had been laid on lands over which they hunted, was a proposal to abandon its April Race Meeting at Punchestown.[73] The farmers called a meeting at Naas which was attended by between 400 and 500 people, and there a list of signatures of 1,500 farmers was displayed, all asking that hunting should be continued and the races held. The loss of business to horse breeders, to farming in general, and to the local traders was feared. A member of *Sinn Féin* who was present at the meeting blamed the Kildare Hunt for the coercion of the country as every member of the county Grand Jury was a member of the hunt, and after the arrests of the *Sinn Féin* members the Jury had asked that firm action should be taken. The culmination of the meeting was that the farmers were prepared to request in writing that the prisoners should be released, but the members of the Hunt Club, who were assembled in another room, said they would not make such a plea as it was a political matter. They proposed that Punchestown be abandoned.[74] A week later in a letter to a local newspaper a writer signing himself 'Tomás' criticised the small and exclusive body who sought to control racing, and he informed St Leger Moore to 'wait and see, colonel, we are only at the beginning of the road, not the end as you seem to think'.[75]

Punchestown was abandoned,[76] and the loss of business not alone in the county, but in Dublin, was expressed in the newspapers. Hotels and boarding houses were closed, the jarveys were idle, and there was none of the traditional decorating of houses in the Naas district. The business people of the county were forcefully made aware that the boom days of the war were gone and a different world was being created.[77]

On the last day of the Curragh races in May the Lancers regiment held a dance in the Curragh Sport's Club, under the patronage of Lady Nutting, the wife of Capt. Sir Harold Stansmore Nutting 17th Lancers,[78] in aid of Mercer's Hospital. But the social life of the camp, to judge from the county newspapers, had not returned to its pre-war gaiety.[79] The condemnation of British rule in Ireland by the Roman Catholic bishops assembled at Maynooth on 24 June 1919 reflected the people's feelings of an administration which the hierarchy described as 'the rule of the sword, utterly unsuited to a civilised nation and extremely provocative of disorder and chronic rebellion'.[80] With such a criticism of their role in the country the military must have realised that their days of social acceptance by the majority of the population were drawing to a close.

Following the signing of the Treaty of Versailles on 28 June 1919 celebrations were held throughout the British Empire, but those in the Curragh camp unfortunately got out of hand. Hundreds of civilians converged on the camp to join the troops there, and from Newbridge and Kildare barracks, at the closing celebrations of the peace festivities. There was dancing in the various ballrooms in the camp and on the greens, to the music of the military bands. When the enjoyment was at its height some of the men fired a number of rockets and explosives which the crowd took to be part of the fun. But the seriousness of the matter was quickly realised when it was seen that a number of people had been burnt or injured. Then rockets which were lying on the ground caught fire and the thousands of people on the green stampeded, knocking men, women and children to the ground. Reporting the débacle the *Kildare Observer* said that 'the rockets went off like machine guns, and passing from one to the other, the blaze spread. A wild scene of confusion followed. Several of the people were burned by the rockets and trampled on, receiving such serious injuries that the call for stretchers was soon given, and a large number were taken away immediately to the Curragh hospital . . . In the midst of the excitement one of the tents became ignited, and the flames spread considerably as the result of men collecting chairs, tables, barrels, etc. and heaping them on the fire'. Fifty-one people were admitted to the hospital, of whom 18 were detained, and two of whom, privates from the North Staffordshire Regiment, were seriously wounded.[81]

How what was intended to be a joyous occasion deteriorated into one of tragedy must surely have been a cause of considerable concern and annoyance to the camp authorities. That the traditional military organisation and discipline had failed to control the situation, and that rockets were fired without adequate safety precautions, as well as the actions of some men in aggravating the conflagration, indicated a break down in army discipline which must have been further questioned following more disturbance in Athy a couple of nights later.

The townspeople of Athy were preparing to welcome home the men who had been demobilised on Saturday 12 July when it was learnt that the *Sinn Féin* prisoners were then also to be released, and that one from the town was to return home that day. The Union Jack had been hoisted over the post office in honour of the soldiers, but when they encountered the *Sinn Féin* supporters there was mayhem. To shouts of 'up the rebels' and 'up the khaki' they set into each other, but no serious injuries were inflicted. On the following day the county Kildare *Féis* was held in the town. It was also disrupted when 40–50 ex-servicemen attacked and gutted the house which had been formerly used as *Sinn Féin* Rooms. Police re-enforcements had to be called.[82]

During the summer months of 1919 sporting activities had continued in the Curragh; in July a team from the county Kildare Club met the Curragh Lawn Tennis Club at the camp,[83] and the County Club defeated the Curragh Garrison cricket eleven in the first match at the county grounds since June 1914.[84] At the end of August the dates for the Punchestown Meeting for the following year were announced, but in the same month a railway signal cabin was torched at Caragh, near Naas,[85] and in September thousands of bales of hay were burnt in Newbridge barracks, despite the efforts of the Curragh fire brigade. The cause of the blaze was not revealed.[86]

An indication of local feelings was given in September when the Naas Urban District Council declined the offer of war trophies as 'they had no statutory authority to accept them'.[87] The council was responding to a letter from the Earl of Drogheda and lord lieutenant of the county offering war trophies, principally canon captured from the Germans or the Austrians, to public bodies for display. While there was some opinion that on account of the thousands of men from the county who had served honourably in the war it would be appropriate to have such mementos, there was little official enthusiasm for the honour. [88]

At the end of the year 1919 the Ordnance Depot Stores at the Curragh held approximately 10,000 seized and surrendered arms, as well as large quantities of explosives and ammunition deposited by dealers or surrendered by individuals. The depot and magazine were in the charge of a regular executive officer, assisted by four temporary officers. There was a civilian staff of 71 men and women, supplemented by a detachment of between 30 and 40 men from the Royal Army Ordnance Corps. The reduction of the numbers of civilian staff was imminent as it was assumed that the garrison was about to be reduced to pre-war strength, and by the end of March 1920 some of the women from the Curragh had been discharged.[89]

The reorganisation of stores, the settling of contracts and the closing of the accounts of demobilised units in the aftermath of the war were major undertakings. The depot at the Curragh was reduced from that of a Mobilization and Camp Depot, and the Island Bridge Depot in Dublin was to be the main Ordnance Depot for the northern half of Ireland. The considerable surplus stores of harness and saddlery in the Curragh were being moved to Dublin when 'rebel action' began to disrupt the operations. The transfer of gun sections from the Curragh was abandoned. Losses from the various ordnance stores were causing concern, and a detective from London was brought in, in the disguise of a private of the R.A.O.C. Though he was recognised as a detective by his fellow workers within a week, he did confirm the official suspicion that 'some of the military and civilian employees were not reliable'. The necessary provision of escorts for stores intransit was an additional demand on personnel.

The depot at the Curragh, owing to its situation, was considered to be 'well protected and not liable to attack by the rebels', and it remained the main issuing depot for the Gun and Machine Gun sections. In Dublin the loss of 169 old-pattern pistols from the Island Bridge Armoury caused grievous alarm, and the theft and destruction of stores, culminating in the burning of the National Shell Factory in June 1921, necessitated the imposition of stricter security on all Arms, and the drafting-in of re-enforcements from England.[90]

Soon the winter sports were revived in county Kildare; cubbing with the Kildares commenced,[91] and the hockey teams took to the field. Naas Club fielded three teams, including one from the Royal Dublin Fusiliers[92]

As social life was resumed so was the union of gentry families with the officers. Frances, the daughter of Major General and Lady Kavanagh of Moorfield House, Newbridge, married Capt. H.O. Wiley 5th Dragoon Guards,[93] and earlier in the year Marguerite de Burgh of Oldtown married an officer of the Black Watch.[94] Evelyn Synnott of Furness, Naas, wed Lieutenant L.F. St Clair 21st Lancers,[95] and Mozra Cassidy, of the Monasterevan distilling family, was given away by her brother Lieutenant James Cassidy, when she married an officer from the Lancers.[96] Even if there were changes in the attitudes of many of the middle classes, the ascendancy still appreciated its social relationships with the army, and regretted the undercurrents of nationalism which disrupted their hitherto pleasant lives. Before Christmas 1919 there was a good turn out for a dance held in the Naas Town Hall in aid of the special fund for the Kildare Hunt Club. The attendance included many officers and their ladies, including Mrs Fane, the wife of the officer commanding the 12th Lancers, who brought a party from her temporary home at Tully House, Kildare. By all accounts it was a most enjoyable night.[97]

But disaffection was widespread, and incidents such as the shooting by the I.R.A. into the home of Inspector Supple R.I.C. at Naas on 14 January 1920, and a couple of weeks later another statement on self government from the hierarchy meeting at Maynooth, caused unease to the loyalists.[98]

Despite widespread disruptions, military routine was observed, as the exigencies of the service allowed, in the various stations in the county Kildare. The existence of the airfield on the Curragh enabled staff officers to fly in there for conferences, and for air mails to be sent out to the other airfields in Ireland.[99] The war time advances in aerial bombardment led to a suggestion from the air secretary in London that the Wicklow mountains should be cleared of all inhabitants to enable bombing to be practiced there.[100] That plan did not come to fruition, although in May 1920 the Secretary to Wicklow County Council promulgated bye-laws which had been made by W.S. Churchill, H.M. Principal Secretary of State for the War Department, for the Glen of Imaal Artillery Range.[101]

While the War Department was finalising its control of the county Wicklow artillery ranges in the spring of 1920 the Irish volunteers in county Kildare were re-organising,[102] and one of their future plans was the burning of the officers' quarters in the Glen of Imaal (as well as the nearby Saunder's Grove, the home of a notorious colonel of 1798).[103] Following the establishment of *Dáil Éireann* in January 1919 the Volunteers were recognised as the army of the Irish Republic, and in that capacity they fought the War of Independence. A North Kildare Battalion had been formed soon after the release of the men interned in Frongoch, and early in 1920 the I.R.A. in north Kildare was organised in two battalions. A training camp was established near Naas, and quickly new companies were formed throughout the area, and a headquarters for the 2nd Kildare battalion was established at Dowdington House at Athgarvan. There the ex-Connaught Ranger soldier William Jones resumed training recruits.[104]

Commandant Michael Smyth, 2nd Kildare battalion, in recording his memories of the period in 1970, described his command: 'There were three military barracks in the battalion area, Naas, Newbridge and the Curragh camp, all strongholds of the enemy, Curragh camp being headquarters of the British military in Ireland. There were also four strongly fortified R.I.C. barracks, Naas being the headquarters in the county of the Black and Tans and Auxiliaries.'[105]

During the War of Independence it was the R.I.C. rather than the military who became the main targets of the I.R.A. In the years of land agitation and civil unrest it had been the police who were to the forefront in implementing the law, and they became the recipients of the distrust and hatred of the affected people.[106] To increase the strength of the police a new force was recruited from among demobilised soldiers in England, and they were to be known as the Black and Tans. In county Kildare a section of Black and Tans was based in a part of Naas barracks which was fenced off from the soldiers' area.[107]

The arrival of the Black and Tans had escalated the violence throughout the country, and when a general strike was called for 13 April 1920, in sympathy with the prisoners in Mountjoy Jail who were on hunger strike, it coincided with the first day of the Punchestown Meeting. As the railways and other transport systems were also disrupted it was decided to postpone the Meeting, but later it was cancelled for the second successive year. Racegoers to the Curragh were inconvenienced in July when the special train from Dublin did not run. The cancellation was due to an attempt by a unit of the Black and Tans to commandeer the train to take them to the Curragh camp for shooting practice on the ranges. However, in those troubled years, the racing at the Curragh was never cancelled, although fixtures in other parts of the country had to be abandoned.[108]

What was described as 'one of the most sensational incidents that has yet occurred in county Kildare' occurred in May when the town hall and courthouse at Maynooth were blown up. A local belief that the town hall was to be occupied by some sort of military outpost was thought to have been the reason for the destruction, even though it was reported that the people generally did not believe that an outpost was planned for the hall. A few months earlier the R.I.C. barracks in the town had been burned down.[109]

An unexpected senior member of that force to visit Naas in November 1920 was Brigadier General F.P. Crozier who was appointed Commandant of the Auxiliary Division of the R.I.C. in Ireland in July 1920. While travelling in a tender near Naas, Crozier and his escort were accidentally thrown from the vehicle, and the general had to be confined to the Curragh Military Hospital for a month. He resigned from his post early in 1921, and was subsequently severely critical of the British conduct of affairs in Ireland.[110]

From their base at Athgarvan on the edge of the Curragh the I.R.A. carried out a campaign of subversion and harassment against the British army. Bridges on the principal roads leading to the camp and Kildare were blown up, roads were trenched, phone lines cut, and soldiers and civilians from the camp and Newbridge barracks were enticed to procure weapons, explosives and ammunition for the I.R.A., in addition to supplying intelligence reports on troop strengths and movements. An attempt late in 1920 to obtain a supply of arms and ammunition, including a machine gun, failed owing to the transfer of the soldiers concerned from Newbridge barracks. Units of *Fianna Éireann* and *Cumann na mBan* attached to some of the companies were employed in carrying dispatches, arms and ammunition.[111]

The main thrust of the I.R.A. activity was directed at the R.I.C. and several police barracks in the county were ignited. That at Lumville, less than a mile from the Curragh camp, with the courthouse nearby, was destroyed. The barracks at

Sallins, Kill, Ballytore, Dunlavin, Donard and Stratford-on-Slaney likewise were burnt. Police patrols were ambushed, and in one such engagement near Kill in the autumn of 1920 a policeman was killed and the others taken prisoner.[112] Subsequently a number of men were arrested in the locality, and the Black and Tans burnt the premises of Boushell's Family Boot Maker and Leather Merchant at South Main street, Naas, in mistake for that of a republican family who lived close by.[113]

The I.R.A. also raided mail trains at Newbridge and Sallins stations, and when the train drivers refused to drive any trains carrying British soldiers or munitions the military themselves boarded the trains at Newbridge each morning. The Monasterevan company I.R.A. captured a consignment of mail bags containing correspondence between the army and Dublin Castle at their local station.[114]

As well as being active in keeping the peace during the Kildare farm labourers' strike, when damage to crops was perpetrated by the labourers who were seeking a minimum wage of 45s per week instead of the then 30s per week scale,[115] the I.R.A. endeavoured to curb cattle stealing in the area of the Curragh and Ballysax and to impose the law in their district. When they were investigating a man from Suncroft on suspicion of robbing in the area he sought refuge in the Curragh camp, but when he set out to return home one night he was apprehended as a spy and executed by the I.R.A.; other people from the locality under similar suspicion also took refuge in the camp.[116]

Michael Smyth's summary of his unit's activity was that: 'The Kildare Volunteers succeeded in making a large number of crown forces stationed in the Curragh very much less effective than they should have been by their constant harassment of them, by implementing the orders of the Volunteers' G.H.Q. and wrecking the enemy lines of communication in every direction.'[117]

Smyth had been elected to Kildare County Council in June 1920 when *Sinn Féin* took the majority of seats; the chairman was the *Sinn Féin* T.D. Donal Buckley, who as *Domhnall O Buachalla* was to be the Governor-General of the Free State from 1932–37.[118] Michael Smyth was amongst the councillors who were arrested on their way to a council meeting on 22 November, and at the same time the office of the County Kildare Insurance Society in Naas was searched. It was reported that during the search the military were most courteous.[119]

The interest of the new members of the county council in the Irish language was demonstrated in the publication of the council's proceedings in both Irish and English in the *Leinster Leader*.[120] Similar interest in the language was shown by Naas Urban District Council and the Newbridge Town Commissioners in the autumn of that year when both bodies, the majority of whom were *Sinn Féin* members, proposed that the names of their towns should be changed to the gaelic forms. The proposals were adopted at the quarterly meeting of the County Council on 22 November 1920, and henceforth the towns were to be known as *Nas na Rí* and *An Droichead Nua*.[121]

Despite the changing political climate and the escalating war situation the British military did not neglect its traditions or its sporting events. When the Royal Horse Artillery held their annual Nery Sports day at Newbridge in September 1920 it served as a commemoration of a daring feat which resulted in the capture of some German guns at Nery in Belgium by L Battery R.H.A. on 1 September 1914.[122] The sports were also reported to have attracted a large audience of military and civilian friends, a reassuring gesture for the embattled soldiers.

As the year had progressed the war intensified, culminating with the events of the Bloody Sunday of 21 November when, in reprisal for the killing of eleven intelligence agents that morning, the Black and Tans fired on the spectators at a match in Croke Park, Dublin, killing 12 and wounding 60 people. By the end of 1920 a total of 165 R.I.C. men had been killed, and 251 wounded, but military casualties were much lower. The British government, in a determined effort to gain control of the situation, declared martial law in December, and the Lord Lieutenant, General French, became 'a virtual dictator' of the country. [123]

In county Kildare violence also continued. In December a party of R.I.C. was fired on at Knockbounce, near Kilcullen, and in the same month a court martial at Kane barracks in the Curragh tried a man from Maryborough and a man from Mountmellick. They were charged with unlawful assembly on the Heath, and assaulting there an ex-soldier named Michael Whelan. He had been stripped, tarred and feathered, with paraffin and Jeyes fluid poured over him. His army ribbons and his pension book were found on him. Whelan succeeded in making his way home where his mother spent two days cleaning the tar from his body. It was said that *Sinn Féin* had been drilling on the Heath, and he may have been suspected of spying on them.[124]

In a move to curtail unlawful activity an internment camp for republicans was opened at the Curragh where the prisoners were guarded by the soldiers. At Christmas-time the internees were permitted to have two masses celebrated there on Christmas Day, and their condition there was reported to be good, with visitors being allowed and parcels received. However, the new year of 1921 brought little joy to the country, or to county Kildare.[125]

In January 1921 Frank and John Corrigan, Blackrath, on the Curragh, were courtmartialled in the Curragh camp on a charge of having seditious literature and possessing a miniature rifle and ammunition.[126] At the same time John Maher from Athy was held in the guard room of Ponsonby barracks also on an arms charge, while in nearby Beresford barracks Michael Grey from Maryborough had an altercation with his guard. Grey was accused of striking and tarring and feathering a one-armed soldier. Under section 55 of the Restoration of Order in Ireland Regulations the photographing of prisoners was permitted, but when an officer of the 2nd King's Shropshire Light Infantry sought to take Grey's photograph he refused and was struck in the face.[127]

On 8 January most of the prisoners in the Curragh were moved to Ballykinlar in county Down, but one man who remained behind had an amusing story to tell. Wearing his coat and hat, he was talking to two visitors outside his hut when the sergeant came along and told them that it was time to go, and moved them towards the fence. Only then did the internee reveal himself to the sergeant, quipping 'I'll go gladly, sergeant, but I am a prisoner'.[128] Less amusing was an incident involving two Black and Tans in Naas about that time. Both ex-soldiers, one a Mexican, the other a Scot, they were drunk and when they discharged their revolvers in the Main street they were arrested.[129]

Another alarming occurrence in the town was the invasion of the ball in the Town Hall organised by the County Kildare Farmers' Union on the night of Friday 4 February. The sudden appearance of military and police 'in the midst of the thronged ballroom, carrying revolvers and rifles with fixed bayonets' made it a night to remember for the dancers, although it was believed that 'much alarm was created amongst some of the ladies'. The forces of the law were said to be looking for a wanted man, but no search was carried out, and after a cursory survey, they left.[130] These incidents demonstrated another aspect of the garrison to the people of the town and neighbourhood, and created strong disagreement between those of loyalist and nationalist views.

The Town Hall was again the location of an unpleasant incident on Tuesday 15 March when a party of Auxiliaries and R.I.C. who were passing through Naas town in half a dozen lorries halted in the centre of the town, and some of them went into the Town Hall. When they noticed a door with a gaelic inscription, and which had formerly been used as a *Sinn Féin* Club, they broke in the door, looted volunteer property, and painted the slogans 'God Save the King' and 'Up the R.I.C.' over the name on the door. When they departed the town some of them were wearing the volunteer caps, and the republican flag was tied to the back of a lorry.[131]

The dismantling of the Remount Depot at the Curragh race-course was under way in February 1921 when the Newbridge auctioneers Steel, Roycroft and Price advertised for auction the huts, stables, hay shed, wagon shed, oats and quarter-master's stores, latrines, 100 forty gallon barrels, and three hangers for housing aeroplanes. At the same time all civilians employed in the military district were asked to submit the name of the trade union to which they belonged, and it was understood that in future all employees should be members of a union.[132]

Rented accommodation was also being vacated as an order had been given to officers that they should move into barracks. Albert Payne, the owner of 'The Nook' in Kildare town, let at £72 a year, feared the loss of rent when his tenant, Lieutenant Samuel Rowell R.A., went temporarily to England. As a precaution he removed the officer's furniture from the house, but at the Kildare Petty Sessions he was ordered to return it, and not to claim for the rent which was due on 5 November 1920.[133]

The *Leinster Leader* reported on 12 March 1921 that 'A camp like Ballykinlar has been opened at Rath, Curragh, and a large number of prisoners have been transferred there from Hare Park'. No visitors were being allowed to the new compound, which was also receiving prisoners from Athlone and elsewhere.[134] The camp was located north-west of the historic Gibbet Rath, and close to the main road from Newbridge to Kildare. It was able to accommodate between 1,200 and 1,500 prisoners and it was to remain in use until after the Treaty was made.[135] Sylvester Delahunt from Tuckmill, Straffan, who was arrested in March 1921 was interned in the camp until December of that year. His autograph book of that time includes a drawing by F. Purcell (782) titled 'a night scene, the escape from the Curragh Internment Camp 1921', and this comment from R. MacDermott (379F) from Athy: 'Rath camp is nice, but there's no place like home.'[136]

During the month of April security operations conducted by the military and the police increased. Courts-martial were held regularly at Beresford barracks of persons found with arms or seditious literature. Matthew Cullen of Hacketstown, the chairman of Baltinglass No. 2 District Council and vice-chairman of the Board of Guardians was tried for being in the possession of a notice issued by D. Coy I.R.A. Brothers Simon and John Nolan from Kilbelan, Newbridge, two young farmers, were charged in connection with a clip of ammunition found in a tree stump on their land. Michael Duffy, Poplar Hall, Inchiquire was given a sentence of two years hard labour for possessing copies of *An tOglagh.*[137] The raids, arrests, trials, curfews, trenching of roads and destruction of bridges, had become every day happenings in the lives of the people.

During the War of Independence the Athy area was an especially disturbed one, and in the first few weeks of May 1921 it was particularly so as widespread activity on the part of both the British forces and the I.R.A. continued in the county. Roads were trenched or blocked with trees, and bridges were damaged. A notice from 'The competent military authorities' was posted in the town warning that in any area in which bridge or road cutting, etc. was done that the holding of fairs and markets would be stopped. Despite the notice the market of the following Tuesday took place in Athy,[138] and the prohibition was not rescinded until April when Col. Skinner, Curragh Command, issued a direction, but curfew was to be continued at 8 p.m.[139] Reconnaissance aircraft from Baldonnell flew over the county from time to time to observe for ambushes or road obstacles, and aircraft were also used to escort convoys of prisoners being moved to Dublin, or to cover troop trains on the lines between Kildare and Dublin.[140]

A joint patrol of military and R.I.C. raided the home of two labouring brothers named Dunne at Barrowhouse near Athy and found seditious ballad sheets. The men were arrested. On Monday 16 May a five man patrol of R.I.C. and Black and Tans was ambushed by a group of eight volunteers at Barrowhouse, and two of the volunteers were killed. The following morning arrests were made in the area, one house was burned, and another demolished. The two volunteers who had been killed were given a military funeral by their comrades, with the Last Post being sounded and a volley of shots fired.[141]

Crown forces surprised a party of I.R.A. men who were attempting to mine the railway at Celbridge early in July, and in the encounter some of the I.R.A. were wounded. Less than a week later when the army supply depot at Newbridge was burned down, and two civilians were killed, it was believed that the fire was accidental.[142]

As reprisals followed incidents the civilian population became more absorbed in the war. On Derby Day at the Curragh the military at *Droichead Nua* [as Newbridge was then sometimes being called in the local press] held up the cars going to the races to look for permits, and at Kildare, after the races, armoured cars surrounded the railway station.[143] Earlier in June at a sports meeting in Athy, in aid of the local *Féis*, 15 army lorries carrying military and police arrived at the sports field. Sentries were posted, and the people were rounded up and questioned. The sports were not allowed to proceed, and three men were taken to the barracks. A cordon was placed around the town and remained in position for an hour.[144] No doubt but an attack on the Athy R.I.C. barracks some time before, and the hold-up of a military car near Castledermot, had aggravated the authorities. The capture of a despatch bag and the burning of the car in the same incident were particularly serious ones.[145] Racing at Kilcock was also disrupted when a mixed posse of soldiers, Auxiliaries and R.I.C. rounded up some of the punters. Even though there were no arrests, the remaining events on the programme were abandoned. The annual sports day at Robertstown was proclaimed by the military, causing the organising committee to complain of the loss of effort and finance. A pig-buyer was arrested at Athy for contravention of the ban on the holding of fairs and markets.[146] In the Curragh area, following a widespread spate of tree-cutting and road blocking, the army commandeered local young men to clear the trees.[147]

Sinn Féin was administering justice through its own courts, and when an ex-soldier was found shot at Kilboggan a military inquiry was held in the Curragh camp. The victim had been tried for larceny by a *Sinn Féin* court and ordered to leave the country for 12 months. An employee with the Royal Engineers in the camp, he had then moved into married quarters there, but when he travelled to Kilboggan one evening he was apprehended and killed.[148]

In a further effort to inhibit aid to the I.R.A. a notice was sent to the Abbeyleix Union that in future the details of persons admitted suffering from wounds from bullets, gunfire or other explosives should be sent to the H.Q. 14th Infantry Brigade at the Curragh. Proceedings were threatened for non-compliance, under the Restoration of Order in Ireland Regulations. The clerk of the union said it was not his problem, and marked the notice as read.[149]

Despite the oppressive role assumed by the army they still managed to maintain some relationships with the civilian population. A football match between the North Staffordshires and the R.F.A. attracted a very large number of spectators at *Droichead Nua*,[150] as Newbridge was then newly named, and in April the Punchestown Meeting was again held. It was the last occasion on which the Irish Grand Military Steeplechase was run under its old conditions. It was won by Capt. A.G. Smith's 'Silk Sail', ridden by Mr R.L. McCreery of the 12th Lancers.[151]

A month before the meeting Percy La Touche, Senior Steward of the Irish National Hunt Steeplechase Committee, had died and he was succeeded in that office by General Sir Bryan Mahon.[152] General Mahon had been general commanding the forces in Ireland for two years after the 1916 Rising, and when he retired from service in 1921 he settled at Mullaboden. General Mahon's confidence in the political future of the country seemed confirmed when he was nominated to the Irish Free State Senate in 1922, though it must have been dented when Mullaboden was burnt by the I.R.A. soon afterwards.[153]

Further evidence of the new political climate was displayed at a meeting of Naas No. 1. Rural District Council when a resolution was adopted protesting against fox hunting taking place. A letter from the clerk of the council to the County Council dated 3 November 1920 explained that the protest was against 'military and upholders of the British law' being allowed to participate in the hunt. On the day on which that letter was read at a county council meeting, 22 November, it was also noted that an application from Lieut. Col. C.L. Graham for permission to erect a hut at Sallins, on the Clane-Sallins road, for the purpose of a soldiers' club, was refused on the recommendation of the county surveyor.

However, the most important resolution considered at that council meeting was the approval of the re-naming of the county town of Naas and the garrison town of Newbridge to the Irish form of their names. The resolution was adopted by the county council.[154]

Chapter 16

BRITISH WITHDRAWAL

When opening the parliament of Northern Ireland on 2 June 1921 King George V appealed for peace between Great Britain and Ireland, and two days later Eamon de Valera, President of *Dáil Éireann*, received an invitation from the British Prime Minister, David Lloyd George, to attend talks. A truce was agreed between de Valera and General Sir Nevil Macready on 9 July, and came into effect two days later, and following a series of talks a Treaty was signed on 6 December.[1]

On the eve of the signing of the truce the Navy and Army Canteen in Newbridge was burnt down, and two soldiers were later commended for their valour in fighting the blaze, though they subsequently had to pay for their uniforms which were damaged in the effort.[2]

The newspapers of the week ending 16 July reported the jubilation which greeted the truce. At Newbridge, Kildare, Athy and Monasterevan there were bonfires, but the joy in the latter town was reduced when a group of children bathing in the river Barrow were frightened when a passing lorry-load of soldiers fired 20 rounds from a machine gun. Shortly before the 'peace moment' several armoured cars were seen driving towards the Curragh, 'fully armed, and rifles were to be seen emerging in the usual way'.[3]

An amusing incident which had happened in the Curragh camp that July was interpreted as a forecast of the future of the place. A decision had been made to paint the Clock Tower, one of the original structures remaining in the camp, and which had not been renovated for a decade. The instructions to the painter were to paint it the same as before. As he cleaned off the old paint he found underneath the colours green, white and yellow, and he asked what should he do. He was told to 'carry on', and he did, and so 'the *Sinn Féin* colours were again painted on the clock, while the other flags were floating about on Peace Day'. It was some time before the authorities noticed the implication of the colours, and as the report in a local paper went, 'who can say but this little mistake on the part of the military by which they displayed the Irish colours in the centre of the greatest camp in Ireland, may not be the happy augury of that peace which is wished for, and a just settlement of the claims of Ireland. The latest news with regard to the matter is that another colour has been added, which will relieve it from any particular significance as far as the painting is concerned'.[4]

In the months before the truce the internees in the Rath Camp, according to the reports which one of them sent to the *Leinster Leader*, kept up their morale with concerts and games of football.[5] By mid-July men were being released at the rate of two a day, but in August new huts were being assembled in the compounds and barbed wire erected.[6] When the Royal Engineers sought to hire men from Naas for the latter task, at the rate of £3.1s.8d a week, there were no takers.[7] The condition of the huts in the internment camps was being criticised as those in Hare Park dated from 1915, and were in poor condition. There was said to be much illness in the camps.[8] Some of the prisoners did not want to wait for parole, and there were frequent escapes. James Staines hid beneath an army lorry taking stores out of the camp, but a comrade who tried the same method was caught.[9]

As the consequences of the truce were considered by the people, speculation started. There was a belief that the many building programmes in progress at the Curragh were to be abandoned, and that 100 men employed at the Sport's Grounds alone were to be left go.[10] As the I.R.A. endeavoured to ensure that the conditions for the truce were observed they prohibited the public singing, especially after dark, of patriotic songs.[11] At Athy a public remonstrance was made to *Sinn Féin* following a raid on Kilmorony House by 30 disguised men. It was the home of the widowed Lady Weldon, of an Anglo-Irish family which had a long military tradition.[12] Field glasses were amongst the few items taken from the house, as her ladyship protested her patriotism.[13] Other military families were leaving the county as the officers returned home to England, and many properties were falling vacant. Brownstown House, where General Gough had lived, was on the market.[14]

Concern for the safety of loyalists, in the event of the negotiations between the British Government and the representatives of *Dáil Éireann* collapsing and hostilities being resumed between the two sides, occupied the attention of members of the General Staff in Dublin that autumn. It was feared that if such was the case that the I.R.A. might take loyalists as hostages, or that as many of the loyalists lived in isolated areas that their property and their lives could be endangered. Colonel J. Brind, at the Dublin District Headquarters, outlined the various possibilities to the Divisional and District commanders, and sought their views.

Brind divided the loyalists into three classes:

1. Those who insist on remaining at home and 'chancing it'.
2. Those willing to be sent out of the country.
3. Those prepared to move into concentration localities, but wishing to remain in Ireland.

The colonel's plan for the latter was that a central town or towns and villages would be selected in each Brigade area and into which the loyalists could move. Suggested towns were Fermoy, Tipperary, Newbridge and Kildare 'where there are plenty of troops . . .' If the truce collapsed it was proposed that persons 'of known rebel sympathies' should be given 24 hours to leave the designated areas, and the accommodation vacated would then be given to loyalists. Patrols of specially raised constables would assist the R.I.C. in protecting the refugees, and any member of the I.R.A. found in the declared areas would be shot as a spy.[15] Major General George Boyd, Commanding the Dublin District, dismissed Brind's plan as 'unpractical', and said that the loyalists who believed themselves to be in danger should leave the country. He believed that clearing 'villages like Fermoy' of rebels would be impossible without the use of force, and even a worse atmosphere would be created for the loyalists. Martial law, Boyd held, was the only solution. The War Office sanctioned contingency plans for military intervention if the republican movement got out of hand.[16] The designation of two towns in county Kildare reflected not only the strength of the military presence, but also the traditional loyal status of the area.

As the negotiations between the British and Irish representatives continued in November 1921 the conditions for the release of internees on parole were issued, and Armistice Day was celebrated in ascendancy locations. In Kildare the artillery 'turned out in strong force for service in the cathedral at 10.30 a.m. . . . and which concluded with the National Anthem. After the service the troops fell in on the Market Square, and while the guns were being fired on the Curragh the Reveille was sounded by massed trumpeters of the Brigade, and the cathedral bell was rung. At 11 o'clock the two minutes was observed in reverent silence by the troops and members of the congregation and then the artillery marched back to barracks'.[17]

The effects of some of the subversion of the I.R.A. was visible at the Kildare Quarter Sessions when awards were made to injured parties. At the end of November the authorities were granted compensation of £246 for damage done to a Crossley tender which had run into a trench on the road near Kildare, and £200 was awarded for the malicious burning of a military motor car. A soldier whose personal bicycle was destroyed was given £18 compensation.[18]

The signing of the Treaty on 6 December generated more rejoicing in the county. The volunteers paraded in Athy, and the internees were welcomed home. On Thursday 9 December 450 men were released from the Curragh, and on the following day 700 more. Processions of hackney cars came to the camp to transport the releasees to Kildare railway station.[19]

In the camp military discipline was seen to be as normal when three sappers were sentenced to 10 years imprisonment for having conveyed letters from internees out of the internment camps for posting.[20] Two men of the Northumberland Fusiliers were arrested for selling military property,[21] and another pair from the 12th Lancers were apprehended for an incident which happened on the road from Sallins to Clane. The Lancers were riding along on their horses when they met a young lady; one of the men dismounted and accosted the girl, but a man came along and they rode off. Later they were arrested by the Republican Police, and handed over to the volunteers, who in turn had them returned to their regiment.[22]

Apart from the law enforcement actions of the nationalists a further confirmation of the changing times was to be seen on the Little Curragh in mid-December when a section of the Kildare Brigade I.R.A. was reviewed there by an officer from G.H.Q. Three battalions, totalling 1,500 men, paraded and made evolutions which were described as 'showing the efficiency and good appearance of the men as they marched past'. Training in signals, engineering and first-aid was also carried out.[23]

The Irish Times of 9 December predicted that it was understood in military circles 'that immediately after the terms of the Peace Agreement have been approved by the British parliament and by the representatives of Southern Ireland, steps will be taken for the removal of all military units now in the country'. The paper then expressed the view that to many people, and not just those of one political party, the withdrawal would be 'a source of profound regret'. The loss of the trade generated by the army to business houses and farmers in, and in the neighbourhood of, garrison towns, of which Newbridge and Naas were mentioned as amongst those which 'owed the greater part of their trade to the military . . . would now be hard hit unless some compensating factor is supplied'.[24]

On 10 December Major General Sir Hugh S. Jeudwine, Commanding the 5th Division at the Curragh, was given information on the withdrawal of troops, and the proposed concentration of units at the Curragh and Dublin before final evacuation. His headquarters passed the details on to the brigades, with the additional notice that 'it was highly improbable that troops will commence to leave Ireland before December 28th'. In a secret document of 24 December the headquarters at the Curragh sent more definite orders to its brigades on the process of the first stage of withdrawal. The number of units which were to remain in certain stations was specified, and in county Kildare four battalions and one cavalry brigade were to be in the Curragh, with the 5th Divisional artillery at Kildare, Newbridge and in Ulster.[25]

Major Thackeray R.M. at Lumville Manse on the Curragh fringe commented in December, 'I think there is no doubt that the Treaty will be signed and that an immediate and drastic change will take place in this country'. Part of that change soon saw his own residence being taken over by a new tenant, the jockey Tommy Burns, and the abandonment of the Deputy Ranger's lodge by Major Percival Henry Compton when he failed to have his five year lease cancelled by the Office of Woods. Compton had pleaded that his 'position in this country would be intolerable, living here after I have been spending the last two years in fighting them'. While there was a fear that the house might be occupied by 'unauthorised persons', the officials in

London believed that while 'the old class of resident' would no longer be on the Curragh it would continue as a racing centre, but that the departure of the army would cause a decline in the demand for horses. Even Col. F.F. McCabe, a Roman Catholic doctor who had served in the Boer War with the South Irish Horse, and who contemplated moving his training establishment to the vacant Deputy Ranger's Lodge, was initially reluctant to move his family to the Curragh.[26]

The Irish Times returned to the consequence of the departure of the British army in its edition of Christmas eve under the heading 'Army withdrawal: how it will hit the Curragh'. A Special Correspondent found that 'it was estimated that from £90,000 to £100,000 a month was circulated by the military in the Curragh district, and that virtually every class of the population in a wide area of Kildare county will suffer loss as the direct result of the withdrawal . . . traders and dwellers in the towns which lie around or near the camp, farmers, market-gardeners, carriers, labourers of every class, and even car-drivers, will be hardly hit'. It was anticipated that land values and house property would suffer marked depreciation. Graziers would no longer be able to afford to rent grass, and dairy farmers would have no outlet for their produce.

The absence of the officers from sporting events was going to be felt, especially in horse breeding areas, and in the great hunting centres, including Kildare. It was feared that the Kildare Hunt could not be kept up if half the subscribers moved away with their regiments. However, it was thought that racing was unlikely to be affected, 'except in the matter of stand-takings at the Curragh and Punchestown'.

On a visit to Newbridge the correspondent learnt that always in the past, 'and even in the troubled times through which we have recently passed, the military in the Curragh district have been on the best of terms with the inhabitants'. The prosperity of Newbridge was as a direct result of liberal military patronage of one sort or another, and in other places such as Brownstown and Suncroft 'almost every family may be said to subsist upon the camp'. Kildare town would also suffer, but not to the extent that would Newbridge.

In Newbridge the man for *The Irish Times* spoke to some of the shopkeepers and learnt that nine-tenths of the traders depended on the military. 'If they did not benefit directly they did so indirectly . . . it is almost entirely a garrison town, and most of the workmen find employment in connection with the camp'. The soldiers of every rank, one shopkeeper said, 'he found pleasant to deal with; they spent their money freely and gave no trouble'. In his opinion the town was a cosmopolitan city in miniature: 'You meet people from everywhere, and there is nothing strange about it, unlike most country towns in Ireland'.

Visiting another establishment the reporter saw a brisk Christmas trade, a regular stream of customers passed in and out, and took as little notice of each other as if they were in a big city emporium. The large staff of assistants was kept busy, and 'it was regrettable to think that, after all this activity and work of the owner in building up the business, a lull might come which would cause the dismissal of half of the staff'. Officers and their wives, said the proprietor, 'were the mainstay of the large shopkeepers in the town, and their departure would naturally make for depression'; still he hoped he would be able to make a living.

The Curragh camp itself (with its normal population of 7,000) the journalist saw as 'virtually a town in which there are no paupers, every individual has a certain amount of money which he is free to use for local expenditure'. As the district was not rated for the upkeep of the camp that money 'could be regarded as all profit'. It was estimated that £7,000 a week was paid in wages in the camp to the civilian clerical and manual workers. The 'huge suburban district, with its many beautiful villas, specially built for the large floating population was expected to be desolate, to become deserted and revert to a state of loneliness and quietude; some of the oldest residents and largest employers, in the absence of congenial company in otherwise pleasant surroundings, may dispose of their properties and go to reside elsewhere'.

The only section of the local community which it was thought might benefit from the exodus was that involved in the horse industry. Fodder would be much cheaper to purchase by the horse-breeders and the big trainers, but even they would lose otherwise in the decline of the market for horses. There was a hope that the British army might continue to purchase remounts in Ireland, and there was even a hope that, if the new Irish army came to train in the Curragh camp that the British army would be given a facility to come there for training. 'There would be plenty of room for both,' observed the Special Correspondent, 'a new link would be established between the two peoples, and large sums of money would still be expended in the neighbourhood.'

Farmers from different parts of the country had asked if it were possible for some detachments of the British army to be left in the country, but the opinion of the newspaper was that such would be alright, if it had been part of the peace offer that Ireland should be recognised as a recruiting area for the army. There was no mention of such an arrangement in the Peace Agreement, and the fact that Ireland was to have a defence force of its own would suggest that the British government would not maintain recruiting depots in Ireland.

While the ratification of the Treaty was welcomed by everyone the reporter spoke to, it was felt in Newbridge that whether the Irish army was to come to the locality or not, that 'whatever the new regime may bring, it will fall short of the force which is now about to disappear'. But it was hoped that there would not be too long a delay before a decision on the future of the area would be made.[27] The euphoria which the ending of hostilities may have generated, and the hopes that the new political situation promised, were quickly to be shattered with the commencement of the Civil War within six months of the approval by *Dáil Éireann* of the Treaty. It was believed by one Anglo-Irish lady of military background that 'the fratricidal Civil War that followed the departure of the British troops' had been directly as a result of 'the methods used against the British during the Troubles'.[28] If that is a theory difficult to comprehend it might be possible to find that the War of Independence did produce guerilla tactics which were again used by both sides in the Civil War, and which were afterwards exported to other colonial countries in their achievement of independence.

Before the end of the year 1921 the British General Headquarters in Ireland issued operation plans for the evacuation of troops which were appropriately given the code name 'Finis'.[29] On the approval of the Treaty by the Dáil on 7 January 1922 the code word 'Finis' was telegraphed to commands, and immediately the machinery for departure was put into operation.[30]

At the Curragh, which was the headquarters of the 5th Division, that meant preparations for the reception of the troops withdrawn from outposts and concentrating at the Curragh camp. There were 12,500 men in the Division which covered counties Carlow, Galway, Sligo, Mayo, Offaly and Kildare. They were withdrawn to the barracks on the Curragh and at Newbridge, Kildare and Naas within three months, and six brigades were disbanded (five infantry, one cavalry, and support elements).[31] The Curragh camp was alive with activity as the units assembled, including battalions from the 6th Division in Cork. Quartermasters and storemen worked seven days a week to process the estimated 700 tons of stores which passed through the camp.[32]

Arrangements had to be made to dispose by local sale of certain effects including huts. If any of the huts remained unsold they were not to be left standing, and if necessary they were to be destroyed. Instructions as to the disposal of the reference library at the Curragh were to be issued later. The general officer commanding 5th Division expressed his view that 'the duties that officers commanding stations will be called upon to carry out in connection with local sales, are difficult and somewhat outside the scope of ordinary military duties'.[33]

If the evacuation was to prove unpleasant for commanding officers and disruptive for those living with their families in the district, it was to be far more traumatic for the traders and workers whose livelihood was directly dependant on the presence of the large and regularly paid British army.

The *Irish Independent* of 11 January 1922 reported that in '*Droichead Nua* artillery barracks 1,500 men are leaving for England. Already large quantities of arms and stores have been dispatched'. But such reports did not entirely dampen the last major public social occasion for the officers of the county Kildare stations. It was the Annual Ball of the Kildare Hunt Club which was held in early January in the Naas Town Hall; when they arrived at the venue the gentlemen and their ladies were introduced to the new Ireland with the presence of Republican police who were regulating the traffic there. The county gentry, with the officers and ladies of the 12th Lancers, 15th Hussars and the 2nd Suffolk Regiment, danced through the night, temporarily forgetting the end of their old world.[34]

At the same time preparations were being made for the evacuation of the Naas Depot of the Royal Dublin Fusiliers; the married quarters were to be abandoned, the families and effects being sent by rail to Dublin. Men in the unit were given the option of transferring to another unit, or being discharged.[35] It was one of the six Irish regiments to be disbanded that year.

On the Curragh feverish activity saw the clearing of military stores with a waggon of rifles dispatched by rail to Dublin. The fragility of the time was demonstrated when Lieutenant Bavin, King's Own Yorkshire Light Infantry, was fired on while travelling on his motor-cycle through Suncroft; the bullet went through his coat. The incident was deplored by the Republican authorities, and the Intelligence Staff of the Irish Republican Army set about investigating the incident.[36] The destruction by fire of the premises of the army forage contractors Messrs Cullen and Allen in Newbridge was another depressing happening in the new year. Fortunately the damage amounting to £1,000 was covered by insurance.[37]

The first of a series of auctions of surplus effects and personal properties had taken place at Naas at the end of January 1922 when Michael Fitzsimons, Auctioneer and Valuer, advertised for sale the household furniture and effects of the Royal Dublin Fusiliers barracks; included were sporting prints, leather covered chairs, harness, and a ten volume encyclopaedia.[38] Soon Robert J. Goff and Co. of Newbridge and Kildare, were publicising the sale of 250 huts in the Curragh, and other hardware. A couple of weeks later service waggons were amongst the variety of items offered for sale. On 4 February Goffs notified the sale of 250 huts, kitchen ranges, boilers, etc. from the Frenchfurze and Rath camps.[39] All through the spring the auctions continued with billiard tables, bedding, antique and modern furniture, cricket nets and poles, garden tools and mowers, and a young strong donkey being offered. The Curragh Sports Club pavilion, the contents of the Church of England Institute, the furniture of the Curragh chapel, and the stores of the Junior Army and Navy Stores came under the hammer. Before Christmas of that year two departing officers were selling their homes when Captain Bald and Col. McCabe respectively put Conyngham Lodge and Osborne Lodge on the market.[40]

Anxiety was also then being expressed in the local press that the considerable employment generated by the presence of the military in the county could not be maintained with the changed conditions; the journalist hoped that the future employment would be 'for purposes that will prove more beneficial for the common welfare of the country'.[41] A couple of weeks later the same newspaper queried the status of the workforce at the *Droichead Nua* Laundry with the new régime in the Curragh and Newbridge. The fact that the military had continued with the building of twelve married quarters at Moorfield in Newbridge, and of a saw-mill, bakery and extension to the military hospital at the Curragh, surprised the media.[42]

In the third week of January the Duke of Wellington's Regiment left the Curragh for England, and in reporting their departure the local paper noted that during the time the regiment had been in charge of security at the Rath Internment Camp in the Curragh, and from which the Irish Volunteer prisoners had been released some weeks before, the volunteers had the kindest feelings towards the Wellingtons (Fig. 36). The 6th Inniskilling Dragoons, many of whom were Irish, were then also being evacuated.[43]

The Provisional Government, which had come into existence on 14 January, was making arrangements for the taking over of 'all barracks and positions, permanent and temporary', as Michael Collins, chairman of the Irish Provisional Government, writing from Dublin City Hall informed the Commander in Chief at Parkgate Street in a letter of 1 February. When inventories of the barracks, stores and supplies had been made and checked by an Irish and a British officer, Collins wrote that 'your troops will then be free to evacuate, and our incoming forces will again take an inventory with the representative of the Provisional Government, and from that time become responsible for the safeguarding of the barracks, stores and supplies'.[44]

An instruction, signed by a colonel on the staff of the Deputy Quartermaster General on 3 February, allowed for the handover of property to accredited representatives of the Provisional Government, on production of their credentials.[45] Two days later, on Sunday 5 February, Richard Mulcahy, Minister of Defence,

Fig. 36. *Republican prisoners, some in a Model T Ford (with two different headlights and a damaged registration plate), after being released from the internment camp in the Curragh in 1922.* Photo: *Joe Cashman* The Irish Times.

General Eoin O'Duffy, Chief of Staff, Gerald O'Sullivan, Adjutant General and Joseph McGrath, Minister of Labour, paid an unofficial visit to the Curragh camp, to quote the local paper, 'to get some idea of the size and suitability of the buildings'. It was reported that the military soon recognised the ministers and that they stood to attention and saluted as they passed-by.[46]

That week the first Republican soldiers arrived in Newbridge to take over the Town Hall, and they were well received by the people; simultaneously a large party of Royal Engineers departed the Curragh for stations in the north of Ireland. It was also rumoured that the 1,500 men in Newbridge barracks were to leave that week, but they did not evacuate until the second week of May; another rumour was that the Curragh would not be completely cleared for a further three months, which proved to be correct.[47] At the end of February General Headquarters in Dublin issued a permit to representatives of the Provisional Government to again visit the camp, when, on arrival, they were to report to the acting officer commanding the 5th Division Headquarters.[48]

If the shooting at Lieutenant Bavin at Suncroft in the second week of January had been simply mischievous, a much more serious incident took place on the morning of Friday 10 February. As Lieutenant John Wogan-Browne R.F.A. was returning to the artillery barracks in Kildare with the regimental pay he was halted by three men who demanded the cash. When he refused to hand it over one of the men shot him in the head and he died instantly. Wogan-Browne was the only son of Colonel Francis Wogan-Browne King's Own Hussars, of Naas; the Wogan-Brownes were one of the oldest and most respected Roman Catholic families in the county, and the murder was widely condemned. The garrison church in the Curragh was packed for the funeral. On the night of the killing there was a hostile demonstration by the artillery men in Kildare town, and the passes of the local traders to the barracks were cancelled. Immediately the traders complained of the possibility of the loss of their livelihood. General Michael Collins, alarmed when Winston Churchill, Secretary for the Colonies, ordered the halting of the withdrawal of the troops from Ireland, sent a telegram to Churchill promising that every effort would be made to bring the culprits to justice. Three men were apprehended, but they were later released without charge.[49]

The barracks at Naas, which had remained the Depot of the Dublin Fusiliers throughout the War of Independence, was also for a time the home of a section of Black and Tans. The building which they occupied was separated with a barbed-wire fence across the square from the rest of the barracks. On 20 January 1922 a large number of Black and Tans were withdrawn. A couple of nights before they had given 'a big farewell dance at the Depot to their friends, and they had a very enjoyable smoking concert' on the following night. The local paper later noted that on the departure of the Tans 'many of the motors passing through Naas towards Dublin during the week bore good humoured inscriptions including "Goodbye to some of the best", "Sorry to leave you", and "Good Luck to Ireland"'. One individual had shrouded himself in a Union Jack for the farewells.[50]

Members of the departing forces were again harassed by irregulars on Monday 20 February when there were three ambushes on military motor lorries on the road from Dublin to the Curragh.[51] This caused the military authorities to reiterate that 'all precautions in practice before 11 July 1921 should still be in practice'.[52] That was the period of the Truce between the British Army and the I.R.A. These attacks, combined with a railway strike, halted the evacuation arrangements for a short time that month. But the transition of responsibility continued with the Head Bailiff of the Curragh, Joe McElveen, assuring the Minister of Defence that he wished to say at once that he would only be too pleased, as an Irishman, to give all the help he possibly could in relation to the area. He added that Captain H. Greer, Director of the Irish National Stud at Tully, Kildare, was also to contact the minister.[53] The Curragh Ranger was then George Wolfe of Forenaghts, Naas, who had served as a major in the Egyptian and Sudanese campaigns, and who was elected to represent Kildare in the Dáil in 1923.[54] Captain Greer and General Sir Bryan Mahon, as well as the Earl of Mayo, were nominated to the Senate in 1922.

The arrival and departure of regiments at the Curragh camp and Newbridge was reported in the newspapers, including the fact that when the Northamptons arrived from Templemore at Newbridge station they had with them a number of greyhounds. That fact prompted the remark that 'at one time the Curragh was a famous coursing centre.'[55] Of more concern now to the public was the fact that nothing was being done about the prospects of unemployment at the camp and at Newbridge. It was known that 'even in these early stages there was much suffering . . . it was absolutely necessary that some practical steps should be taken to save homes being broken up, idle men are now walking the streets'. Those men, it was said, 'were hard working men who had daily walked from Newbridge and other places to the camp to start work at seven o'clock in summer and eight in winter; they passed through Newbridge before dawn'. The urgency for action was reflected in a representative meeting held in the town which sent delegates to the *Dáil* to plead their case.[56]

But it was not all gloom and doom; the sergeants of the artillery during the last week of February gave a Fancy Dress Ball in their mess in Newbridge Barracks which was very successful,[57] while preparations for departure accelerated in the barracks. Orders were issued for the disposal of equipment and stores and specific instructions were given on the items to be retained by units, and those to be lodged in stores. The service pattern furniture and the crested china and glass from the officers' messes were to be returned to the ordnance depots. Two of the libraries in the Curragh Camp were to be handed over to the new army, while quantities of surplus stores were to be sold at public auction. Headquarters in Dublin had issued an order that no property was to be left to the depredations of the local inhabitants, and that anything left unsold at the auctions, with the exception of latrine buckets, would be destroyed.[58] The searchlight on the Curragh Tower was ordered to be removed and sent elsewhere, and notification was given that the Horse Transport Company was to be disbanded.[59]

A further instruction concerning the final evacuation of the Curragh issued from 6th Division Headquarters on 21 March. It included the requirement that 'all married families, both officers, non-commissioned officers and men [were to be] evacuated from the Curragh, Kildare, Newbridge and Naas by 15 May'. Naas barracks was to be evacuated about 22 May, the troops there being withdrawn to the Curragh, and the final evacuation of the Curragh and Newbridge was to take place within 48 hours about one week after the 22 May. At the conclusion of the evacuation no officers or men of any unit were to remain at the Curragh or Newbridge.[60]

General Richard Mulcahy, the Minister of Defence of the Provisional Government was reminding the Evacuation Officer, Commandant Dalton, on 26 March that 'in connection with the take over of the camp it is necessary to have our officers in constant attendance to inspect in great detail the plant and buildings to be received', and a list of requirements was attached.[61]

Men employed in the camp were endeavouring to clarify their status, and many of them sought re-engagement with the new army; 43 workers notified their willingness to remain with the engineers in the camp and at Newbridge, and 10 working at the hospital offered to stay on as caretakers for as long as was necessary. A list was compiled of the foremen who were recommended and willing to take over and act as representatives of the Free State Government, on the conditions of a stated remuneration, and of receiving three months notice of termination of employment.[62]

The shepherds on the common sward of Greenlands were also conscious of the changing times, and on 15 April 1922 it was announced that it had been decided that the sheep on the Curragh would in future be marked with a shamrock instead of a crown.[63] However, that chauvinistic innovation did not dispel Mr McElveen's fears for the future of the plain and he again wrote to the Minister of Defence on 9 May to say that there was a considerable amount of harm being done there, and that it was desirable 'that steps are taken to keep the Greenlands under some control'.[64]

The date of the final evacuation of the barracks remained uncertain, and was changed as circumstances required, but plans for the hasty evacuation of the Curragh Camp in the event of an emergency had been made. They ordered that, if things came to the worst, the Destruction Parties were to go into action; they would destroy surplus military stores, and blow up the magazine.[65] Fortunately that situation did not arise, and, finally, Tuesday 16 May 1922 was decided on for the evacuation of the Curragh. Already the strength of the garrison had been reduced to a minimum, and certain property, such as Ballyfair House, in the countryside one mile south of the camp and the residence of the General Officer Commanding, had been handed over to the provisional government. It was to be used as a base for the assembly of men to take over the camp at the appointed time.[66] Ballyfair had been emptied of its contents, and the new occupants had to purchase bedding from the store of Messrs Todd Burns in the Curragh Camp when they had formally taken over the property.[67]

The orders issued from General Headquarters in Dublin on 4 April for the evacuation of the Curragh confirmed the earlier ones, and were very detailed.[68] They specified that all Ordnance, Barrack, Royal Engineer and Royal Army Service Corps stores were to be moved to either Ulster or England, and that some 700 tons of stores and 132 vehicles were to go as soon as possible. The families of all ranks were to be evacuated out of the country by the 15th of the month. A date was fixed for the evacuation of the Royal Scots Fusiliers from the Hare Park Camp, the mess furniture was to go north, and a caretaker was installed. If it was found necessary that personnel of the Provisional Government had to be accommodated, the Hare park would be available. Plans for the evacuation of the Kildare barracks were that the 4.5" Howitzer Battery would move to Newbridge, and other dismounted personnel from Kildare and Newbridge would go to the Curragh. The furniture from the officers' mess and officers' married quarters was to be sent to Ulster, and the remainder of the equipment, the Barrack and Ordnance stores, with the exception of Regimental equipment, was to be handed over to the representatives of the new order. The barracks at Naas was to be the last in the county to be evacuated. It was then occupied by a company of the Leicesters, and they were to rejoin their battalion wherever it was stationed. The arrangements there for the disposal of the furniture and stores was the same as for Kildare barracks.

It was expected that at the end of April advance parties of all of the units remaining on the Curragh, with all the heavy baggage and mobilization equipment, would be despatched to England or elsewhere; the furniture of the officers' messes was again to go to Ulster. Both the general and families' hospitals in the camp were to be closed, and caretakers appointed. The supplies of fuel, food and forage in the camp were to be worked off to the amount required to feed the troops up to the day of departure, and allowing for two days preserved rations for their journey.

The conditions of employment of the selected civilians who were to take over the stores, etc. on behalf of the Irish government were specified; this arrangement was believed to be to the advantage of the Provisional Government in that they would get experienced men, but if the government wished to interview the employees, arrangements would be made. It was also appreciated that the continuation in office of the foremen, etc. would simplify the take-over for both the outgoing and incoming incumbents, the War Department and the Provisional Government.[69]

Further detailed arrangements for the transition of responsibility were given under such headings as Barracks, Ordnance, Royal Engineers, R.A.S.C., R.A.M.C., and fire-brigade. Up-to-date ledgers of stores, married families furniture and ordnance were to be available for the hand-over, and in the R.E. installations allowance was made for the manning of electrical, sewage and pumping stations;

it was planned that for those works the engineers which were to be provided by the Provisional Government would arrive in the camp at least one week before evacuation. The R.A.S.C. was to close its Motor Transport Shops, bakery and abattoir as decided on, and the R.A.M.C. was to transfer such patients as was necessary to the King George V Hospital in Dublin. At the fire station the horsed and hand manual fire engines were to be handed over, while the motor and steam brigades were to go to Dublin. Thought was given to the circumstances of the civilian employees who were to transfer to the new employers; it was hoped that they would be allowed to continue in occupation of their War Department quarters in the camp, and for which they paid rent. Five detailed schedules with the orders listed the civilian employees who were recommended to be retained, as well as the recommended military personnel necessary for the maintenance of the vital installations.[70]

By 15 April the Curragh married quarters were evacuated, with 200 women and children departing for England from Newbridge railway station. The military hospital and Kildare barracks were also evacuated.[71] Another month was to elapse before the Curragh Camp itself was finally cleared, but evacuation continued with a full battalion of the Scottish Rifles, with their total baggage, departing for Collinstown Camp in county Dublin.[72]

The transference of authority and personnel was proceeding against the ominous background of irregular and criminal activity. While the artillerymen in Kildare Barracks were having a last fling with a dance at the end of March the safe was stolen from the canteen, later to be found unopened in a field nearby.[73] On 1 April General Dudley had his car stolen from him while travelling on the road from Newbridge to Naas,[74] and two staff cars were stolen from the Curragh camp in the second week of April.[75] Even the famous racing festival at Punchestown was not respected when two armed men held up the driver and stole the Crossley Saloon of the lord lieutenant. In that same week it was planned that 800 men would be quartered in the barracks at Kildare to commence training as civic guards. They were hardly installed there when they came under attack from irregulars, and the guard on the barracks had to be strengthened.[76]

As the time of departure of the British forces approached the civilian residents of the Curragh area, and especially those of Newbridge, became more apprehensive, fearing a great loss of employment and a decline in trade. Various proposals were submitted, and deputations sent, to the Provisional Government. It was proposed that Newbridge should be made the headquarters for the army, and that the British government should be asked to retain all civilian workers pending the take-over. No less than 1,500 men were said to be then employed in preparing sporting grounds for the British troops. The Minister of Defence promised to investigate where the funds for that project came from, with a view to having the work continued. There was a fear locally that 1,100 men were to be discharged, and also that local enterprise, such as market gardening, dairy farming and carting would be badly affected.[77]

When the parish priest of Newbridge was asked how the withdrawal would affect him, he replied succinctly that it meant that he would be down £50 a week [from the loss of income as an officiating clergyman].[78]

County Councillor J. J. Fitzgerald said that there were 3,000 men unemployed in north Kildare, but the deputies admitted that they were 'somewhat barren on practical suggestions, but urged that it was an immediate problem for the government'. The establishment of a local industry was proposed, with the electric plant and machinery at the Curragh possibly being useful. However, it was feared that some of the departing troops might cause vandalism there, and an immediate inspection of the Curragh 'by heads of the I.R.A.' was recommended. It was thought that 25% of those people likely to be unemployed had no binding ties with the district, but that a far greater number were married with houses and families in the area.[79]

The movements of the British and Irish military were being closely followed by the journalists of the provincial press, and in the 6 May issue of the *Leinster Leader* the arrival of the 500 R.H.A. men from Kildare at Newbridge, and the departure of the 71st R.F.A. from there was noted. The reporter observed that 'there was considerable interest taken at *Droichead Nua* on Saturday evening when motors arrived with I.R.A. officers and men from G.H.Q., Beggars Bush, to occupy different barracks at the Curragh Camp. The motors remained in the main street for a considerable time while provisions were purchased by some of the men detailed for the purpose. The men seemed in excellent form and whiled away the time by waiting in the motors in a pleasant manner'. But the composure of the men may have hidden some apprehension as their colleagues at Ballyfair House had again been attacked, and reinforcements had to be brought from Dublin for the garrison there.[80] With the house secured, the officers and men stationed at Ballyfair gave a dancing party to guests from all over the county to celebrate the departure of the British.[81] In another incident, but directed at the departing forces, shots had been fired from a train after it had passed through Newbridge, no doubt intended to alarm the British soldiers on guard behind sandbags at the station and on the railway bridge there.[82]

The Final Orders for the evacuation of the Curragh were issued on 10 May; they specified that the evacuation would be worked from a zero hour of 10.30 a.m. on Tuesday 16 May. At 3 p.m. on Monday 15th Hare Park Pumping Station was to be occupied by Free State Troops, when the British guard would withdraw; from that hour Hare Park would be out of bounds to the latter troops, and the Curragh camp was to be placed out of bounds to the incoming army until half an hour before zero hour on the following day. Three guards for the vital installations (the magazine, electric light station and the pumping station) and three patrols on the roads within the camp were detailed by the Headquarters 14th Infantry Brigade for the morning of the 16th, and they were to remain in position until relieved by the Irish army. The possibility of any disturbance during the hand-over was to be offset by the availability of a picquet of one officer and 20 men and an armoured car 'in the event of reinforcements being required before zero hour on that day'. Civilians were to be banned from entering the camp until the British garrison had departed.[83]

On Thursday and Friday 11 and 12 May, components of the artillery held their sports at the fine sportsfield in Newbridge barracks, and on the first evening the R.A. of the Curragh and Newbridge hosted 'a very pleasant time to a large number of friends. The meeting was regarded as a farewell one on all sides'. On Friday the 142nd Battery R.F.A. attracted a large attendance also to their events. Describing the meetings the man from the *Leinster Leader* wrote 'it must be said that the artillery in this area have been a body whose relations with the people have always been of the most friendly nature. Both officers and men speak of their relations very highly and state that they carry with them the very kindliest recollections of their long connection with the town and district'.[84]

There was again speculation in the press on 13 May; while it was believed that the R.F.A. were under orders to leave Newbridge, it was rumoured that they might be kept there until after the coming elections for *Dáil Éireann.* In the meantime men from the locality of the town were joining the Free State Army and undergoing training in Beggar's Bush barracks, Dublin. Indeed it was even suggested that G.H.Q. might move from that barracks to Newbridge, a move which 'would be welcomed as the departure of the British troops was a great loss to business circles, and doubtless, with the army in residence in the Curragh trade will revive'.[85]

The arrival of the Free State troops to relieve the British soldiers at the appointed locations was arranged so that the latter men would march away as the former approached, and if for some reason the Irish soldiers did not arrive, then all of the British guards and patrols would be withdrawn at 10.30 a.m. Compliments would not be paid.[86] The incoming troops were to enter the camp along the main Hare Park to Ponsonby barracks road to establish detachments at specified places. Lieutenant Colonel Sir F. Dalrymple, the Commanding Officer, was to meet the General Officer Commanding Free State Troops (as he termed them) at Staff House, on the western end of the camp, and to remain at the Free State Headquarters until the evacuation of the remnants of the garrison, two platoons of the 1st Battalion Northamptonshire Regiment, was completed. The evacuation procedures had been discussed, and agreed to, with the G.O.C. Free State Troops.[87]

The occupation of Hare Park by an advance party of 80 Free State troops took place on the afternoon of the 15 May as planned; according to a later report in a local paper, when the men were about to raise the tricolour over the camp, they were prevented from doing so by the British. Also present in the camp that night was Lieut. General J. J. (Ginger) O'Connell, preparing for the historic encounter on the next morning.[88]

Finally zero hour came for the hand-over of the Curragh Camp; the British and Irish officers met at the Staff House as planned, and in the pouring rain proceeded to formalise the transfer of the camp from the Crown to the Provisional Government. There was no exchange of compliments between the guards, and the British soldiers marched off as the Irish approached. Groups of Irish and British officers proceeded to inspect the perimeters of the camp; then the two remaining platoons of the Northamptonshires formed up, the Last Post was sounded, and the hand-over was officially completed. There was delay before the colours of the new state could be hoisted over the camp as, before departing, the British had ordered the cutting down of the three flag poles in the camp. When two Board of Works men set about repairing them they were arrested. It was midday on 16 May before a pole was found to fly the flag which was raised by General O'Connell on the water tower while fellow soldiers saluted (Fig. 37). The only spectators present were Desmond Fitzgerald T.D., representing the cabinet, and a number of Irish and foreign journalists.[89]

The British evacuation of the camp, and the co-ordinated evacuation of the barracks at Newbridge and Naas, was effected by road in over 300 vehicles, and by rail, with special trains departing from the Curragh Siding and Newbridge. Newbridge barrack was formally taken over at 10.30 a.m. on 16 May also, following which 200 trainee civic guards from Kildare barracks were billeted there. Again there was no ceremony, except the change of guard, and everything 'passed off in a business-like manner'. A company of the Leicesters evacuated Naas barracks at the same time, without ceremony, but having removed the flagstaff, thus preventing the national flag from being immediately raised.[90] However, before they vacated the barracks in the Curragh that morning, a National Army lieutenant later recalled, the British soldiers blackened the fire grates, ironed the billiard tables and tipped the cues. He believed that the cutting down of the flag poles was an army tradition[91] (during the hand over at Athlone General Sean MacEoin also found that the flagstaff had been cut down, and the British contingent there did not return the compliments paid by his men).[92] That flagpoles were also cut down in Cork and Dublin was reported in contemporary newspapers. Brigid Tracy, the daughter of the Canteen Manager in Keane Barracks in the Curragh camp, recalled in 1994 watching from near her home on the Green Road for the lowering of the Union Jack on the camp water tower on 16 May 1922. She remembered that there was a delay before the tricolour was raised, and later she learnt that it was because the flagpole had been removed.[93]

The departing troops were to be accommodated in a number of locations in the Phoenix Park side of Dublin prior to embarking for England.[94]

The evacuation of the camp was fully covered in the national and provincial press. On 17 May 1922 the *Freeman's Journal* reported that as in the passing-over of Dublin Castle, there was no spectacular display at the Curragh: 'The significance of the event needed no emphasizing . . . there was no ceremony to mark the surrender of this, the greatest of England's military strongholds in Ireland.'[95] *The Irish Times* carried the story under the heading 'New Order at the Curragh', and a photograph of the presiding officers (Fig. 38). The camp was described as

Fig. 37. *A group of Irish officers saluting the newly raised tricolour in the Curragh camp on 16 May 1922.* Photo: *Joe Cashman* The Irish Times.

Fig. 38. *Approaching the Water Tower during the handover of the Curragh Camp, 16 May 1922 are Lieut. Col. Sir F. Dalrymple on the right, with a walking stick; behind, left with gauntlets, Lieut. General J.J. O'Connell, Deputy Chief of Staff of the Free State Army. The Dublin-registered car is an Overland Tourer.* Photo: *Joe Cashman* The Irish Times.

having accommodation for 12,000 troops 'but often many more thousands pitched their tents upon the ground. These now are gone, and what the loss involved by the removal of so many human beings means for the commercial interests of the district is a problem which the Irish Government will have to solve. The remedy is one that will require time to effect'.[96]

The ominous warning from *The Irish Times* was echoed in the *Kildare Observer* under the heading 'A Loss to the District', but in fact it was the withdrawal of the Curragh fire-engine which was lamented. It had often been called to fires throughout the county, including one at Naas in the previous January when the *Kildare Observer* had remarked that 'the townspeople are solely dependant on the military fire-brigade from the Curragh', before speculating as to what would happen on the withdrawal of the British troops.[97] In the *Leinster Leader* it was reported that the departing military convoy took an hour to pass through the town of Naas, and it was noticed that 'there were several machine guns at the ready position and many light armoured cars also in evidence'. Crowds of people had turned out along the road to Dublin to witness the passing of the British.[98]

The first collapse of prices in the county was reported in the local papers on 20 May 1922 when the price of forage had fallen. Hay was fetching only £9 a ton instead of the usual £14. Mr O'Callaghan of Cork, the chief buyer of hay for a number of years in the Curragh military area and who resided in Newbridge, had now no military contracts to fill.[99] The huge reduction in the number of horses would not alone harm the breeders and suppliers of fodder, but the arrangement by which the farmers could purchase the manure from the barrack stables would also cease. It was drawn away to all parts of the county, and a grower of potatoes at Two-Mile-House, near Naas, still frequently finds army buttons and clay pipes in his fields.[100]

Unemployment was being felt keenly in the district and the business fraternity of Newbridge let it be known that they felt that 'it was but right that the new Irish Forces in the barracks and Curragh camp should procure their provisions and necessaries as far as possible from the traders within the military area'. A local journalist believed that as the new army was then replacing the old, and more settled conditions were promised, 'it is to be hoped that an early attention to the needs of those affected will be forthcoming.'[101]

In the Curragh camp there were a number of unpleasant incidents; on Tuesday and Wednesday 23 and 24 May no civilians, even those in possession of passes, were allowed through the lines. They were being stopped at the Water Tower on the main county road, something which the British forces had never attempted, protested Mr H. Colohan, a member of the Kildare County Council. The blame was laid on the troops who had arrived from Beggar's Bush barracks in Dublin, and who had already caused offence when passing through Naas. Then they were seen to be rowdy, drunk, and shouting 'Up Collins', causing them to be described in the local paper as 'uniformed terrorists, who terrified the women and children, and a credit to the Black and Tans'. The County Council was to contact the Minister of Defence on the outrage, and Capt. Joyce, O.C. Naas Barracks, said that some men were already under arrest, and that military discipline was being taken against others.[102]

The Civil War caused a major increase in the strength of the units in the Curragh, and by 3 June it was reported that '300 I.R.A. men', with officers, from the south had arrived for training.[103] Three battalions (29th, 43rd, 54th) were formed at the Curragh.[104] As the training of the men in the camp progressed, military order was also being introduced. By the middle of July the hospital was functioning, under the command of 'a well known Irishman, lately a colonel in the R.A.M.C.'.[105] The welfare of the troops was not neglected, and in August the management of the Sandes Soldiers' Home, which was to continue its benevolent work, appealed in the local paper for illustrated papers, magazines and story books for the troops, adding that 'we have a large reading room crowded with soldiers who are glad of something interesting to read.'[106]

But there were problems also. A refugee family from Belfast had occupied quarters at number one South Road in the Curragh camp, and even though the tenant was said to be employed as a clerk of works, there was official doubt of his status. He eventually vacated the quarters.[107] As late as August 1922 two unauthorised females were in residence in quarters in Gough Barracks, and when they had departed the house was being kept under the surveillance of the military police.[108] Nor was the accounting for furniture and other effects handed-over on inventories by the British proving to be satisfactory; again at Gough barracks, furniture was missing from the serjeant's mess and from the married quarters, and 'practically all the sheets, blankets, and messing and kitchen utensils had disappeared'. The quartermaster admitted that he did not know where they had gone.[109] A few months later it was recommended that the camp should be designated a military area, and that nobody should have access to it unless they were holders of special permits, and which would be renewed weekly.[110]

By August 1922 the employment of civilians in the Curragh and Newbridge barracks was proceeding. A story in a local paper on 12 August was that a large number of the new military forces on the Curragh were members of the Irish Transport and General Workers' Union, and an unemployed man said that every member of the union available was now in good employment either in the Curragh or at *Droichead Nua* barracks, the majority in connection with the military. The loyalty of the workforce to the army was demonstrated on Wednesday 23rd when a large number of them ceased working for two hours to mourn the death of Michael Collins.[111]

Before departing the Curragh the officer commanding the Royal Engineer Depot had given instructions that all works on hand were to be completed, and those included 30 houses on the outskirts of the Curragh, at Newtown, Kildare, Kilcullen, Kill and near Newbridge. The houses were described as 'large double two storey houses' and they were completed and occupied by 17 June. A number of houses built to contract at Artillery Place, Newbridge, had, as the local paper put it, 'since the departure of the military been occupied by civilian families with the result that to some extent the great congestion is being relieved in the town and neighbourhood as far as housing is concerned'.[112]

From her Home in the Curragh Elsie Sandes oversaw the closure of 13 of her institutions throughout the country, with only three in the Free State remaining open, including the Curragh. When the new army had moved in there the commander in chief had requested that it should remain, and so it did. For the one evening when soldiers from both the British and Irish armies were stationed there together Miss Sandes, from her window, saw a British drummer boy approaching, and coming towards him, a Free State drummer boy. As there had been anxious days waiting for the change-over she feared what might happen when the boys met: 'For a few seconds the boys faced each other. Then the British boy, with a friendly grin, held out his hand and the lad who had recently been his enemy came forward and took it.' Miss Sandes turned to a friend and prophesied 'all will be well, the two armies are going to be friends'.[113] But the quality of the men in the new army was being questioned by a journalist in the *News of the World* of 18 March 1923. It carried a report that 'a lunatic who escaped from Mullingar Asylum was discovered in surprising circumstances. After much searching he was traced as having joined the Free State Army at Birr. He was found there, drilling a squad of some 30 men in the barrack square. These facts were reported by Dr Gavin to the Asylum committee'.[114]

Naas was also suffering from a decline in trade, and in March 1922 the Urban District Council asked the Minister for Labour to receive a deputation in connection with the number of persons disemployed in the district on account of the evacuation of the military barracks in the town. Further appeals elicited some monies for public works in the urban area.[115] Again, due to the Civil War, a new battalion was formed at the town barracks. It was the 33rd Battalion, 350 strong.[116] The influx of men again benefited the economy, but the ending of hostilities necessitated the reduction in numbers of the army, and in September 1923 began the process of bringing the force down to a peace-time strength; by 1927 the total strength had fallen from 48,000 in 1923 to 11,500.[117]

That the hoped for increase in business in Newbridge did not materialise is evident from accounts of the serious disquiet of the Town Commissioners and other residents in 1924. The ultimate fate of the barracks had been postponed in August 1922 when, during the Civil War, an internment camp for about 1,200 Republican prisoners was opened there. The camp was closed before November 1923 and the barracks was left in the care of the Board of Works.[118] On 17 April 1923 the chairman of the Town Commissioners sent a case to the Minister of Defence expressing the opinion that the barracks should be utilised, and reminding him of the deputation which was received by the government in February 1922. The Commissioners' submission emphasised that the Curragh, which accommodated from 7–8,000 men, and the barracks at Newbridge and Kildare, had led to the development of numerous trades and industries. Building and allied trades, shop-keeping, dairying, mineral water manufacture, market-gardening, laundry works, bakeries, and practically everything connected with the business of supplies came into existence and did a remunerative trade. The population of the town had grown to some 3,000, to supply the wants of the army and to fill the numerous civilian appointment in the camp. Now hundreds of the working-class, skilled and unskilled, were unemployed, and there were no local industries in which they might find work. The position of farm labourers was especially bad, and their housing conditions poor. Immediate action was being sought, and the removal of army headquarters to the Curragh camp recommended.[119]

A table attached to the Town Commissioners' submission showed that in the Curragh camp 353 men were normally employed by the Royal Engineers, and up to as many as 700 at maximum. Their wages averaged £2,500 a week. A further 50 to 70 men were employed by the Army Service Corps, and the same numbers by the Ordinance; another 50 were employed by the Mechanical Transport, totalling, in all, between 500 and 900 men [this last figure was the number employed there during the Great War]. Another 300 were employed by the Board of Works, with about 280 in the Curragh and the remainder between Newbridge and Kildare.[120] On 5 May 1923 the *Freeman's Journal* took up the matter, reporting that 'notices of service no longer required' had been issued to between 300 and 400 men. Public meetings were being organised to protest against the government's actions, which were also causing dismay in other military stations throughout the country.[121]

By Christmas 1924 nothing had been done to relieve the situation in Newbridge, and again the chairman of the Town Commissioners wrote to the Minister of Defence, attaching a letter from the parish priest, Fr Laurence Brophy. A memo with the submission emphasised that the town was entirely dependent on the troops, and that the business of the town was now gone. Supplies for the military were being brought in from outside the area, and local caterers were not being given an opportunity to tender. Over 600 people were consequently unemployed, with 150 on Home Assistance. In former times, it was claimed, local conditions were always considered in the allocation of troops, and the Newbridge barracks was always occupied even when there were very few troops on the Curragh. 'If something was not done immediately,' the chairman of the Town Commissioners warned, 'the government will be faced with a starving people and a derelict town.'[122]

The parish priest also expressed grave indignation at the treatment of the town by the new government. He pleaded that 'the honest traders of the town were left idle, and almost reduced to pauperism, while a few men in Dublin had suddenly become immensely rich. Surely,' he asked, 'it should be possible for Kildare farmers to compete with those suppliers from the city who sent vegetables to the camp?' The deteriorating condition of the empty barracks was also a cause of annoyance, and Fr Brophy made the point that if the barracks was good enough for the British Army, why was it unsuitable for the National Army: 'it may not be up-to-date in every modern accommodation, but is an army entitled to the latest luxury? If you take the rank and file of the army, they are taken from very humble homes, and I venture to say, that if you take our middle class over the country, the housing accommodation in Newbridge barrack is far superior to theirs. Good drainage, gas and water from the town supplies, nothing wanted.'[123]

In January of the new year 1925, the quartermaster-general was briefing the Minister of Defence on the contracting system for the Curragh: he held that 'practically all our requirements of milk and vegetables were supplied by local residents'. The tender system was explained, especially the fact that if local traders were offering terms equal to those of outsiders, the former were given preference. That the quartermaster was indignant at the priest's allegations of favouritism was obvious, and he hinted that political factors were involved. Eventually the Minister of Defence made an explanation to the Executive Council, refuting the various complaints. The names of the suppliers of milk and vegetables were given, and they were all living locally. Again the tender system was explained and defended. As to the future of the Newbridge barracks, it was known that the internees had badly damaged the buildings while incarcerated there, and that essential repairs would cost £6,000; in any case the Defence Department did not require the barracks, and the Board of Works had been requested to make whatever arrangements were possible for the use of it, or for its disposal. The status of the married quarters was also the responsibility of the Board. As to a proposal that employment might be given in repairing the roads on the Curragh, the situation was that, with few exceptions, they were the responsibility of the Kildare County Council, though it was admitted that there was some doubt about the section of the main road through the camp, and the Attorney-General was to advise on that matter.[124]

By 1926, 28 of the houses in the married quarters in Newbridge barracks had been taken over by the Commissioners and plans were being made to acquire another 30.[125]

That the new state was already competent in ignoring or shelving problems, and that the once prosperous population in the vicinity of the Curragh did not merit sympathy, is apparent from a reply which the Minister of Defence gave to a Parliamentary Question on 3 March 1927; asked if he was aware that a large number of plot holders and labourers with gardens had derived a livelihood from the sale of vegetables and fruit in the Curragh District, and if he would consider the possibility of establishing a market at the Curragh for such commodities before any new contracts were arranged, he replied that (a) he was not aware of the extent to which the persons in question formerly obtained a livelihood from their trade in the camp area, and (b) that as the contract referred to had already been arranged, it was not possible to establish a market there.[126]

The depression of those post-evacuation years is still remembered by the older townspeople of Newbridge. The fact that there were 'a lot less soldiers' in the Curragh and a 'big fall off in trade and business in the district' was recalled by a Curragh resident and journalist in the centenary supplement of the *Leinster Leader* in 1980, and a few years later in a similar supplement to the *Nationalist and Leinster Times* (1983) the devastating affects of the British withdrawal were noted, as well as a futile attempt by some local people to keep a British army based on the Curragh. The fact that business in Naas was likewise affected, and that the U.D.C. sought a meeting with the Minister of Labour was also mentioned.

In recent years a Newbridge business man reminisced on the boom years of the Great War when there were huge numbers of troops in the town and the camp, with constant marching to and from the Newbridge railway station. Newbridge, he recalled 'was a grand busy little town when the British Army was there. One lived off the army; no matter what one had to sell the army bought it, from a horse to a chicken. One could poach a couple of salmon or shoot pheasant, anything, the army would buy it, hay, straw, logs. There was no fear in the town or never any violence, except an occasional brawl from over-drinking. But there was no sectarianism, they were part of society and they were welcome.'[127]

An old countryman remarked that 'many people thought it a sad day they [the British army] left the Curragh . . . many people knocked a good do out of them . . . soldiers were easy codded, not well up. Newbridge missed them very much'.[128]

An aged widow's memory was that Newbridge:

> was a lively place in those times, as my late husband could relate, he being born in Canning Place. He used often to deliver vegetables and rabbits and poultry to the military, and told many a tale of Newbridge on Saturday nights. The soldiers would come out to town for a bit of 'diversion', and many from the Curragh Camp. They were okay if they stayed in the former King's Arms, the military pub, but space was limited so many proceeded down to the pubs at Lower Main Street [public houses were sometimes placed 'out of bounds' to the troops] and many a fine fight would ensue with the bogmen, quite often someone would end up in the river, but they must have pulled each other out because no one was drowned, but sometimes a Highlander would go home without his kilt. But there was no great malice. There would be a large audience at the wall across the street watching for the fights to spill out or be evicted on to the pavement, and much cheering for our stalwarts from below the college who were equal to the largest Scotsman.[129]

A former Town Commissioner of Newbridge described the town in the 1930s as 'a rather poor place and lowly populated, due to the departure of the British troops from the barracks ten years previously', but he also acknowledged that while livelihoods were affected, 'the British left bitter memories for most with their departure from this country in 1922, and a popular saying on people's lips was "burn everything English except their coal"'.[130] If the commissioner's views were the popular ones they are not reflected in other opinions expressed then, nor were they evident in other reminiscences of the British occupation as already quoted in this book. Indeed, another old resident who could remember the evacuation of the barracks said that it had 'made a ghost town of it. The population was about 2,000 and we had no industry and a lot of unemployment'.[131] The widow already quoted recalled that 'the town was in low water when the military left in '22, no supplies being needed for the barracks from the farms, cattle, fodder for the horses and mules, a great deal of coal and coke and goods from the local shops, also milk, these could not be sold. The jarveys with their side-cars were also badly hit, they had done much trade plying to and from the station which was very busy, then also to the camp, the race-course and the barracks in Naas.'[132]

That even the better-off members of the community suffered from the evacuation was noticed by the sporting journalist Stanislaus Lynch decades afterwards when he wrote 'in Kildare, as in many other parts of Ireland, hunting had always been supported by army officers who were members of the huge British garrison stationed in Ireland. In addition to the numerous cavalry and artillery regiments stationed throughout the country, many officers of infantry regiments also kept their own horses and hunted regularly. When the war started, these were called up and many never returned'. *Irish Life*, in its edition of 17 February 1922, predicted that 'the fine golf course on the Curragh will be something of a white elephant now that the British troops are leaving'. As to cricket, a devotee of that game believed that 'the withdrawal of the British army in 1922 was to have a devastating impact on cricket in Kildare from which it never fully recovered'.[133]

Lionel Fleming, a distinguished journalist of Anglo-Irish background, expressed a sympathetic view of the garrison: 'They [the soldiers] regarded the pacification of Ireland as a job to be done; they had no interest in the country and had none of the hate for their opponents which their opponents had for them'.[134] Indeed, evidence from the memoirs of soldiers who served in Ireland quoted in this book would confirm that as far as the gentlemen were concerned it was the sporting and social aspects of the country which they most understood and remembered. As William St Leger Alcock, Royal Welsh Fusiliers, had confided to his diary while stationed in Naas in 1832, Naas was 'the stupidest place imaginable, no one but Lord Mayo living in the neighbourhood':[135] the social difference between the officers and the people created just as much of an oblivion of the ordinary folk and their concerns for the officers as it did at home in England. The perception of the rank and file of the army of Ireland and the Irish is less easily defined; from the evidence presented here it is clear that most of them adapted to life in the garrison towns just as they would in any other country. They sometimes married local girls and settled here, but the majority soon moved on to yet another station. Their interest in the political, economic or social life of the the country would have been minimal.

Nora Robertson, the daughter of an Anglo-Irish Royal Artillery officer, remained with her husband in the new State. This is how she recalled the perception by her class of the new army: 'The Anglo-Irish vividly remembered the Mecca of the pre-war Curragh, and the smart old days. Unfair and unkind comparisons would have been made [with the Irish army].'[136] How right she was! One smart quip circulating in certain circles reflected on the social standing of the new occupants of the officers' messes on the Curragh. In the knowledge that a small number of ex-British officers had joined the new army[137] it was said that they had been left behind to introduce the Irish officers to behaviour and procedures in the Curragh messes.

That there would have been disappointment amongst some sections of the community at the loss of the social aspects of the British army persisted into the 1950s when, as a young officer enjoying a party in Ceannt Officers' Mess (the former Beresford barracks) in the Curragh camp, this writer was ingeniously told by a middle aged lady from the locality that 'the camp was never the same since the military left'. But the great lines of barracks on the Curragh remain as testimony to the 70 years of British occupation, while on the outskirts of Kildare town red brick villas with such names as Simla and Lucknow are faint reminders of colonial times, and of residents who had served in such exotic places as well as on the hallowed plain of Kildare.

Epilogue

Following the departure of the British army, and the ending of the civil war, the population adjusted gradually to self-rule. In the former garrison areas where troops of the Free State army were now quartered there was a short period of financial respite before the strength of the force was reduced from 60,000 to half that figure. The census of 1926 noted that 'the changes in the population of what were formerly British garrison towns will also be observed. The decline in the populations of Fermoy (34%), Newbridge (34%), Kildare (20%) and Tipperary (16%) were due principally to the exodus of the British military.[1] Naas on the other hand, showed an increase in population, due to the temporary presence of Free State troops there. The closure of barracks and the loss of revenue was to change drastically the status of some towns, and in county Kildare this was especially so.

At Naas barracks in July 1922 there was a brief burst of excitement when it came under fire from anti-treaty forces. However, prospects for the future of the barracks improved after the civil war when it became the station of the 33rd Battalion, which had a strength of 520.[2] In May 1923 the officers entertained 100 guests at a dance in the Town Hall, with music by the Domino Jazz Band and catering by Mrs Lawlor.[3] However, with the re-organisation of the army in 1926 that battalion was disbanded, and a small signal unit from the Curragh was then the only occupier of the barracks. Two years later it was handed over to the Office of Public Works.

From that time until the Emergency parts of the barracks were leased to various industrial firms, and the officers' mess became a vocational school. In 1934 the Naas Urban District Council acquired the other ranks' married quarters as local authority housing, while the commanding officer's house became the dispensary doctor's residence.

During the Emergency small detachments from the Curragh again occupied the barracks, and in 1945 it became the base of No. 1 Depot of the Construction Corps, and so remained until the Corps was disbanded in 1948. A caretaker detachment from the 3rd battalion on the Curragh then remained there until the Army

Apprentice School opened in the partly refurbished buildings in 1956. It was named Devoy Barracks, after the locally born Fenian.[4] Today the school has a strength of 140, and ten civilian teachers.

At *Droichead Nua*, following evacuation, the barracks was occupied by the new civic guards, the Garda Siochana, until August when it was converted into an internment camp for Republican internees. In May 1923, with the Civil War over, the warrant officers and sergeants of the 54th Infantry Garrison Battalion which was based there hosted a dance in the Picture Palace, 'the first', it was noted in the press, 'since the town passed to the soldiers of Ireland'.[5]

In 1926 the married quarters were leased to the Town Commissioners for local authority housing, and in the following year the garrison church assumed the role of a Town Hall, which it still retains. An appropriate reminder of the origins of the building and of the town is an inscription over the door recording that on 9 March 1859 Lieut. Gen. Sir James Chatterton, Commanding the Cavalry Brigade in the District, had laid the foundation stone for the church.

By 1939 the barrack sports ground was transformed into a gaelic football pitch, reportedly with many of the old British army cannon being buried beneath the sodding. During the 1930s also parts of the barracks were converted to industrial use, thus again making it an economic bonus to the townspeople.[6] Today the site remains a business and industrial one, with most of the barracks gone, except for the Reading and Recreation Rooms building of 1880 which has become the administration block of *Bord na Móna* Headquarters.

In Kildare the barracks was pressed into service immediately following evacuation with the despatch there by the Provisional Government of several hundred men to be trained as civic guards. They were transferred to Newbridge barracks in the summer of 1922, following which the hutment barracks remained vacant until January 1925 when the 1st Artillery Battery transferred there, from Marlborough barracks in Dublin. By 1942 most of the old huts had been dismantled, and new barrack buildings completed. Named Magee barracks in 1939, to honour Sergeant Magee, a hero of the battle of Ballinamuck in 1798, it continues as the principal base of the artillery corps.[7] During the Emergency up to 1,000 men were billeted there, but today the strength of the garrison is 245, with 81 residents in married quarters.

In 1922 the situation and scale of the newly reconstructed training camp on the Curragh dictated that its future function should be an important one for the new army. The Army Census taken in November of that year returned a military strength of 3,608 for the camp, which was officially designated a Training Camp in 1924.[8]

Writing in the *Handbook of the Curragh Camp* in February 1930 Major General Aodh MacNeill, General Officer Commanding Curragh Military District, said that from its take-over from the evacuating forces in May 1922, 'the camp was forthwith established as the main training centre for the Irish Army. As such it has a played a vital part in the development of the army, and it is destined to play an even more important part in the future'; the camp was to be 'the service home of the regular army . . . it is here that these traditions of service, of efficiency, and of discipline and loyalty will be maintained and developed . . . it is here that the standards for the whole army will be set'.[9]

However, before the traditions which General MacNeill had expressed were established, several serious events occurred in the camp. As the Civil War continued seven men were executed there just before Christmas 1922,[10] and in April of the following year, 70 internees tunnelled their way out of the Tintown compound.[11] Nevertheless, early in February 1923, what was described as 'the first marriage of importance since the Free State was established took place in the R.C. garrison church'. Sergeant Major Kelly, a native of the Curragh, married a local girl of the same name, with a guard of honour of non-commissioned officers and men; the honeymoon was to be spent in Paris.[12]

On 17 March 1923 the Commander-in-Chief, General Richard Mulcahy, came to the Curragh to review a march past of battalions on the sward. Afterwards he addressed the men, emphasising that 'it was the first St Patrick's Day that the Curragh of Kildare was in the hands of the Irish army'. Even the *Kildare Observer* was impressed by the ceremonial, reporting that 'a large crowd turned out for the celebrations . . . the parade was carried out in every detail with almost amazing efficiency on the part of an army which has been so recently created'.[13]

By September it was considered safe to reissue passes for the camp to the traders who had been excluded since the take-over.[14] Then, later in the autumn, what has been described as 'the only true mutiny at the Curragh' occurred.[15] It was the ultimatum, from the old I.R.A. element within the army, demanding that the demobilisation which was intended to reduce the strength of the force should be discontinued. The fact that some ex-British army officers were to be retained in service caused further annoyance. In April 1922 Emmet Dalton had invited veterans of the Great War to report to the Curragh to raise and train an army for the new Free State. Dalton, who had been awarded a Military Cross for service as a major in the Dublin Fusiliers during the war, had subsequently joined the Irish Republican Army. He was appointed General Officer Commanding Cork in July 1922.[16]

Officers who had completed the officers' training programme in the Curragh camp were amongst the first group to be demobilised, and on 9 November seven officers refused to accept their demobilisation papers. They were arrested and court-martialled, causing the unrest to spread to other officers at the camp. The mutiny really began early in March 1924 when two officers presented an ultimatum to the government demanding reorganisation and the removal of the Army Council. Before the mutiny had ended 55 officers had absconded with their weapons and ammunition, and 49 others, including three major-generals and five colonels, had resigned.[17]

Just a month after the commencement of the officers' insubordination in the Curragh another bizarre incident occurred. There was still a number of republican internees in Tintown camp, and it was suspected that a military policeman named Corporal Joseph Bergin was carrying information to sympathisers outside. In the second week of December Bergin's body, which had five bullet wounds, was found in the canal at Milltown Bridge. Subsequently an army Intelligence Officer was tried for the killing and imprisoned, where he died.[18] A memorial to Bergin was erected on the bridge in 1938 by the National Graves Association.[19]

The departure of the officers and gentlemen coincided with the decline of the 'nobility and gentry' in the county. Colonel Richard St Leger Moore of Killashee had died by his own hand in October 1921 at the age of 73. In reporting 'the tragic suddenness' of his death the *Leinster Leader* said that it was 'the passing of a valuable link between the old county family traditions and the more democratic thought of today'.[20] Killashee became a convent, as soon also did Moore Abbey, the mansion of the earls of Drogheda at Monasterevan. The ducal house of Leinster fell on bad times in 1922, and from which it never recovered, when the young sixth duke died, and his heir was confined in a psychiatric hospital.

As to the 'unfortunate' women, there is no mention of them to be found in the pages of the *Leinster Leader* for the troubled year of 1923, but their presence in Newbridge about that time was remarked on by a Free State officer. The changed circumstances of the military, from that of hardened colonial soldiers to idealistic young men, mainly of rural background, would have reduced such fraternisation. Nevertheless, as elsewhere throughout the country, as the authority of the Roman Catholic church increased and the façade of a high moral tone was adopted, the existence of prostitutes would have been denied. But it is certain that just as the ancient profession of soldiering prospered, neither did what is often termed the 'world's oldest profession' disappear. As late as the 1960s, two unauthorised females were evicted from an Emergency vintage pillbox on the plain, where they had taken up residence.

By 1928 the strength of the camp had fallen to 2,019, and the names of the Anglo-Irish soldiers which had graced the various barracks in the Curragh had been removed. Also obliterated were most of the decorative emblems of empire and regimental insignias, though the armorial shield bearing the British royal arms remains on the Water Tower.

The barracks were renamed and all of the men thus honoured were the signatories on behalf of the Provisional Government of the 1916 Proclamation.[21]

Keane barracks was re-named Pearse barracks.
Gough barracks as MacDonagh barracks.
Engineer barracks as MacDermott barracks.
A.S.C. barracks as Clarke barracks.
Stewart barracks as Connolly barracks.
Beresford barracks as Ceannt barracks.

By 1939 the strength of the army had been reduced to 19,791, but during the following Emergency years it was greatly increased. In 1942 it was 38,787, with 4,405 men in the Curragh camp, and inevitably there was a resurgence of trade in the neighbouring towns and villages. [22] Again, the internment camps in the Curragh were re-opened, this time to hold not only Republicans, but also Allied and Axis soldiers who were fortunate enough, as a result of shipwreck or aeroplane crash, to find themselves in a neutral haven.[23] During times of I.R.A. activity in post-war years the confinement of individuals in the camp continued to be resorted to.

Territorial reorganisation in 1977 designated the camp as the base for the headquarters of the Curragh Command, which consists of the counties Wexford and Carlow, and parts of counties Kildare, Wicklow, Laois, Offaly, Kilkenny and Waterford. The military college and the majority of the army schools are also located in the camp.[24]

The military strength of the camp in February 1996 was 1,738. Civilian employees numbered 139, and there were 885 residents of married quarters, giving a total day-time population of 2,762. Just as the camp in its heyday experienced a population increase seasonally, so it does still, but on a more modest scale, when twice each year about 350 men proceeding overseas on United Nations service muster there for a months training (Figs. 39 and 40).[25] The old problem of safety on the Ballysax to Donnelly's Hollow road during the range practices has yet to be resolved to everyone's satisfaction.

The fabric of the camp had received its first major rent in April 1923 when the wooden Roman Catholic church was accidentally burnt.[26] Subsequently services were held in the Protestant church, while the much reduced congregation from there was accommodated in the smaller red brick Wesleyan chapel. A new magazine was enclosed on the western fringe of the camp, beside the fabled Gibbet Rath, in 1937, and three years later the old outdoor swimming pool was roofed. But the camp retained its Victorian character, even if it grew shabby, until the 1950s. Though the Roman Catholic Soldiers' Home was closed and demolished about that time, an elaborate Marian Shrine was made, and a large new Catholic Church was erected, obscuring the once dominant clock tower. Also constructed were dining halls, a primary school and a gymnasium. Other ranks' married quarters were built outside the perimeter of the camp. A small number of billets were built to house the female soldiers recruited in 1980.

One of the last two remaining institutes of 'the good old days', Sandes Home, closed its doors in 1986, but the Wesleyan Home continues to cater for the troops. So does Darling's Hairdressing Rooms, the only pre-evacuation commercial premises left in the camp, following the demise of Messrs Todd Burns and Messrs Easons (see Fig. 26) in recent years.

As ranges of buildings, such as bath houses, laundries, stables and married quarters became redundant they were closed, but for many years allowed to

Fig. 39. *N.C.O.s of the 3rd Infantry Battalion outside the Orderly Room of Connolly Barracks in the Curragh Camp, March 1996. In the centre Sgt. Major C. Flynn, with on his right Sgt. N. O'Driscoll, and on his left, C.Q.M.S.T. James. The latter's paternal grandfather served in a Welsh regiment, and after 1922 in the Military Police in the Irish Army. His maternal grandfather served in the Connaught Rangers, and afterwards in the Irish Army where he was known as 'Rajah'.* Photo: *Pat Foley.*

Fig. 40. *The 72nd Cadet Class, composed of 30 Irish and 10 Zambian students, drilling on the square of the Military College, Pearse Barracks, Curragh Camp, March 1996.* Photo: *Paddy Walsh.*

moulder. Now they are being either demolished or refurbished. In a former Recreation Hall a library has been installed, while a staff officer's married quarters has been converted into the United Nations School. The clock tower, the only original building of consequence remaining, has been restored and again the chimes are heard throughout the camp. If the cricket and tennis clubs have closed the golf and rugby clubs are prospering in new premises.

In the current re-organisation of the Defence Forces it is planned that the regional Commands will be abolished, and the Curragh camp is designated as the Defence Forces Training Centre with a strength of over 700. As the plans come to fruition, it is expected that considerable resources will be made available for the improvement of this unique and historic place, and that the Curragh of Kildare will continue to welcome the soldiers, as it has done for centuries past.

However, the plain itself has suffered considerably from development and over-use in the last 70 years. Until 1961 it was in the care of the office of Public Works, and since then, under the Curragh of Kildare Act of that year, it has been in the stewardship of the Department of Defence. That Act repealed the 1868 Act and parts of the 1870 Act, and the Bye-laws of 1964 superseded the laws of a century earlier.

Even before the making of the motorway (M7) across the sward, and which it was feared could damage the aquifer which feeds Pollardstown Fen, much change had come on the plain. The unprecedented enclosure of a large area at the racecourse, and the tree planting within, has not alone affected the open aspect of the landscape, but also the quality of the sward. The gradual increase in the numbers of horses (over 1,200 in 1995) in training on the plain has led to damage to the turf, as has the series of long artificial gallops with their accompanying roadways. Similarly, in other areas, the surface of the plain has been destroyed by the tracks of military machines and the intrusion of other vehicles. Over-grazing by sheep, and in particular the planting of trees in areas other than in the camp itself, has further contributed to a deterioration of the landscape. The common problems of litter and dumping are widespread. The cemetery too has suffered; the mortuary section of the lychgate was dismantled in the 1930s, and confusion over responsibility for maintenance has led to its neglect.

The classification of the entire Curragh of Kildare as a National Monument in 1995, and the proposal to designate it a National Heritage Area, must surely herald a more concerned approach to the use of the celebrated sward. With the good will of the sheep-owners, the horse owners and the military, combined with the interest of the County Council, local residents and the public at large, it is to be hoped that the official confirmation of the scientific and archaeological importance of the plain will ensure that what remains of St Bridget's open pastureland will be respected, and its uniqueness safeguarded for generations yet to come.

NOTES

Chapter 1

1. Griffith, Richard. *Valuation, County Kildare. Union of Naas.* (1853), p. 177–8. 182. Griffith gives the area of the plain as 4,885 acres, 2 roods, 3 perches.
2. Petty, Sir William. *The County of Kildare.* (London 1685) (Reprinted Newcastle-upon-Tyne 1968).
3. Noble and Keenan. *Map of County Kildare,* 1752. Edition Irish Georgian Society No date.
4. Walker, Henry. *Map of Curragh of Kildare,* (Dublin 1807).
5. 31 George 3, *c.* 38 April 1791.
6. O'Flanagan, Rev. M. (Editor) *Letters collected during the progress of the Ordnance Survey. County Kildare. 1837.* (Part II. Bray 1930). Part II. p. 73.
7. Ibid., p. 72.
8. Brewer, J.N. *The beauties of Ireland.* (London 1825), p. 37.
9. Mitchell, Frank. *The Irish landscape.* (Dublin 1976), p. 163.
10. *Kildare County Development Plan,* Third Revision. Part I. (1985). p. 44.
11. Feehan, John and McHugh, Roland. 'The Curragh of Kildare as a hygrocybe grassland', In *The Irish Naturalists' Journal* Vol. 24, No.1. (1992), p. 13.
12. Lewis, Samuel. *A topographical dictionary of Ireland. Part II.* (London 1837), p. 81.
13. Rawson, T.J. *Statistical Survey of County Kildare.* (Dublin 1807), p. 121.
14. 27 Edward 1. 1299.
15. Curragh of Kildare Act, 1868. Award of Commissioners. (Dublin 1869). p. 1.
16. Ordnance Survey Townland Survey of the County of Kildare. 6" = 1 statute mile. Ordnance Survey. Sheets No. 22, 23, 28. 1909. Sites and Monuments Record County Kildare. Archaeological Survey of Ireland. O.P.W. (1988).
17. Ordnance Survey, Sheet 18.
18. Ordnance Survey, Sheet 17.
19. O Riordáin, S.P. 'Excavation of some earthworks on the Curragh, County Kildare', In *Proceedings of the Royal Irish Acadamy.* Vol. LIII, Sec. C. No.2. (1950), p. 272.
20. Ibid., p. 250.
21. Ibid., p. 269.
22. Ibid., p. 258–72.
23. Wailes, Bernard. 'Excavations at Dún Ailinne', In *Journal of the County Kildare Archaeological Society.* Vol. XV, (1962), p. 353.
24. Smyth, A.P. *Celtic Leinster.* (Dublin 1982), p. 3.
25. O'Flanagan Rev. M. (Ed). *Letters collected during the progress of the Ordnance Survey, County Kildare.* 1837. Part II. (Bray 1930), p. 73.
26. Ibid., p. 77.

27. Ibid., p. 65.
28. Ibid., p. 66.
29. Mac Lysaght, Edward. *Irish Life in the 17th century.* (Cork 1950), p. 360.
30. Pochin Mould, D.D.C. *Saint Brigid.* (Dublin 1964), p. 41.
31. In *All the year round.* (London 25 May 1867), p. 521.
32. Andrews, J.H. *Irish Historical Towns Atlas. No.1. Kildare.* Royal Irish Academy (1986), p. 1.
33. Fitzgerald, Lord Walter. 'The Curragh: its history and traditions', In *Journal of the County Kildare Archaeological Society.* Vol. III, (1902), p. 3.
34. Joyce, P.W. *Irish names of places. Vol. 1.* (Dublin 1895), p. 463.
35. O'Grady, J.S. 'History and antiquities of the Hill of Allen', In *Journal of the County Kildare Archaeological Society.* Vol. IV, (1905), p. 455.
36. De Courcy-Wheeler, Capt. H.E. 'The tower on the Hill of Allen', In *Journal of the County Kildare Archaeological Society.* Vol. VII (1914), p. 413.
37. Joyce, Ibid., p. 115.
38. Connolly, Sean, and Picard. J.M. 'Cogitosus: Life of St Brigit', In *Journal of the Royal Society of Antiquaries of Ireland* No. 117 (1987), p. 5–27.
    Connolly, Sean. '*Vita Prima Sanctae Brigitae*', In *Journal of the Royal Society of Antiquaries of Ireland* No. 119 (1989), p. 5–49.
    Andrews, 1986, p. 1.
39. Condren, Mary. *The serpent and the goddess.* (San Francisco 1989), p. 55.
40. Pochin Mould, Ibid., p. 41
41. Andrews, Ibid., p. 7.
42. Curragh of Kildare; transcript notes of the Commission of Inquiry, held at Newbridge. 1866. Part II. D.O.D. 16597/66, p. 394.
43. Fitzgerald, Ibid., p. 5.
44. In *All the year round.* (London 25 May 1867), p. 520.
45. Otway-Ruthven, Jocelyn. 'The medieval County of Kildare', In *Irish Historical Studies.* Vol. XI, No. 43 (1959), p. 181.
    Otway-Ruthven, A.J. *A history of medieval Ireland.* (London 1968). p. 77–78.
46. Fitzgerald, Ibid., p. 32.
47. Curragh Commission,1866. Part I. p. 141.
48. Berry, H.F. (editor). *Statutes and Ordinances and Acts of Parliament in Ireland: King John to Henry V.* (Dublin 1907), p. 217.
49. D'Arcy, Fergus. 'The Ranger of the Curragh of Kildare', In *Journal of the Royal Society of Antiquaries of Ireland* Vol. 122 (1992), p. 4.
50. *Annals of the Four Masters.* Part I. (Dublin 1990), p. 382.
51. Gilbert, Sir John. *History of the Viceroys of Ireland.* (Dublin 1865), p. 96.
52. *Annals of the Four Masters.* Part III. (Dublin 1990), p. 272.
53. Fitzgerald, Ibid., p. 7.
54. Lawless, Hon. Emily. *With Essex in Ireland.* (London 1902), p. 49–67.
55. Gilbert, Sir John. *History of the Irish Confederation and the War in Ireland. Vol. II.* (Dublin 1882–85), p. 249.
56. Monk, Thomas, 1682. quoted in Fitzgerald Ibid., p. 9.
57. *Calendar of Mss of the Marquess of Ormonde,* New Series. Vol. VIII. (London 1920), p. 350–353, 387.
58. Boulger, D.C. *Battle of the Boyne.* (London 1911), p. 121–2.
59. Bryan, Col. Dan. 'The Curragh training camp', In *Irish Sword.* Vol. 1 (1953), p. 347.
60. Cited in Mac Lysaght. Ibid., p 360–1.
61. *Calendar of Mss of the Marquess of Ormonde.* New Series. Vol. III. (London 1920), p. 115, 163, 217, 228–9, 235, 240, 257. Jackson, Maj. E. S. *The Inniskilling Dragoons.* (London 1909), p. 32.

62. Moody, T.W. and Martin F. X. (editors). *The Course of Irish History.* (Cork 1987), p. 232.
63. Delany, Ruth. *The Grand Canal of Ireland.* (Newton Abbot 1972), p. 28.
64. In *Faulkner's Dublin Journal* 8–10 July 1783.
65. Crooks, Major J.J. *History of the Royal Irish Regiment of Artillery.* (Dublin 1914), p. 294.
66. Fitzgerald, Ibid., p. 11. According to evidence given at the Curragh Commission of 1866 (Part II, p 374) there was a camp on the Curragh in either 1796 or 1797.
67. In *Finn's Leinster Journal,* 30.4.1796.
68. In *Freeman's Journal,* 2.10.1843.
69. Fitzgerald, Ibid., p. 3.
70. Delany, Ruth, Ibid., p. 105.
71. Curragh Commission, 1866. Part II, p. 374.
72. McAnally, Sir Henry. *The Irish Militia 1793–1816.* (Dublin 1949), p. 187.
73. In *Faulkner's Dublin Journal.* 20.12.1804.
74. Ibid., 9.3.1805.
75. McAnally, Ibid., p. 191.
76. Ibid., p. 187.
77. Ibid., p.191/2.
78. Ibid., p. 192.
79. Morrin, Noel. Postage collection. (1992).
80. McAnally Ibid. p 229.
81. Ibid., p. 228–9.
82. Curragh Commission 1866. Part II, p. 374.
83. Swan, Comdt. D.A. 'The Curragh of Kildare', In *An Cosantoir. The Irish Defence Journal.* (May 1972), p. 14.
84. Lugard, Lieut. Col. H.W. *Narrative of operations in the arrangement and formation of a camp for 10,000 infantry on the Curragh of Kildare.* (Dublin 1858), p. 14.
85. Curragh Commission 1866. Part I. p. A 15.
86. Delany, Ibid., p. 132–3.
87. Cooke, John. *Bord na Móna, Peat Research Centre.* (Newbridge 1991), p. 28.
88. Lugard, Ibid., p. 7.
89 Thom's Directories 1855–1923.
90. Hennessey, W.M. 'The Curragh of Kildare', In *Proceedings of the Royal Irish Academy* Vol. 9 (1867), p. 346–7.
91. Fitzgerald, Ibid., p. 9.
92. Ibid., p. 10.
93. Kelly, James, (editor). *Letters of Lord Chief Baron Edward Willes to the Earl of Harwick 1757–1762.* (Aberystwyth 1990), p. 69–70.
94. Atkinson, A. *The christian tourist.* (Dublin 1815), p. 216.
95. D'Arcy, Fergus. *Horses, lords and racing men.* (The Curragh), p. 93.
96. Brewer, Ibid., p. 37.
97. In *Freeman's Journal,* 8.5.1843.
98. Ibid., 8.5.1843, 9.5.1843.
99. Hennessey, Ibid., p. 352.
100. In *Freeman's Journal*, 8.5.1843, 9.5.1843.
101. D'Arcy, Ibid., p. 139, 143.
102. Murray, K.A. and McNeill, D.B. *The Great Southern and Western Railway.* (Dublin 1976), p. 173.
103. Noble and Keenan. *Map of the County Kildare. 1752* Edition. Irish Georgian Society. No date.
104. Taylor, Lieut. Alexander. *A Map of the County Kildare 1783.*
105. Walker, Henry. *Map of the Curragh of Kildare. 1807.*

106. D'Arcy, Ibid., p. 355.
107. Ibid. p. 127.
108. An Act to make better provision for the management and use of the Curragh of Kildare. 16.7.1868. 31 and 32 Victoriae, Cap. 60. p. 551. Fitzgerald, Lord Walter. 'The Rangers of the Curragh of Kildare', In *Journal of the Royal Society of Antiquaries of Ireland.* Vol. 7, pt.4 (1897), p. 371.
109. D'Arcy, Fergus. 'The Ranger of the Curragh of Kildare', In *Journal of the Royal Society of Antiquaries of Ireland.* Vol. 122 (1992), p 7.
110. Ibid., p. 8.
111. *The Irish Times.* 11.2.1889.
112. Ibid.
113. Myler, Patrick. *Regency Rogue.* (Dublin 1976), p. 54.
114. Fitzgerald, Ibid., p. 12.
115. Young, Arthur. *A tour in Ireland.* (Cambridge 1925), p. 145.
116. Ní Chinneide, Sheila. 'An 18th century French traveller in Kildare', In *Journal of the County Kildare Archaeological Society* Vol. XV (1976), p. 380.
117. Lewis, Samuel. *A topographical dictionary of Ireland.* (London 1837), Part II, p. 81.
118. Hall, Mr and Mrs S.C. *Ireland, its scenery, character etc.* (London 1841–43). Part II, p. 259.
119. Curragh Commission 1866. Part I c. 5/78. D.O.D. 16597/66.
120. Feehan, John and McHugh, Roland. 'The Curragh of Kildare as a hygrocybe grassland', In *The Irish Naturalists' Journal* Vol. 24, No. 1 (1992), p. 16, 17.

Chapter 2

1. Ferguson, Kenneth. 'The development of a standing army in Ireland', In *Irish Sword.* Vol. XV (1983), p. 153.
2. Hayes-McCoy, G. A. *Irish Battles,* (Dublin 1969), p. 58.
3. Trenchard, John. An argument showing that a standing army is inconsistent with a free government. (London 1697), p. 7, quoted by Ferguson, Ibid. p. 153.
4. Ferguson, Ibid.. p. 153.
5. Ibid., p. 157.
6. Ibid., p. 154.
7. De hOir, Siobhain 'Guns in medieval and tudor Ireland', In *Irish Sword,* Vol. XV (1983), p. 80.
8. Fox, Capt. S.E. 'First use of gun powder in Irish history', In *Irish Sword,* Vol. XI (1974), p. 193.
9. Kerrigan, P.M. 'Seventeenth-Century Fortifications, forts and garrisons in Ireland', In *Irish Sword,* Vol. XIV (1981), p. 3.
10. Ibid., p. 3, p. 135.
11. Bryan, Col. Dan, 'Ballyshannon Fort, Co. Kildare,1642–1650', In *Irish Sword* Vol. IV (1960), p. 93, 97.
12. Johnston, Maj. S.H.F., 'The Irish Establishment'. In *Irish Sword* Vol. I (1953), p. 34.
13. Heffron, Col. M. 'Collins Barracks, Dublin', In *An Cosantoir. The Irish Defence Journal* (May 1968), p. 132.
14. Loeber, Ralph. *A Biographical Dictionary of Architects in Ireland 1600–1720.* (London 1981), p. 37.
15. Simms, J.G. 'Collins Barracks, Dublin', In *Irish Sword,* Vol. III (1958), p 73.
16. Wyse Jackson, R. 'Queen Anne's Irish Army Establishment in 1704', In *Irish Sword,* Vol. I (1953), p. 134.
17. Pereira, H.P.E. 'Barracks in Ireland, 1729, 1769', In *Irish Sword,* Vol. 1 (1953) p. 142.

18. Gibson, W.H. 'The North and South Moats, Naas', In *Journal of the County Kildare Archaeological Society* Vol. XVII (1991), p. 51.
19. Mooney, Fr. Canice, O.F.M. 'Quarters of the Army in Ireland 1769', In *Irish Sword* Vol. II (1956), p. 230/1.
20. Kerrigan, P.M. 'A military map of Ireland in the late 1790s', In *Irish Sword*, Vol. XII (1976), p. 247–251.
21. Sutcliffe, Sheila. *Martello Towers.* (Newton Abbot 1972), p. 123.
22. Kerrigan, P.M. 'The Defences of Ireland', In *An Cosantoir. The Irish Defence Journal* (August 1974), p. 287–290.
23. Cooke, John. *Bord na Mona, Peat Research Centre.* (Newbridge 1991), p. 27.
24. Begley, Lieut. Col. Michael. MS draft of a history of Devoy barracks, Naas. p. 7.
25. Cooke, Ibid., p. 31.
26. Kerrigan, P.M. 'A return of barracks in Ireland, 1811', In *Irish Sword*, Vol. XV (1983), p. 277–283.
27. Begley, Ibid., p. 8.
28. Gibson, Ibid., p. 51.
29. Berman, David and Jill. 'Journal of an officer stationed at Naas and Baltinglass, 1832–3', In *Journal of the County Kildare Archeological Society.* vol. XV (1976), p. 268–278.
30. Cunliffe, Marcus. *The Royal Irish Fusiliers 1793–1950.* (London 1952), p. 199.
31. Thackeray, W.M. *The Paris Sketch Book; The Irish Sketch Book etc.* (London 1872), p. 279.
32. In *Leinster Express.* 18.8.1855.
33. Lewis, Samuel. *A topographical dictionary of Ireland. Part II.* (1837), p. 424.
34. Lugard, Lieut. Col. *Narrative of operations in the arrangement and formation of a camp for 10,000 infantry on the Curragh of Kildare* (Dublin 1858). Larkin, William. *Map of Co. Kildare 1811.*
35. *Thom's Irish Almanac and Official Directory.* (1846), p. 238.
36. Muenger, E.A. *The British dilemma in Ireland.* (Dublin 1991), p. 2.
37. *Thom's Irish Almanac and Official Directory.* (1846), p. 131.
38. Pelly Patricia, and Tod Andrew. (editors) *Grant, Elizabeth. The Highland Lady in Ireland.* (Edinburgh 1991), p. 472.
39. Ibid., p. 294.
40. Johnston, Major S.H.F. 'The Irish Establishment', In *Irish Sword,* Vol.1 (1953), p. 34.
41. Ibid., p. 35.
42. Brady, John. *Catholics and catholicism in the eighteenth-century press.* (Maynooth 1965), p. 294–5.
43. Cooke Ibid., p. 43.
44. Lugard, Ibid., p. 7.
45. In *Leinster Express* 7.4.1855.
46. Kilmainham Papers, MS 1266–8. NL. Vol. 268.22.1.1855.
47. In *Leinster Express* 7.4.1855.
48. In *Leinster Express* 17.2.1855.
49. In *Leinster Express* 24.2.1855.
50. In *Leinster Express* 7.4.1855.
51. Fitzgerald, Lord Walter 'The Curragh: its history and traditions', In *Journal of the County Kildare Archeological Society* Vol. III (1902), p. 14.
52. Lugard, Ibid., p. 10.
53. In *Leinster Express* 3.3.1855.
54. Ordnance Survey Townland Survey of the County Kildare.Sheets Nos. 22, 23, 28. 6" = 1 Statute Mile. 1837.
55. Lugard, Ibid., p. 18.

56. Lugard, Ibid., p. 14.
57. James, N.D.G. *Plain Soldiering. A history of the Armed Forces on Salisbury Plain.* (Salisbury 1987), p. 10.
58. Lugard, Ibid., p. 9.
59. War Office 44/611. XC/A 034335. 12 .3.1855.
60. Ibid., 16.3.1855.
61. Lugard, Ibid., p. 9.
62. War Office 44/611 XC/A 034335. 14.4.1855.
63. Lugard, Ibid., p. 9.
64. Lugard, Ibid., p. 17, 24.
65. Lugard, Ibid., Curragh Sheet. Ordnance Survey.
66. Lugard, Ibid., Enlarged Plan of Squares C and G.
67. Lugard, Ibid., p. 12.
68. Lugard, Ibid., Curragh Sheet. Ordnance Survey.
69. Lugard, Ibid., p. 12–13.
70. Kilmainham Papers. Vol. 161, 4 Jan. 1856. p. 377.
71. Lugard, Ibid., p. 13, 18.
72. Kilmainham Papers. Vol. 268. 16 August 1855. Vol. 161. 13 July and 10 October 1856, p. 410, 437.
73 At the railway station he was greeted by Robert Browne, the Curragh ranger, and officers from the cavalry depot in Newbridge, with a guard of honour from the 11th Hussars. Also in the reception party were the County High Sheriff, E. Cane Esq., Sir Thomas Burke, Lord Dunkellin, Sir E. M'Donnell, and Messrs Millar and Ilberry of the railway depot.
74. In *Leinster Express* 10 March 1855.
75. Lugard, Ibid. p 10.
76. Kilmainham Papers Vol.161, 24 March 1855, p. 344.
77. Ibid., 23 March 1855, p. 344.
78. Bryan, Col. Dan 'The Curragh Training Camp', In *Irish Sword* Vol. I (1953), p. 346.
79. In *Leinster Express* 7 April 1855.
80. Ibid.
81. Lugard, Ibid., p. 9.
82. In *Leinster Express* 5 May 1855, 16 June 1855.
83. War Office 44/611. XC/A034335. 8 June 1855.
84. Ibid.
85. In *Leinster Express* 5 May 1855, 19 May 1855.
86. In *Leinster Express* 4 August 1855.
87. Kilmainham Papers. Vol. 268. 31 May 1855, 1 June 1855, 4 June 1855, 8 June 1855.
88. In *Leinster Express.* 16 June 1855.
89. Wall, Thomas 'Two Fenians in Kildare', In *Journal of the County Kildare Archeological Society* Vol. XIV (1970), p. 322–9.
90. In *Leinster Express* 5.5.1855.
91. In *Leinster Express* 16.6.1855.
92. Kilmainham Papers Vol. 161. 1 June 1855, p. 355.
93. Ibid., 8 June 1855, p. 356.
94. Ibid., 23 June 1855, p. 359.
95. *Burke's Irish family records.* (London 1976). p. 338; *Thom's almanac and official directory.* (1852), p. 274, 412.
96. In *Leinster Express.*7 July 1855.
97. Kilmainham Papers. Vol. 161. 24 July 1855, p. 363.
98. In *Leinster Express.* 28 July 1855.

99. In *Leinster Express.* 1 September 1855.
100. Kilmainham Papers. Vol. 268. 22 August 1855.
101. *Warrant for the regulation of the barracks in Great Britain and Ireland and the colonies.* H.M.S.O. (London 1824).
102. Ibid.
103. Kilmainham Papers. Vol. 268. July 1855.
104. Ibid., Vol 268. August 1855.
105. Kilmainham Papers. Vol.161. 31 December 1855. p. 376.
106. Ibid., Vol. 268. 26 July 1855.
107. Ibid., Vol. 161. 10 July 1855, p. 361.
108. Lugard, Ibid., p. 25–26
109. Ibid., p. 11.
110. Kilmainham Papers Vol. 161. 16 April 1866.
111. Ibid., Vol. 161. 5 January 1866. p. 378.
112. Ibid., Vol. 161. 18 March 1866. p. 390.
113. Ibid., Vol. 161. 18 September 1866.
114. In *Leinster Express* 24 Nov. 1855.
115. Kilmainham Papers. Vol. 268. 6 April 1867.
116. Ibid., Vol. 162. 21 October 1859.
117. Ibid., Vol. 161. 21 May 1857.
118. Ibid., Vol. 270. 19 May 1868.
119. Photograph. Eddie Chandler Collection 1868. 27/28 Irish Architectural Archive.
120. Noel Morrin Collection. 1992.
121. Kilmainham Papers. Vol. 268. 6 August 1855.
122. Ibid., Vol. 161.31 October 1855, p, 371.
123. Ibid., Vol. 268. 12 and 17 July 1856.
124. Ibid., Vol. 268. 31 October 1856.
125. Kilmainham Papers. Vol. 162. 29 July, 15 and 30 August 1859.
126 Ibid., and Vol. 162. 13 Sept. 1859.
127. Ibid., Vol. 268. 11 June 1857.
128. Ibid., Vol. 162. 9 Nov. 1858.
129. In *Leinster Express.* 1 Sept. 1855.
130. Lacy, Thomas *Sights and scenes in our fatherland.* (London 1863).
131. Ibid., p 69–70,176–7.
132. In *Leinster Express.* 20.10.1855.
133. Ibid., 6.10.1855.
134. Lugard, Ibid., p. 19.
135. Curragh Commission 1866. Part II, p. 516.
136. Return to House of Commons on the building of the Curragh and Aldershot. 25 June 1858. p. 63, p. 316.
137. Return to House of Commons on all the Expenses on the Curragh etc. 17 May 1861. p. 67, p. 302.
138. Lugard, Ibid., p. 29.
139. Ibid., p. 30.
140. Ibid., p. 34.
141. Memorandum of Agreement for the Forming of an Encampment on the Curragh of Kildare. 13 Jan. 1859. D.O.D. Curragh 97.
142. Curragh Commission 1866. Part 1, p. 61.
143. An Act to make better Provision for the Management and Use of the Curragh of Kildare. 16 July 1868. 31 and 32 Vict. *c.* 60, 1868.

CHAPTER 3

1. Return . . . to the House of Commons 12.5.1865, with extracts of any correspondence that has taken place with reference to the granting of a Lease of the Curragh of Kildare by the Crown to the War Department 5.7.1864. War Office 22.5.1865. Parliamentary papers, 1865 (301), xxxii. 339.
2. In *Leinster Express.* 13.5.1865.
3. Ibid., 13.5.1865.
4. Ibid., 20.5.1865.
5. Ibid., 17.6.1865.
6. Ibid., 1.7.1865.
7. In *Leinster Express.* 11.11.1865.
8. Return . . . to House of Commons. 1865. (301) xxxii. 339.
9. Fitzgerald, Lord Walter. The Curragh and its history and traditions. In *Journal of the County Kildare Archaeological Society.* Vol III (1902), p. 14.
10. A Bill to make better provision for the management and use of the Curragh of Kildare. Parliamentary papers, 1866 (136), ii. 383; Parliamentary papers 1867–68 (134), ii. 123; Parliamentary papers, 1867–68 (192), 11.143.
11. The commissioners were Maj. Gen. The Hon. A. H. Gordon, the Officer Commanding at the Curragh (Chairman) (and living at Knocknagarm on the Curragh), Maj. Edmund Mansfield of Morristown Lattin, A. Wetherell B.L., Mr T. P. Wilkinson (Secretary). Gentlemen from the county who attended the hearings included John La Touche, Major Borrowes, G. P. L. Mansfield, Eyre Powell, J. E. Medlicott, G.C.G. May, and Edward More O'Ferrall with William Lewis, solicitor. The county surveyor, Thomas Brasill, and Keith Hallowes Esq., of the firm of Hallowes and Hamilton, solicitors to the Woods and Forest Commissioners in Ireland, and Horace Watson Esq, of London, solicitor in England of the Commissioners of Woods and Forests, appeared for the Woods and Forests Department. Many other landowners, tenants and residents of the neighbourhood, as well as representatives of the military, Turf Club and the Grand Jury, were there. A total of 56 tenants gave evidence, and a further 15 were represented by solicitors, with a total in all of 71 witnesses. Curragh of Kildare. Commission of Inquiry. Transcript of Notes. Parts 1 and 2. Minutes of Proceedings at Public Sittings. Newbridge 1866. Property Section, D.O.D. Dublin. Part 1, p. A2, A3.
12. *A Bill to make better provision for the management and use of the Curragh of Kildare* 1866. p. 5.
13. *Commission,* 1866. Part 1. p. A 31–2, A 3, A 8–9.
14. In *All the year round.* 25.5.1867, p. 521.
15. Commission, part 2, p. 401.
16. Mayo Papers.NL MS. 1.22 (1).
17. Commission, part 1, p. 138.
18. Ibid., part II, p. 491.
19. Ibid., part I, p. 179, 193, 285.
20 Ibid., part I, p. 308–9.
21. Ibid., part I, p. 188–9.
22. Ibid., part I, p. A11, A12, A22, C 10–83, C 16–89, 211; part II, p. 364–5.
23. Ibid., part I, p. 183. Part II, 426, 497–8, 500–1, 509–10, 554–5.
24. Ibid., part II, p. 487–8, 494.
25. Ibid., part II, p. 491–2, 547.
26. Ibid., part I, p. 139.
27. Ibid., part II, p. 576.
28. Ibid., part II, p. 384, 522–35, 563–76.
29. Ibid., part II, p. 529, 565.
30. Ibid., part II, p. 379, 384, 529–30, 563–4, 567–8, 575–7.

31. Ibid., part II, p. 556–8.
32. Ibid., part I, p. 146, 190; part II, p. 357–64, 559–63.
33. Ibid., part I, p. 245–6; part II, p. 593–4.
34. Ibid., part II, p. 420.
35. Ibid., part II, p. 421.
36. Ibid., part II, p. 421, 600–4.
37. Ibid., part II, p. 421.
38. Ibid., part II, p. 421.
39. Ibid., part II, p. 421.
40. Ibid., part I, p. 42.
41. Ibid., part I, p. 277.
42. Ibid., part I, p. 211.
43. Ibid., part I, p. 40.
44. Ibid., part I, p. 142.
45. Ibid., part I, p. 179–80.
46. Ibid., part I, p. 246–7.
47. Ibid., part II, p. 585.
48. Ibid., part II, p. 586–7.
49. Ibid., part II, p. 600–1.
50. Ibid., part II, p. 601–2.
51. Ibid., part II, p. 602–3.
52. Mayo Papers. NL MS. 11.222 (6).
    Report of Commissioners who met in Newbridge on 14, 15, 17, 18, 19, 20, 21, and 24 September 1866, to the Lords Commissioners of H.M. Treasury. Parliamentary papers. 1867–68 (329) lv. 743.1, 28.6.1886, p.5, para. 56.
53. *Commission*, part II, p. 604.
54. Ibid., part II, p. 604–5.
55. Ibid., part II, p. 605, 613.
56. Ibid., part II, p. 603, 607, 610–11.
57. Ibid., part II, p. 610.
58. Ibid., part II, p. 608–9.
59. Information from the Russell librarian, Maynooth College, 1994.
60. *Report of Commissioners who met in Newbridge on 14, 15, 17, 18, 19, 20, 21, and 24 September 1866, to the Lords Commissioners of H.M. Treasury.* Parliamentary papers. 1867–68 (329) lv. 743.
61. Ibid., p. 2, para. 9.
62. Ibid., p. 2, para.. 9, 10, 14; p. 4, para. 35–36.
63. Ibid., p. 5, para. 49–55.
64. Ibid., p. 11, para. 123.
65. Ibid., p. 11, para. 122–4; p. 12, para. 125–8.
66. Ibid., p. 12, para. 129–132.
67. *Report from the select committee on the Curragh of Kildare Bill.* 1868. Parliamentary papers 1867–68 (404), x. l. Minutes of Evidence. p. 12.
68. Ibid., p. 7, 12.
69. Ibid., p. 13.
70. Ibid., p. 21.
71. Ibid., p. 27.
72. Ibid., p. 30.
73. Ibid., p. 19.
74. *Report from the select committee on the Curragh of Kildare Bill. 1868, together with the In Proceedings of the Committee, Minutes of Evidence and Appendix.* Parliamentary papers 1867–68 (404), x. l.

75. *An act to make better provision for the management and use of the Curragh of Kildare.* 31 and 32 Victoria, *c.* 60. 1868.
76. Act 1868. 555–19.558–49, 560–56.
77. Act, 550–1, 4, 5, 6.
78. Act, 551–7.
79. Act, 553–16.
80. Act, 562–63.
81. Act, 553–14.
82. D'Arcy, Fergus. *Horses, lords and racing men. (The Curragh,* 1991), p. 172.
83. Ibid.
84. Ibid., p. 173.
85. Ibid., p. 172.
86. *Curragh of Kildare Act, 1868, Award of Commissioners.* Dublin. 30.6.1869.
87. *Bye-Laws for the Regulation of the Curragh, by the Lord Lieutenant of Ireland in Council.* Dublin. 4.12.1868.
88. *An act to confirm the award under the Curragh of Kildare Act, 1868, and for other purposes relating thereto.* 33 and 34 Victoria. Ch. 74, 1870.
89. In *Leinster Leader* 25.4.1896.
    Curragh Rifle Ranges. Bye-laws as to Donnelly's Hollow road, by order of the Secretary of State for War. Copy signed by Colonel Commandant 14th Infantry Brigade amd Troops Curragh. 12 October 1921.
90. Draft Award of Arbitrator, Edmund Murphy, 81 Pembroke road, Dublin. In the matter of Bye-Laws made by H.M. Principal Secretary of War Department for regulating the Curragh Range ground and the closing of the Donnelley's Hollow road in county Kildare. 3.7.1896. Copy in office of the County Registrar, Naas.

Chapter 4

1. In *Leinster Express.* 17.11.1855.
2. Ibid., 22.12.1855.
3. Kilmainham Papers. Vol. 161. 10 November 1855, p. 372.
4. Ibid., Vol. 110, 12 January 1856.
5. Ibid., Vol. 161. 8 January 1856, p. 379.
6. Ibid., Vol. 161. 5 November 1856, p. 449.
7. In *Leinster Express.* 29.12.1855.
8. Kilmainham Papers. Vol. 161, 6 and 8 May 1856, p. 398–9.
9. Lugard. *Narrative of Operations in the arrangement and formation of a camp for 10,000 Infantry on the Curragh of Kildare.* (Dublin 1858). p. 10.
10. Tulloch, Maj. Gen. Sir Alexander Bruce. *Recollections of Forty years' Service.* (London 1903).
11. Ibid., p. 34–35.
12. D'Arcy, Fergus *Horses, lords and racing men.* (The Curragh 1991), p. 157.
13. Lugard, Ibid., p. 8.
14. Kilmainham Papers. Vol. 162. 29 January 1859. p. 224.
15. Lugard, Ibid., p. 7.
16. Notes. 'The Irish Crimean Banquet, 1856', In *Irish Sword.* IV (1960), p. 73.
17. Lugard, Ibid., p. 7.
18. Fortescue, Hon. J.W. *A history of the British army. Vol. XIII 1852–70.* (London 1930), p. 547.
19. Verner, Col. Willoughby, late Rifle Brigade. *The military life of H.R.H. George, Duke of Cambridge.* (London 1905), p. 124.
20. Ibid., p. 124–6.

21. *Thom's Almanac and Official Directory,* Dublin (1858), p. 274, 412.
22. Kilmainham Papers. Vol. 161, 11 September 1857, p. 556. Ibid., Vol. 162. 21 March 1858, p. 76. Ibid., Vol. 162. 24 March 1859, p. 278. Ibid., Vol. 162. 17 September 1859, p. 373.
23. Creagh, General Sir O'Moore and Humphris, H.M. *The V.C. and D.S.O.* Vol. I. London (1924). p. 41.
    Clark, B.D.H. 'The Victoria Cross, a register of awards to Irish-born officers and men', In *Irish Sword* Vol. XVI (1986), p. 189.
24. *The Irish Times.* 21.5.1994.
25. Drawing in the collection of the Knight of Glin.
26. In *All the year round.* May 1867. p. 522.
27. Verner, Ibid., p. 125.
28. In *Illustrated London News.* 7.11.1861, p. 249.
29. Royal Commission. Appx. LX p. 105. 1904.
30. Ibid., Appx. LV. p. 98. 1904.
31. In *Leinster Express.* 1.9.1855.
32. In *Leinster Leader.* 6.6.1896, 1.8.1896.
    In *Kildare Observer.* 8.4.1899.
33. Kilmainham Papers. Vol. 162. 16 November 1858.
34. Ibid., Vol. 162. 18 February 1860.
35. Monument on plain to Captain Collins.
36. *Curragh Commission, 1866.* Part. II. p. 431.
37. Award of Arbitrator 3.7.1896. Land Registrar of Titles, county Kildare.
38. Kilmainham Papers. Vol. 269. 22 April 1863.
39. In *Leinster Express.* 24.6.1865.
40. Kilmainham Papers Vol. III. 6 September 1856, p. 338.
41. Ibid., Vol. 60, 3 July 1867.
42. Ibid., Vol. 112, 8 October 1867, p. 363.
43. Wyndham, Horace *The Queen's Service.* (London 1899), p. 40.
44. D'Arcy, Fergus *Horses, lords and racing men.* (The Curragh 1991), p. 159.
45. In *Leinster Express.* 18.8.1855, 1.9.1855.
46. Ibid., 1.9.1855.
47. Ibid., 6.10.1855
48. Cooke, John. *Bord na Móna, Peat Research Centre.* (Newbridge 1991), p. 47; In *Leinster Express,* 13.10.1855.
49. In *Leinster Express,* 20.10.1855.
50. In *Leinster Express,* 17.11.1855, 29.12.1855.
51. Moore, Capt. A.M. Quoted in Proceedings, in the *Journal of the Royal Society of Antiquaries of Ireland* Vol. V (1859), p. 443/4.
52. Kilmainham Papers. Vol. 346, 22 August 1860.
53. Ibid., Vol. 163, 18 July 1862.
54. Ibid., Vol. 268, 19 December 1860.
55. Ibid., Vol. 161, 11 September 1857.
56. Ibid., Vol. 269, 16 January 1866.
57. Ibid., Vol. 268, 3 January 1859.
58. Ibid., Vol. 268, 26 September 1857; Vol. 162. 18 December 1858.
59. *Curragh Commission 1866.* Part II, p. 508.
60. Ibid., Part II, p. 475, 488, 494.
61. Kilmainham Papers. Vol.162, 19 January and 10 February 1860.
62. Ibid., Vol. 163, 30 July 1860; Vol. 268, 31 October 1860.
63. Kilmainham Papers. Vol. 268, 9 April 1857.

---

64. Ibid., Vol. 162, 4 May 1859.
65. Ibid., Vol. 268, 19 June 1860.
66. Ibid., Vol. 3, 29 July 1865.
67. Haire D.N. In aid of the Civil Power, 1868–90, p. 123–5 in *Ireland under the Union. Varieties of Tension. Essays in Honour of T. W. Moody.* (ed). F. S. L. Lyons and R.A.J. Hawkins. (Oxford 1980).
68. Kilmainham Papers. Vol. 16, 24 June 1860.
69. Ibid., Vol. 163, 21 April 1865.
70. Ibid., Vol. 269, 1 May 1866.
71. Devoy, John *Recollections of an Irish Rebel.* (New York 1929), p. 131, 143.
72. Kilmainham Papers. Vol. III, 24 July 1865.
73. *An Act to make Better Provision for the Management and Use of the Curragh of Kildare. 1868.*
74. *Bye-Laws for the Regulation of the Curragh, by the Lord Lieutenant of Ireland in Council.* Dublin. 4.12.1868.
75. Haire, D.N. In *Ireland Under the Union.* (Oxford 1980), p. 117; Turner, Maj. Gen., Sir A.E. *Sixty Years of a Soldier's Life.* (London 1912), p. 39.
76. Haire, D N. *British Army in Ireland, 1868–1890.* M. Litt. Thesis, T.C.D., 1973, p. 50.
77. Hogan, Brigadier General, Tables of Service of British regiments in Ireland.
78. Kilmainham Papers. Vol. 270, 2.9.1871.
79. Ibid., Vol. 271, 3.6.1872.
80. Mercer, Major J.D. *Record of the North Cork Regiment of the Militia 1793–1880.* (Dublin 1886).
81. Wyndham, Horace. *The Queen's Service.* (London 1899), p. 83.
82. Kilmainham Papers. Vol. 271, 26.7.1872.
83. Speirs, E.M. *The late Victorian army. 1868–1902.* (Manchester 1992), p. 222.
84. *Everyman's Encyclopaedia.* (London 1961), Vol. 1, p. 538; Vol. 3, p. 95.
85. In *Leinster Express.* 13.5.1876, 27.5.1876, 5.8.1876.
86. Ibid., 10.6.1876, 5.8.1876.
87. In *Leinster Express.* 9.9.1876.
88. Ibid., 23.9.1876.
89. Kilmainham Papers. Vol. 271, 5.2.1879.
90. Ibid., Vol. 271, 3.11.1880.
91. In *Leinster Express.* 11.11.1876.
92. *Kildare Observer.* 23.5.1885.
93. Ibid., 19.9.1885.
94. Ibid., 12.9.1885, 26.9.1885.
95. *The monthly official directory of the Curragh Camp and Newbridge.* (Dublin 1887), p. 18.
96. Ibid., p. 4, p. 15–17.
97. Godley, General Sir Alexander *Life of An Irish Soldier.* (London 1939), p. 22.
98. Ibid., p. 24.
99. Ibid., p. 23.
100. Wyndham, Horace *The Queen's Service.* p. 43.
101. Drawing in Collection of Hubert Hamilton, Moyne, Co. Laois.
102. *The Monthly Official Directory of the Curragh Camp and Newbridge.* (Dublin 1887), p. 9.
103. Hogan, Ibid., p. 92.
104. In *Kildare Observer.* 4.6.1887.
105. Ibid., 20.8.1887.
106. MacCauley, J.A. 'The Dublin Fusiliers', In *Irish Sword.* Vol. VI (1964), p. 266.
107. In *Kildare Observer.* 25.6.1887.
108. In *Kildare Observer.* 23.7.1887.
109. Ibid., 16.7.1887.
110. Cooke, John *Bord na Móna, Peat Research Centre.* (Newbridge 1991), p. 47.

111. Wyndham, Ibid., p. 81.
112. Ibid., p. 43–53.
113. Godley, Ibid., p. 28.
114. O'Toole, Jimmy *The Carlow Gentry.* (Carlow 1993), p. 211.
115. Godley, Ibid., p. 30.
116. Verner, Col. W. *The military life of H.R.H. George, Duke of Cambridge.* (London 1905), p. 378.
117. McCance, Capt. S. *History of the Royal Munster Fusiliers.* (Aldershot 1927), p. 18.
118. Ibid., p. 18–19.
119. *The navy and army illustrated.* (London 9.7.1897), p. 120.
120. Programme of 1895 tournament.
121. Poster of the 1895 meeting.
122. In *Leinster Leader.* 25.1.1896.
123. Ibid., 8.2.1896.
124. Ibid., 15.2.1896.
125. Ibid., 15.2.1896.
126. Ibid., 7.3.1896.
127. Ibid., 25.1.1896.
128. Ibid., 29.2.1896.
129. Ibid., 14.3.1896.
130. Ibid., 21.3.1896.
131. Ibid., 28.3.1896.
132. Ibid., 14.3.1896, 2.5.1896.
133. Ibid., 25.1.1896.
134. Ibid., 25.4.1896.
135. Ibid., 5.9.1896.
136. Ibid., 4.4.1896.
137. Ibid., 25.4.1896.
138. In *The Regiment. An Illustrated Journal for Everybody.* (London. 23.5.1896). p. 115.
139. In *Leinster Leader.* 23.5.1896, 11.6.1896.
140. Ibid., 2.5.1896.
141. Ibid., 23.5.1896.
142. Ibid., 9.5.1896.
143. Ibid., 20.6.1896.
144. Ibid., 13.6.1896.
145. Ibid., 6.6.1896.
146. Ibid., 13.6.1896.
147. Ibid., 25.7.1896.
148. Ibid., 20.6.1896.
149. Ibid., 27.6.1896, 11.7. 1896.
150. bid., 27.6.1896.
151. Ibid., 11.7.1896.
152. Ibid., 4.7.1896.
153. Ibid., 25.7.1896.
154. Ibid., 21.3.1896.
155. In *Kildare Observer.* 20 4.1901.
156. Speirs, E.M. *The late Victorian army, 1868–1902.* (Manchester 1992), p. 222.
157. In *Leinster Leader* 18.7.1896.
158. Ibid., 1.8.1896.
159. Ibid., 8.8.1896.
160. Ibid., 1.8.1896.

161. Ibid., 12.9.1896.
162. Ibid., 14.11.1896..
163. *Country Gentleman.* Quoted in Muenger, E.A. *The British Military Dilemma in Ireland 1886–1914.* (Dublin 1991) p. 77.
164. In *Leinster Leader.* 8.8.1896.
165. Ibid., 5.9.1896.
166. Ibid., 19.9.1896.
167. Ibid., 24.10.1896.
168 Ibid., 14.11.1896.
169. Ibid., 10.10.1896, 31.10.1896.
170. Ibid., 7.11.1896.
171. D.O.D., Property Management Section. Crown Lands. File 1406.

CHAPTER 5

1. Census of Ireland, 1861. Leinster Vol.1, p. 555.
2. In *Illustrated London News*, 7.11.1861. p. 249; *Curragh Commission*, Part I, p. 54; In *Leinster Express.* 21.1.1865.
3. St Aubyn, Giles *Edward VII. Prince and King.* (London 1979). p. 50.
4. In *Freeman's Journal.* 26 August 1861.
5. St Aubyn, Ibid., p. 50.
6. St Aubyn, Ibid., p. 50.
7. In *Freeman's Journal* . 26 August 1861.
8. *The Annual Register 1861.* p. 154.
9. *Illustrated London News.* 7.11. 1861. p. 249.
10. *The Annual Register 1861.* p. 153.
11. In *Freeman's Journal.* 26 August 1861.
12. In *Leinster Leader.* 30.12.1916.
13. D'Arcy, Fergus. *Horses, lords and racing men.* (The Curragh 1991), p. 159–160.
14. Curragh Commission. Part II, p. 576.
15. DeCourcy-Wheeler, Capt. H.E. The tower on the Hill of Allen. In *Journal of the County Kildare Archaeological Society.* Vol. VII (1914), p. 414–5.
16. St Aubyn, Ibid., p. 51–2.
17. Ibid., p. 52, 69.
18. Ibid., p. 52–5.
19. In *Kildare Observer.* 16.2.1901.
20. In *Illustrated London* News. 7.11.1861. p. 322.
21. Ibid., p. 322.
22. In *Green Howard's Gazette.* July 1899.
23. From Daniel Gillman. 19.11.1992.
24. Gorry Gallery, Dublin. Exhibition catalogue 18th, 19th, 20th century Irish Painters. 28 April 1989.
25. In *Illustrated London* News. 7.11.1861. p. 322.
26. Souvenirs of Soldiering at Camp. (Curragh 1861). Royal Archive, Windsor Castle.
27. D'Arcy, Ibid., p. 163.
28. In *Leinster Express.* 29.10.1864.
29. In *Leinster Express.* 15.4.1865.
30. Ibid., 28.10.1865.
31. Devoy, John. *Recollections of an Irish rebel.* (New York 1929), p. 62.
32. *Dictionary of National Biography.* (London 1917).

33. Hogan, Brig. Gen. P.D. Tables of Service of British regiments in Ireland. No date.
34. Semple, A.J. *The Fenian infiltration of the British army in Ireland 1864–7* Unpublished M. Litt. thesis, T.C.D., 1971, p. 58.
35. Devoy, Ibid., p. 131, 143.
36. Ibid., p. 60.
37. In *Leinster Express* 28.10.1865.
38. In *All the year round.* May 1867. p. 524.
39. Devoy, Ibid., p. 142.
40. O'Broin, Leon. *Fenian fever. An anglo-American dilemma.* (London 1971), p. 32.
41. Semple, A.J. 'The Fenian inflitration of the British army' In *Journal of the society for army historical research.* Vol. 52, No. 211 (London 1974), p. 133; Semple, Ibid., p. 162.
42. Semple, Ibid., p. 188.
43. In *All the year round.* (May 1867). p. 524.
44. Semple, Ibid., p. 31.
45. Comerford, R.V. *The Fenians in context. 1848–82.* (Dublin 1985), p. 125.
46. Semple, Ibid., p. 97, 101.
47. In *All the year round.* (May 1867). p. 524.
48. Cunliffe, M.F. *The Royal Irish Fusiliers.* (London 1952), p. 227.
49. Kilmainham Papers. Vol. 270, 1.1.1868.
50. In *All the year round.* May 1867. p. 523.
51. Kilmainham Papers. Vol. 270, 16.3.1868.
52. Morris, M. O'Connor *Hibernia Venatica.* (London 1878), p. 168.
53. Nelson, Thomas *The Land War in County Kildare.* Maynooth Historical Series No. 3 (Maynooth 1985.), p. 5, 15, 20.
54. Hawkins, Richard 'An army on police work, 1881–2', In *Irish Sword* Vol. XI (1974), p. 78.
55. *United Ireland.* 26 November 1881.
56. O'Neill, Maire *From Parnell to de Valera. Jennie Wyse Power.* (Dublin 1991), p. 19–20.
57. Haire, D.N. 'In Aid of the Civil Power, 1868–90', In *Ireland Under the Union. Essays in honour of T.W. Moody.* (editors) F.S.L. Lyons and R.A.J. Hawkins. (Oxford 1980), p. 142.
58. Haire, Ibid., p. 122–3.
59. In *The Times.* 21.11.1881.
60. Hawkins, Ibid., p. 82.
61. Hawkins, Ibid., p. 96.
62. Muenger, E.A. *The British military dilemma in Ireland.* (Dublin 1991), p. 118, 120.
63. Puirseal, Padraig *The G.A.A. in its time.* (Dublin 1982), p. 72.
64. O'Connor, Tommy (ed) *Ardfert: a hurling history.* (Ardfert 1988), p. 18; *G.A.A. Official Guide 1914–15* (Wexford 1915), p. 22.

Chapter 6

1. Wyndham, Horace *The Queen's Service.* (London 1899), p. 64.
2. Kilmainham Papers. Vol. 161, 24.3.1855.
3. In *Leinster Express.* 22.12.1855.
4. Return to House of Commons. Divine Service 1856.
5. Kilmainham Papers. Vol. 268. 12 May 1856.
6. Ibid., Vol. 162. 6 and 20 July 1859, p. 339, 349.
7. Return to House of Commons. Divine Service 1856.
8. Kilmainham Papers. Vol. 161. 5 October 1856.
9. Ibid., Vol. 162. 13 October 1859, p. 381.
10. *Thom's almanac and official directory.* (Dublin 1858), p. 693.

11. Curragh R.C. Garrison Church Register of Baptisms.
12. Thom, Ibid., 1865. p. 1045.
13. Thom, Ibid., 1878. p. 950; Mc Evoy, John, *Carlow College 1793–1993*, ((Carlow 1993), p. 150,189
    Comerford, Rev. M. *Collections relating to the Diocese of Kildare and Leighlin.* Vol.1, p. 216 (Dublin 1883).
14. Curragh Commission, 1866. Part I, p. 64.
15. Curragh Plan: Revised Ordnance Survey map. of 1875, 6" : 1 Statute mile. In connection with the matter of Bye Laws made by H.M. Principal Secretary of State for War Department (1894) for the regulation of the Curragh Rifle Range ground and the closing of the Donnelly's Hollow road in county Kildare.
16. Act 31 and 32. Vic. *c.* 60. 1868. Second Schedule. p. 567.
17. In *Leinster Express.* 1.7.1876; *Curragh Camp, and district, illustrated and described.* (Dublin *c.* 1908), p. 14.
18. Kilmainham Papers. Vol. 269. 22 February 1866.
19. Ibid., Vol. 269. 12 March 1867.
20. Ibid., Vol. 269. 26 March 1867.
21. Ibid., Vol. 269. 30 May 1867.
22. H.C. Parl. Papers 1865. Vol. XXXII. Return to an Address of the Hon. The House of Commons. 17.3.1865.
    W.O. 3.4.1865. Allowance to Officiating Clergymen to Troops. p. 14.
23. In *Kildare Observer.* 14.2.1885.
24. In *Leinster Leader.* 8.2.1896.
25. Ibid., 9.5.1896.
26. Ibid., 25.4.1896
27. Ibid., 4.7.1896.
28. Ibid., 24.10.1896.
29. Ibid., 30.12.1916.
30. Kilmainham Papers. Vol. 161. 2 January 1856, p. 376.
31. Lugard, Lieut. Col. H.W. *Narrative of Operations in the arrangement and formation of a camp. for 10,000 Infantry on the Curragh of Kildare.* (Dublin 1868); Ordnance Survey Sheets 22, 23, 27, 28; Mayo papers. NL MS 11, 222(1).
32. D.O.D. Dublin. Property Management Section. Box 293. No. 26.
33. Paterson, Rev. John *Meath and Kildare, an historical guide.* (Kingscourt 1981), p. 37.
34. H.C. Parl. Papers 1865. Vol. XXXII. Return to an Address of the Hon. The House of Commons. 17.3.1865; W.O. 3.4.1865. Allowance to Officiating Clergymen to Troops. p. 14.
35. St David's church, Naas. Minutes of Select Vestry 1805–1884. Parish register 1679–1830.
36. Berman, David and Jill 'Journal of an officer stationed in Naas and Baltinglass', In *Journal of the County Kildare Archaeological Society.* Vol. XV (1976), p. 268.
37. St David's church, Naas Ibid.; Letter in *The Irish Times*, 23.6.1984.
38. In *Kildare Observer* 2.11.1889, 30.6.1900.
39. Oral. Mr Kenneth. Dunne, Kildare.
40. Dunlop, Rev. Robert. *Plantation of renown.* (Naas 1970), p. 15.
41. Young, M.F. *The letters of a noble woman.* (London 1908), p. 135.
42. Kiely, K., Newman, M., and Ruddy, J. *Tracing your ancestors in County Kildare.* (Naas 1992), p. 14.
43. In *Leinster Leader.* 12.9.1987.
44. In *Leinster Express.* 1.9.1855.
45. Oral. Molloy. James 1994.
46. *An Baile Seo Ghainne.* (Kilcullen 1977), p. 25–27; *Bulletin of the Mill Hill Fathers.* (Summer 1986). p. 18–21.

---

47. *Annals of the Sisters of Mercy, by a member of the Order.* (New York 1881), p. 360.
48. Fitzpatrick, W.J. *Times and Correspondence of Rt. Rev. Dr. Doyle. Vol. II.* (Dublin 1861), p. 436–437.
49. Haly Papers. No. 11. Archive Diocese of Kildare and Leighlin. Carlow
50. Ibid., No. 10. 6.8.1855.
51. Mac Suibhne, Peadar. *Paul Cullen and his Contemporaries.* (Naas 1962), p. 119.
52. Curragh Command Headquarters. File Q. M. Ch-10287. Lettings Sandes Home. 1949–1984. Military Archives. File C.42. Sandes Soldiers' Home.
53. In *All the year round.* 26.11.1864. p. 369.
54. Kilmainham Papers. Vol. 162. 13.10.1859. p. 381.
55. Ibid., Vol. 271. 8.1.1878.
56. J.H. Andrews and A. Simms (editors) *Irish Historic Towns Atlas. No. 1. Kildare.* R.I.A. (Dublin 1986), p. 6; National Education Report. p. 128.
57. In *Kildare Observer.* 10.12.1886.
58. Jeffrey, M.H. *The trumpet call obey.* (London 1968), p. 29, 30, 82.
59. Carson, Patricia. Sandes Soldiers' Home. In *Journal of the County Kildare Archaeological Society* Vol. XIV (1970), p. 473.
60. Jeffrey, Ibid., p. 88.
61. *Curragh Camp and District. Illustrated and Described.* (Dublin c. 1908), p. 14. *Soldier* (London, August 1954).
62. In *The Irish Times.* 30.11.1911.
63. Ibid., 1.6.1911.
64. Military Archives, Dublin.
65. Wyndham, Horace. *The Queen's Service.* (London 1899), p. 84, 85.
66. In *Leinster Leader.* 4.7. 1896.
67. *Curragh Camp and District,* p. 15.
68. In *Leinster Leader.* 4.7.1896.
69. *Curragh Camp and District,* p. 14.
70. *The Regiment. An Illustrated Journal for Everybody.* (London 23.5.1896).
71. In *Leinster Leader.* 25.4.1896.
72. *Curragh Camp and District,* p. 15–16.
73. Stone in situ. *Curragh Camp and District,* p. 16.
74. Kilmainham Papers, Vol. 271. 3 April 1878.
75. Volumes in Military Archives, Dublin.
76. Kilmainham Papers. Vol. 271. 3 April 1878.
77. In *Leinster Leader* 2.5.1896.
78. Robins, Joseph *The Lost Children.* (Dublin 1980), p. 256.
79. Shaw, Bernard *Man and Superman. Act III.* (Odhams Press, London N.D.), p. 379.
80. *Warrant for the Regulation of Barracks.* (HMSO, London 1824).
81. In *Leinster Express.* 1.4.1855. In *All the year round.* 7.9.1867.
82. In *Leinster Express.* 1.4.1855.
83. Lugard, Lieut. Col. H.W. *Narrative of Operations in the arrangement and formation of a camp. for 10,000 Infantry on the Curragh of Kildare.* (Dublin 1858). Plans 10, 13, 14.
84. Kilmainham Papers. Vol. 268. 4.4.1857.
85. Ibid., Vol. 162. 5.1.1858.
86. Ibid., Vol. 110. 13.8. 1859.
87. Ibid., Vol. 163. 13.8.1860.
88. In *All the year round.* 28.12.1867. p. 58.
89. Ibid., p. 59–60.

90. Kilmainham Papers. Vol. 68. 18.11.1876. p. 351.
91. *The monthly official directory of the Curragh Camp. and Newbridge. December 1887.* (Dublin 1887), p. 19.
92. Kilmainham Papers. Vol. 162. 10.2.1860.
93. Ibid., Vol. 268. 2.6.1860.
94. Ibid., Vol. 162. 10.2.1860.
95. Ibid., Vol. 163. 1.6.1861; Ibid., Vol. 163. 24.6.1861.
96. Ibid., Vol. 269. 15.11.1867.
97. Note on Soldiers' and Sailors' Help Society, in the archive of the Soldiers' Sailors' and Airmens' Families Association and Help Society, Republic of Ireland. Dublin.
98. Kilmainham Papers. Vol. 163. 7.8.1860.
99. In *All the year round.* 28.12.1867. p. 60.
100. *The Irish Times.* 4.1.1932; Loeber, Rolf. Biographical dictionary of engineers in Ireland 1600–1730. In *Irish Sword.* Vol. XIII (1979), p. 236.
101. Oral. Cummins, Conleth. Newbridge. 1992.
102. In *Leinster Express.* 11.4.1855.
103. Lugard, Ibid., Plan. No. 1.
104. Watercolour sketch by Lady Fremantle. *c.* 1866. Collection. The Knight of Glin.
105. Kilmainham Papers. Vol. 268. 8.5.1857.
106. In *The Irish Times.* 28.6.1866.
107. Lacy, Thomas *Sights and scenes in our Fatherland.* (London 1863), p. 177.
108. Lugard, Ibid., p. 31–33.
109. *Return of number of persons flogged in Great Britain and Ireland, 1856.* House of Commons, British Parliamentary papers. 1857–8. XXX VII 307.
110. In *Leinster Leader.* 18.4.1896.
111. Ibid., 26.9.1896.
112. In *Leinster Leader.* 13.6.1896.
113. Kilmainham Papers. Vol. 271. 1.8.1878.
114. In *Leinster Leader.* 21.3.1896.
115. Ibid., 21.3.1896.
116. Ibid., 21.3.1896.
117. In *Leinster Express.* 18.8.1855, 26.8.1876.
118. In *Leinster Leader.* 28.3.1896.
119. In *Kildare Observer.* 15.8.1885.
120. Ibid., 14.11.1885.
121. Ibid., 31.7.1886.
122. In *Kildare Observer.* 24.12.1887.
123. In *Kildare Observer.* 21.4.1900, 23.6.1900, 20.10.1900.
124. In *Leinster Leader.* 27.6.1896.
125. Ibid., 12.9.1896.
126. Ibid., 7.11.1896.
127. In *Kildare Observer.* 17.7.1886.
128. In *Leinster Leader.* 21.3.1896, 28.3.1896, 27.6.1896.
129. *Army statistical, sanitary and medical report of Army Medical Dept. for the year 1860.* Vol. XXXIII. 1. (H.M.S.O., London 1862), p. 12.
130. Ibid., 1869. Vol. XXXVIII.375. (HMSO, London 1871), p. 11, 40–41.
131. Haire, D.N. *British Army in Ireland 1868–1890.* M. Litt. T.C.D., 1973, p. 264–5.
132. *Army Statistical Report, 1887.* Vol. XLIX.169. (H.M.S.O., London 1889), p. 48.
133. *Army Statistical Report, 1869.* Vol. XXX VIII 375. (H.M.S.O., London 1871), p. 11, 40–41.
134. Ibid., 1887. Vol. XLIX.149, p. 48.

135. Ibid., 1896. Vol. LIV. 557, (H.M.S.O. London 1897), p. 64; Murray, James 'The early formative years in Irish Radiology', in J.C. Carr (ed.) *A Century of Medical Radiation in Ireland. An Anthology.* (Dublin 1995).
136. In *Leinster Express.* 2.12.1876.

CHAPTER 7

1. In *Leinster Express.* 1.9.1855.
2. Ibid., 8.9.1855.
3. Ibid., 6.10.1855.
4. Ibid., 24.11.1855.
5. Ibid., 21.7.1855.
6. Lugard, Lieut. Col. H.W. *Narrative of Operations in the arrangement and formation of a camp. for 10,000 Infantry on the Curragh of Kildare.* (Dublin 1858), p. 24. Plans 49, 50, 51.
7. In *Leinster Express.* 27.10 1855.
8. House of Commons. *Return of Number of Convictions and Arrests for Trespass on the Curragh of Kildare, before the Camp. Court and Petty Sessions at Newbridge or Kildare, 1860 and 1861.* Parliamentary papers (141) XXXII.299.
9. Kilmainham Papers. Vol. 163. 22.7.1864.
10. Ibid., Vol. 268. 26.6.1862; Vol. 269. 7.5.1865; *Thom's Almanac and Official Directory.* (Dublin 1862). p. 785.
11. In *Leinster Express.* 15.7.1865; Kilmainham Papers. Vol. III. 13.7.1865; *Thom's Almanac and Official Directory.* (1865). p. 877.
12. In *Leinster Express.* 25.11.1865.
13. Ibid., 8.4.1865.
14. Ibid., 17.6.1865.
15. Crown Lands file. 1406. 1866–1910. Property Management Section. D.O.D.
16. In *Leinster Express.* 24.6.1855.
17. Crown Lands file. Ibid.
18. Bye-Laws, Curragh of Kildare. 4.12.1868. D.O.D.
19. In *Leinster Express.* 20.5.1876.
20. Ibid., 26.2.1876.
21. Ibid., 6.5.1876.
22. Ibid., 22.7.1876.
23. Ibid., 28.10.1876.
24. Ibid., 18.11.1876.
25. Ibid., 8.7.1876.
26. Ibid., 5.8.1876.
27. Ibid., 22.7. 1876.
28. In *Kildare Observer.* 29.5.1886.
29. Ibid., 3.7.1886.
30. Ibid., 19.9.1855.
31. Ibid., 27.3.1886.
32. Ibid., 13.6.1885.
33. Ibid., 6.2.1886.
34. Ibid., 29.5.1886.
35. Ibid., 2.10.1886.
36. Ibid., 7.5.1887.
37. Hamilton, I.B.M. *The happy warrior, a life of Sir Ian Hamilton.* (London 1966), p. 17.
38. In *Kildare Observer.* 25.12.1886.

39. In *Leinster Leader*. 25.7.1896.
40. Ibid., 22.7.1876, 5.8.1876.
41. Ibid., 5.8.1876.
42. Ibid., 2.3.1896.
43. Ibid., 20.6.1896.
44. Ibid., 14.11.1896.
45. In *Kildare Observer*. 12.8.1899.
46. Ibid., 11.3.1899.
47. Ibid., 15.7.1899.
48. Spiers, E.M. *The late Victorian army*. (Manchester 1992), p. 132.
49. Wyndham, Horace. *The Queen's Service*. (London 1899).
50. Army Medical Department report 1884. p. 9.
51. Curragh Commission. 1866. p. 602.
52. Shaw, A.G.L. *Convicts and the Colonies*. (London 1966), p. 205, 302.
53. *Curragh Commission 1866*. p. 205.
54. For Dublin see Purcell, Mary. *Matt Talbot*. (Dublin 1954), p. 22–4.
55. In *Leinster Express*. 1.12.1855.
56. Lugard, Ibid., p. 33.
57. Kilmainham Papers. Vol. 268. 22.10. 1859; Vol. 162, 26.10. 1859.
58. Ibid., Vol. 163. 7.8. 1860.
59. Kilmainham Papers. Vol. 268. 30.7.1860.
60. Board of Guardians, Naas Union. Minutes of Meetings. 20.5.1865, 4.8.1866.
61. Ibid., Minute book. 28.7.1860.
62. Ibid., Minute book. 4.8.1860.
63. *Army Medical Department Report for the year 1860*. Parliamentary papers. XXXIII.I. 1862.
64. Census of Ireland 1861. Part V. p. 154, 555, 556, 692.
65. D'Arcy, Fergus *Horses, lords and racing men*. (The Curragh 1991), p. 156.
66. Board Guardians. Naas Union. Minute book 12.1.1861.
67. Burke, Helen *The people and the Poor Law in 19th century Ireland*. (Littlehampton 1987), p. 164–171.
68. In *The Irish Times*. 24.11.1863.
69. In *All the year round*. 26.11.1864. p. 369–372.
70. In *The Irish Times*. 19.10.1863.
71. In *All the year round*. 26.11. 1864. p. 369–372.
72. In *Leinster Express*. 6.2.1864.
73. Ibid., 2.4.1864.
74. Ibid., 26.3.1864.
75. *The Contagious Diseases Prevention Act, 1864*; Board of Guardians, Naas Union. Minutes of Meeting. 2.7.1864; Kilmainham Papers. Vol. 163. 23.2.1865.
    Blanco, R.L. 'The attempted control of V.D. in the army of mid-Victorian England', In *Journal of the Society for Army Historical Research*, Vol. 45, No. 184 (London 1967), p. 234.
76. Return to House of Commons. 9.6.1869; Crown Leases. 1852–1866.
77. In *Leinster Express*. 8.4.1865.
78. Ibid., 15.4.1865.
79. Ibid., 24.6.1865.
80. Ibid., 5.8.1865.
81. Ibid., 18.11.1865.
82. Ibid., 18.11.1865.
83. Lowndes, F.W. *Lock hospitals and Lock wards in general hospitals*. (London 1882).
84. In *Leinster Express*. 18.2.1865.

85. Ibid., 21.2.1865.
86. Ibid., 21.2.1865.
87. *Kildare County Infirmary*. (Carlow No date), p. 8.
88. Kilmainham Papers. Vol. 163. 23.2.1865.
89. Board of Guardians. Naas Union. Minute book. 13.5.1865, 20.5.1865.
90. Ibid., 4.8.1886.
91. In *Leinster Express*. 4.3.1865.
92. Ibid., 18.3.1865.
93. *Curragh Commission, 1866*. p. 602.
94. Ibid., p. 557–8.
95. Kilmainham Papers. Vol. 59. 19.12.1866, p. 101.
96. Kilmainham Papers. Vol. 59. 19.2.1867, p. 207.
97. Ibid., Vol. 60. 8.4.1867, p. 4.
98. Ibid., 21.11.1867, p. 361.
99. Board of Guardians. Naas Union. Minute book. 14.8.1869.
100. *The Wren of the Curragh*. Reprinted from the *Pall Mall Gazette*. (London 1867).
101. Ibid., p. 1.
102. Ibid., p. 6.
103. Ibid., p. 10–20.
104. Luddy M. and Murphy C. (ed) *Women Surviving*. (Dublin 1990), p. 60; Luddy Maria, Women and the contagious diseases acts 1864–1886, in *History Ireland*. (Dublin, Spring 1993), p. 32.
105. *The Wren of the Curragh*, Ibid., p. 44–47.
106. Ibid., p. 17.
107. Ibid., p. 35.
108. Ibid., p. XI.
109. Ibid., p. XII.
110. Ibid., p. 32, 51.
111. Ibid., p. 32.
112. Ibid., p. 17, 52.
113. Board of Guardians. Naas Union. Minute book. 8.2.1868.
114. Ibid., 13.4.1867.
115. Ibid., 11.1.1868.
116. Ibid., 13.6.1868.
117. *Act for the Better Prevention of Venereal Diseases. 1866*. Victoria 28 and 30.l, *c*. XXXV. 1868. *Report of the Committee on the Pathology and Treatment of V. D*. p. 241.
118. *Curragh Commission, 1866*. p. 146, 188, 191–193, 244–5, 357–365, 426, 511–512, 555–563.
119. Oral. Lieut. Col. C.M.L. Clements and Mr. L. Moran. 1994.
120. Record site plan Kildare Lock Hospital. Scale 50' : 1". 3.5.1876. Map. No. 17–9. No. 4. Kildare Lock Hospital. Plan of well and tank tower. 5' = 1". 1.1.1869. Ref. 17–9. Military Archives, Dublin.
121. Andrews, J.H. and Simms. A. *Historic towns atlas, Kildare*. (R.I.A., Dublin 1986), p. 10.
122. Map of Lock Hospital, County of Kildare. 1882. Scale. 50' : 1".
Kildare barracks: Record of reappropriation of Lock Hospital. 28.3.1906. Sheets No. 2, 3, 4. Scale. 8' = 1". Ref. 16–9. Military Archives, Dublin.
123. Map of Kildare hut barracks. County Kildare Sheets. XXII, 12–16. Military Archives, Dublin R.I.A., *Historic Towns Atlas*. p. 10.
124. Kilmainham Papers. Vol. 270, 20.12.1869, 22.12.1869.
125. Ibid., Vol. 270, 24.12.1869.
126. Ibid., 27.12.1869.

127. *Army Medical Report for the year 1869.* Vol.XXXVIII.379. (London 1871).
128. McGee, William. *The moral utility of a Lock hospital.* (Dublin 1872), p. 1.
129. Ibid., p. 2.
130. Ibid., p. 3.
131. Ibid., p. 3.
132. Ibid., p. 5.
133. Ibid., p. 8.
134. *Report to the War Office by the Inspector of Certified Hospitals on moral affects of the contagious diseases Act.* (1873) (209) XL. 433. p. 4.
135. Ibid., p. 3.
136. Ibid., p. 5.
137. In *The Dublin Gazette.* 10.1.1873. p. 25; Curragh Bye-laws. 3.1.1873.
138. In *Leinster Express.* 5.8. 1876.
139. Ibid., 2. 7.1876.
140. Board of Guardians. Naas Union. Minute book. 12.1.1878.
141. Kilmainham Papers. Vol. 71. 19.11.1879, p. 30.
142. Ibid., 10.2.1880. p. 98.
143. Comerford, Rev. M. *Dioceses of Kildare and Leighlin.* (Dublin 1883), Vol. II, p. 51.
144. Kilmainham Papers. Vol. 71, 11.2.1880. p. 105–6.
145. Ibid., 1.3.1880, p. 115.
146. Ibid., 1.3.1880, p. 115.
147. Ibid., 26.2.1880, p. 115: 1.3.1880, p. 116.
148. Ibid., Vol. 71, 6.4.1880, p. 150.
149. Ibid., 13.4.1880, p. 159.
150. Board of Guardians. Naas Union. Minute book. 17.4.1880.
151. Ibid., 26.1.1881.
152. Ballymore Development Group. *The History of Ballymore Eustace.* (1992), p. 45.
153. In *Kildare Observer.* 11.7.1885.
154. Ibid., 16.10.1886.
155. Ibid., 30.10.1886.
156. In *The Lancet.* Vol. I, 25.6.1887, p. 1303.
157. In *Kildare Observer.* 8.1.1887.
158. In *Leinster Leader.* 14.11.1896.
159. Oral. Tom Garrett. 1992.
160. Oral. Lieut. Col. J. Daly, Naas. 1980.
161. Oral. Garda Con Bradley, Newbridge.

Chapter 8

1. *The monthly official directory of the Curragh Camp and Newbridge, December 1887.* (Dublin 1887), p. 30–32.
2. *Burke's peerage and baronetage.* (London 1887), p. 834.
   Bateman, J. *The great landowners of Great Britain and Ireland.* (London 1883), p. 265.
3. *Burke's,* Ibid., p. 985; Bateman, Ibid., p. 306.
4. *Burke's Irish Family Records.* p. 245, 355, 692, 784; *Burke's peerage and baronetage,* p. 306.
   Bateman, Ibid., p. 97.
5. *Burke's peerage and baronetage,* p. 443; Bateman, Ibid., p. 138.
6. *Burke's peerage and baronetage,* p. 306; Bateman, Ibid., p. 97.
7. Boylan, Lena 'Celbridge in Vanessa's time', In *Journal of the County Kildare Archaeological Society* Vol. XVI (1986), p. 404; Boylan, Lena *Castletown and its owners.* (Celbridge 1968), p. 63

8. Wolfe, George 'The Wolfe family of County Kildare' In *Journal of the County Kildare Archaeological Society* Vol. III (1902), p. 361, 389.
9. Wyndham, Horace *The Queen's Service.* (London 1899), p. 84.
10. Ibid., p. 104.
11. Oral. Con Cummins, Newbridge. 1992.
12. Oral. Lieut. Col. C. M. L. Clements. 1992.
13. Kilmainham Papers. Vol. 269. 10.10.1866; In *Leinster Leader.* 9.5.1896.
14. D'Arcy, Fergus *Horses, Lords and Racing Men.* (The Curragh 1991), p. 155–6.
15. *Baily's Hunting Directory 1906–7.* (London 1906), p. 213.
16. Kildare Hunt Club. Accounts 1913.
17. Mayo, Earl of and Boulton, W.B. *History of the Kildare Hunt.* (London 1913), p. 116, 117, 119, 120.
18. Kildare Hunt Club. Accounts 1866.
19. Mayo, Earl of and Boulton, Ibid., p. 330.
20. Kildare Hunt Club. Accounts 1899.
21. Ibid., 1901.
22. Ibid., 1913.
23. O'Connor Morris, M. *Hibernia Venatica.* (London 1878), p. 2.
24. Ibid., p. 37–38.
25. Ibid., p. 45.
26. Ibid., p. 102–3.
27. Ibid., p. 109.
28. Ibid., p. 197
29. Ibid., p. 230–1.
30. Ibid., p. 419.
31. Ibid., p. 141, 412.
32. Norton. James E. *History of the South County Dublin Harriers and some neighbouring packs.* (Dublin 1991), p. 180.
33. Ibid., 185.
34. *The Irish Military Guide.* (Dublin 1909), p. 22.
35. In *Leinster Express.* 7.1.1865.
36. Gonne MacBride, Maud *A Servant of the Queen.* (London 1938), p. 13.
37. *United Ireland.* 26.11.1881.
38. In *Leinster Leader.* 14.10.1882; ibid. 24.5.1986.
39. In *Kildare Observer.* 7.11.1885.
40. Ibid., 5.12.1885.
41. Ibid., 6.11.1886.
42. Ibid., 7.11.1885.
43. Ibid., 1.4.1899.
44. In *Kildare Observer.* 25.2.1899.
45. In *Kildare Observer.* 4.11.1899.
46. In *Leinster Leader.* 7.11.1914, 28.11.1914, 5.12 1914,
47. Mayo, Earl of and Boulton, Ibid., p. 353.
48. *Burke's Irish family records.* p. 745.
49. Watson, S.J. *Between the flags.* (Dublin 1969), p. 77.
50. In *The Irish Times.* 17.4.1868.
51. Welcome, John. *Irish Horse-racing.* (Dublin 1982), p. 47.
52. Repington, Lieut. Col. Charles a Court. *Vestigia.* (London 1919), p. 69–70.
53. Haire, D. N. *The British Army in Ireland 1868–1890.* Unpublished M. Litt. Thesis No. 109, T.C.D., 1973, p. 244.

54. In *Kildare Observer* 4.4.1885, 25.4.1885.
55. Ibid., 15.8.1885.
56. Ibid., 7.8.1886.
57. Ibid., 6.3.1886.
58. Godley, General Sir Alexander *Life of an Irish soldier.* (London 1939), p. 20.
59. In *Kildare Observer.* 10.10.1885.
60. In *Leinster Leader.* Centenary Supplement. 1980. p. 35.
61. O Muineog, M. *Kilcock G.A.A. history.* (Clane 1987), p. 26, 29.
62. Gonne MacBride, Maud *A Servant of the Queen.* (London 1938), p. 40.
63. In *Kildare Observer.* 26.3.1887.
64. Godley, Ibid., p. 28–9.
65. In *Leinster Leader.* 14.3 1896, 11.4.1896, 19.9.1896, 15.5.1886.
66. In *Leinster Leader.* 25.4.1896.
67. In *Kildare Observer.* 1.3.1899.
68. Ibid., 25.2.1899.
69. Ibid., 25.3.1899.
70. Ibid., 15.4.1899.
71. Godley, Ibid., p. 17.
72. Ibid., p. 18.
73. Ibid., p. 18–20.
74. Ibid., p. 21–22.
75. Ibid., p. 20–21.
76. Ibid., p. 27.
77. In *Leinster Leader.* 28.11.1896.
78. Ibid., 25.1.1896.
79. In *Leinster Express.* 22.1.1876.
80. *Thom's Almanac and Official Directory.* (Dublin 1880), p. 305.
81. In *Kildare Observer.* 20.11.1886.
82. In *Kildare Observer.* 4.9.1886.
83. Godley, Ibid., p. 18, 28.
84. Clark, Brian 'The Victoria Cross: a register of awards to Irish-born officers and men', In *Irish Sword.* Vol. XVI (1986), p. 190.
85. In *Kildare Observer.* 13.8.1887.
86 Haldane, General Sir Aylmer A Soldier's Saga. (London 1948), p. 58.

CHAPTER 9

1. In *Kildare Observer.* 21.2.1885.
2. Ibid., 12.9.1885.
3. Ibid., 7.3.1885.
4. Ibid., 15.8.1885.
5. Ibid., 12.12.1885.
6. Ibid., 26.9.1885, 6.2.1886.
7. Ibid., 9.2.1889.
8. Registers of churches, copies in Kildare County Library.
9. Wyndham, Horace. *The Queen's Service.* (London 1899), p. 205–6.
10. In *Leinster Leader.* 25.1. 1896, 1.2.1896.
11. Ibid., 29.2.1896.
12. Ibid., 29.2.1896.
13. Ibid., 14.3.1896.

14. Ibid., 2.3.1896.
15. In *Kildare Observer.* 4.6.1887.
16. Ibid., 4.4.1896.
17. Ibid., 3.10.1896.
18. Ibid., 4.4.1896.
19. Ibid., 9.5.1896.
20. Ibid., 13.6.1896.
21. Ibid., 27.6.1896.
22. Ibid., 8.8.1896.
23. Ibid., 1.8.1896.
24. Ibid., 4.7.1896.
25. Godley, General Sir Alexander *Life of an Irish soldier.* (London 1939), p. 21.
26. In *Leinster Leader.* 5.12.1896.
27. Ibid., 11.4.1896.
28. Ibid., 8.4.1897.
29. In *Kildare Observer.* 6.1.1900.
30 Bell, Pat *Long shies and slow twisters. 150 years of cricket in Kildare.* (Celbridge 1993), p. 16.
31. Ibid., p. 52.
32. Ibid., p. 14.
33. Ibid., p. 61.
34. In *Kildare Observer.* 18.6.1887.
35. Ibid., 15.5.1886.
36. Ibid., 29.5.1886, 26.6.1886.
37. Godley, Ibid., p. 17.
38. In *Kildare Observer.* 13.8.1887.
39. Ibid., 24.7.1886.
40. Ibid., 7.8.1886.
41. Hamilton, I.B.M. *The happy warrior.* (London 1966), p. 17.
42. In *Leinster Express.* 26.8.1876.
43. In *Leinster Leader.* 20.6.1896.
44. Bell, Ibid., p. 64.
45. Ibid., p. 16.
46. In *Leinster Express.* 15.4.1876.
47. In *Kildare Observer.* 22.8.1885.
48. Ibid., 9.10.1886.
49. In *Leinster Leader.* 12.9.1896.
50. Ibid., 18.7.1896, 25.7.1896.
51. In *Leinster Express.* 7.10.1876.
52. In *Leinster Leader.* 2.3.1896, 14.3.1896, 28.3.1896, 2.4.1896, 11.4.1896, 18.4.1896, 2.5.1896, 20.6.1896, 11.7.1896, 11.7.1896, 18.7.1896, 25.7.1896, 1.8.1896, 12.9.1896, 17.10.1896, 5.12.1896, 12.12.1896.
53. In *Leinster Leader.* 5.12.1896.
54. Ibid., 12.12.1896.
55. Ibid., 17.10.1896.
56. Gibson, W.H. *Early Irish golf.* (Naas 1988), p. 154–5.
57. *The Regiment. An Illustrated Journal for Everybody.* (London, 23.5.1896) p. 15.
58. In *Kildare Observer.* 28.1.1899.
59. Ibid., 18.2.1899.
60. Ibid., 4.2.1899.
61. Robertson, Nora *Crowned Harp.* (Dublin 1960). p.101.

62. Spiers, E.M. *The late Victorian army.* (Manchester 1992), p. 98.
    Muenger, E.A. *The British military dilemma in Ireland. 1886–1914.* (Dublin 1991), p. 18.
63. Robertson, Ibid., p. 26.
64. Muenger, Ibid., p. 63.
65. Ibid., p. 64.
66. Cooper, Bryan *The Tenth (Irish) Division in Gallipoli.* (Dublin 1993).
67. Lyons, M. R. *Naas Lawn Tennis Club, a short history 1881–1992.* (Naas 1992), p. 5.
68. Ibid., p. 6.
69. Barton, Derick. *Memories of ninety years.* (Dublin 1989), p. 19, 20.
70. Postcard. Camp life in Newbridge 1908. In collection of Mr. P. Bell. Celbridge.
71. Parkinson, Bro. R.E. *The lodge of research.* (Dublin 1959), p. 119.
72. Ibid., p. 120.
73. Ibid., p. 119.
74. Ibid., p. 119.
75. Crossle, Philip *Irish Masonic records.* (Dublin 1973), p. 48.
76. Ibid., p. 54.
77. Ibid., p. 90.
    Oral, The Masons in Newbridge. Social Employment Scheme Folklore Project. Kildare County Library 1988–9. p. 2.
78. Crossle, Ibid., p. 56.
79. Ibid., p. 143.
80. Ibid., p. 88–9.
81. Curragh Lodge (No. 397). Minute book 1917–1920; Seal of Lodge in Masonic Archive, Dublin.
82. Grand Lodge Registers. Vol. 4. 4th Series. Curragh Lodge (No. 397), Newbridge, County Kildare. 1906–1908. p. 97.
83. Grand Lodge Registers. Vol. 2. 4th Series. United Services Lodge (No. 215) Newbridge 1906–1914. p. 611–2.
84. Oral. Mr. Kenneth Dunne. 1992.
85. Grand Lodge Registers. Vol. 3. 4th Series. Travelling Warrants. Glittering Star (Lodge No. 322), Worcestershire Regiment. 1878–1921. p. 465–6. St Patrick's (Lodge No. 295), 4th Dragoons. 1888–1908. P. 359–360.
86. In *Kildare Observer.* 17.6.1899.
87. Oral. The Masons in Newbridge. SES Folklore Project. Kildare County Library 1988–9. p. 2.
88. Burke, Sir Bernard *Burke's peerage and baronetage.* (London 1887), p. 838.
89. In *Kildare Observer.* 29.7.1899.
90. Verner, Col. W. *The Military Life of H.R.H. George, Duke of Cambridge.* (London 1905), p. 126.

Chapter 10

1. Lugard, Lieut. Col. H.W. *Narrative of Operations in the arrangement and formation of a camp. for 10,000 Infantry on the Curragh of Kildare.* (Dublin 1858). p. 23–24. Plan 37.
2. Ibid., p. 25–26.
3. In *Leinster Express.* 11.4.1855.
4. Kilmainham Papers. Vol. 162. 26.1.1860.
5. In *Leinster Express.* 22.9.1855.
6. Ibid., 8.12.1855.
7. Kilmainham Papers. Vol. 268. 6.10.1859; Vol. 162. 6.9.1859, 4.10.1859.
8. Ibid., Vol. 270. 9.4.1870.
9. Lugard, Ibid., Curragh Sheet. Ordnance Survey.
10. Lamprey, in Gorry Catalogue, Dublin. April 1989. p. 6.

11. Lugard, Ibid., p. 14.
12. Lugard, Ibid., p. 20.
13. Lynch, P. and Vazey, J. *Guinnesses' brewery in the Irish economy 1759–1876.* (Cambridge 1960), p. 215.
14. Lugard, Ibid., p. 25.
15. In *Leinster Express.* 24.6.1865.
16. Crown Lands. File 1406. 1866–1910. Property Section, D.O.D., Dublin.
17. Mansion House Fund: Local committee no. 565. Kildare, county Kildare. Dublin Corporation Archives.
18. Kilmainham Papers. Vol. 161. 6.7.1855, p. 360.
    Ibid., Vol. 268. 9.1.1856.
19. *Thom's Almanac and Official Directory* (Dublin 1856). p. 213.
20. Kilmainham Papers. Vol. 110. 11.5.1859, 20.5.1859, 14.10.1859.
21. Ibid., 1.3.1860.
22. Ibid., Vol. III. 14.7.1866.
23. Ibid., Vol. 268. 14.5.1857.
24. Ibid., Vol. 162. 29.7.1859, 15.8.1859, 30.8.1859.
25. In *Leinster Leader.* 26.9.1896.
26. Kilmainham Papers. Vol. 268. 12.7.1861.
27. In *Leinster Express.* 21.1.1865.
28. Ibid., 3.6.1865.
29. Ibid., 15.7.1865
30. Ibid., 21.1.1865.
31. Ibid., 13.5.1865.
32. *An Act to make better Provision for the Management and Use of the Curragh of Kildare. 1868.* Schedules, p. 567.
33. D'Arcy, Fergus *Horses, lords and racing men.* (Curragh 1991), p. 156.
34. In *Kildare Observer.* 3.1.1885.
35. *Porter's Post Office guide and directory for the counties Carlow and Kildare.* (Dublin 1910), p. 82.
36. Kilmainham Papers. Vol. 162. 25.5.1858, p. 109.
37. Ibid., Vol. 162.9.2. 1859, p. 236.
38. Ibid., Vol. 269. 31. 10. 1866.
39. In *Leinster Express.* 4.3.1865.
40. Ibid., 22.4. 1865; Army Circular; Camp Curragh, fuel, light and straw supplied by Barrack Dept. 1867, clause 52.
41. In *Leinster Express.* 14.12.1865, 21.12.1865, 10.12.1865, 30.12.1865.
42. Oral. Kenneth Dunne. South Green, Kildare. 22.6.1994.
43. Local tradition. 1992.
44. In *Kildare Observer.* 3.1.1885.
45. Ibid., 14.2.1885.
46. Ibid., 21.2.1885.
47. Ibid., 7.2.1885.
48. Ibid., 4.7.1885.
49. Ibid., 7.11.1885.
50. The contract prices for bread, flour and fresh meat supplied to the Curragh in 1875 were higher than for supplies to other stations. Parliamentary papers. lxxx. 334–5. 1875.
51. In *Kildare Observer.* 27.3.1886.
52. Ibid., 30.1.1886.
53. *Thom's Irish Almanac and official directory.* (Dublin 1868). p. 592.
54. Oral. Dick Brophy. 11.3.1992. Brophy family documents in McGlynn papers. 1992.

---

55. McGlynn papers. Cash books: 1886, p. 308–9; 1887, p. 334; 1891, p. 726; 1894, p. 870.
56. Ibid., 1883, p. 1–14; 1885, p. 273–280; 1893, p. 824–830; 1893, p. 879–886.
57. Oral. James Molloy. 1994; Oral. Conleth Cummins. 1993.
58. Th*e monthly official directory of the Curragh Camp. and Newbridge.* (Dublin 1887), p. 26–9.
59. In *Leinster Leader.* 6.6.1896.
60. Ibid., 10.10.1896, 24.10.1896.
61. McGlynn papers.
62. In *Kildare Observer.* 16.5.1885.
63. Ibid., 3.10.1885, 26.12.1885.
64. Ibid., 9.2.1889.
65. Ibid., 3.1.1885.
66. Ibid., 6.2.1886.
67. In *Leinster Leader.* 31.7.1896.
68. Ibid., 18.7.1896.
69. In *Kildare Observer.* 12.2.1887.
70. Ibid., 10. 12.1887.
71. Hickey, D.J. and Doherty, J.E. *A dictionary of Irish history. 1800–1980.* (Dublin 1987), p. 413.
72. In *Kildare Observer.* 21.1.1899, 11.2.1899.
73. Ibid., 8.4.1899.
74. Ibid., 16.9.1899, 9.12.1899.
75. Ibid., 16.2.1901.
76. Wyndham, Horace *The Queen's hussar.* (London 1899) p. 256, 259–260.
77. Ibid., p. 256, 271.
78. In *Kildare Observer.* 8.4.1899.
79. Larkin, William *Map. of Kildare. 1811.* National Library Maps Vol. 15a 3–16. Map. No. 50.
80. Lugard, Ibid., Map. Curragh sheet.
81. Valuation lists, County Kildare. Valuation Office, Dublin.
82. Tully, Col. F.J. Letter to the editor. In *An Cosantoir. The Irish Defence Journal* Vol. XXX. No. 9. (September 1965), p. 481.
83. Computation by Mr Tony Kinsella, statistician, from 1911 Army Medical Department reports and annual estimates.
84. Army estimates. Appendix 2. 1911.
85. British Labour Statistics. Historian abstracts. 1886–1968. (H.M.S.O. 1971), Table 97, p. 111. Also information from Central Statistics Office Dublin. 8.7.1994.
86. Census of Ireland 1911. Vol. 1. Leinster. Miscellaneous tables. p. 44.
87. Ibid.
88. Military archives. File A13709.
89. Census of Ireland 1911. Vol. 1. Leinster. p. 44, 71.
90. In *Leinster Leader.* 7.3.1914.
91. Ibid., 7.3.1914.
92. Ibid., 18.4.1914.
93. Ibid., 4.4.1914.
94. Ibid., 8.8.1914, 15.8.1914, 17.10.1914, 12.12.1914, 26.12.1914.
95. Ibid., 21.1.1914, 11.11.1916.
96. Ibid., 15.8.1914.
97. In *The Irish Times.* 24.12.1921; See also Financial Estimates in Haire, D.N. British Army in Ireland. 1868–1890. Unpublished M. Litt. Thesis, No. 109. T.C.D., 1973, p. 311–320.

CHAPTER 11

1. Kilmainham Papers. Vol. 270. 17.8.1871.
2. Ibid., Vol. 271, 12.10.1878; Vol. 272, 14.4.1879.
3. Map. 9-A3. Corps of Engineers, Curragh Camp.
4. Map. 16–14, Ibid.
5. In *Kildare Observer.* 12.12. 1885.
6. Kilmainham Papers. Vol. 271. 30.10.1873.
7. Ibid., Vol. 271. 30.11.1877.
8. Ibid., Vol. 271. 11.2.1878.
9. Ibid., Vol. 271. 25.2.1878.
10. Ibid., Vol. 271. 3.8.1878.
11. Ibid., 11.12.1877.
12. Army Medical Department Report for the year 1878; Parliamentary papers. Vol. XLIV. 331. 1878–79; Army Medical Department Report for the year 1882; Parliamentary papers. Vol. XLIX.299. 1884.
13. Kilmainham Papers. Vol. 272. 8.3.1880.
14. Ibid., Vol. 271. 1.10.1878.
15. Ibid., Vol. 272. 1.4.1880.
16. Map. 24–24 No. 3. Military Archives, Dublin.
17. In *Kildare Observer.* 12.12. 1885.
18. Map. 8–28 No. 1. Military Archives, Dublin; in *Leinster Leader*, 24.5.1986.
19. In *Kildare Observer.* 12.12. 1885.
20 Adam, James *Catalogue Irish Art Sale 27 March 1996.* Dublin, p. 19.
21. *The monthly official directory of the Curragh Camp and Newbridge. December, 1887.* (Dublin 1887), p. 15, 20–24, 26–27.
22. Spiers, E.M. *The late Victorian army.* (London 1992), p. 221.
23. Wyndham, Horace. *The Queen's Service.* (London 1899), p. 81–5.
24. Standard Plan. No. 23. Capt. Stockley R.E. for Director of Fortifications and Works. 1905. C.O.E. Curragh.
25. Conversation between Brigadier Denis Fitzgerald and Lieut. Col. C.M.L. Clements. 27.4.1991.
26. Maps 164–13 No.2; 125–13; 81–13; 109–13. Military Archives, Dublin.
27. Maps. 12–6; 3–17; 17–17; 3–6. Military Archives Dublin; Army Department Medical Report. 1887. p. 48.
28. Maps. 15–21; 5–23; 2–3. C.O.E. Curragh Camp.
29. Maps. 16–14; 42–15; 34–15 No. 2; 73–21; 33–15; 54–18. C.O.E. Curragh Camp.
30. In *Leinster Leader.* 28.11.1896.
31. Swan, Comdt. D.A. 'The Curragh of Kildare', In *An Cosantoir. The Irish Defence Journal.* (May 1972). p. 54. Originals displayed in Ceannt Officers' Mess, Curragh Camp.
32. Date on Curragh Water Tower.
33. In *Leinster Leader.* 25.1.1896.
34. Ibid., 3.3.1896.
35. Ibid., 12.9.1896.
36. Ibid., 10.10.1896.
37. Ibid., 18.4.1896.
38. Ibid., 25.4.1896.
39. Folklore Project, S.E.S. 1988–89. Kildare County Library. p. 12.
40. Swan, Ibid., p. 52.
41. In *Leinster Leader.* 20.1.1994.
42. Army Medical Report. 1896. p. 64.
43. In *Kildare Observer.* 15.7.1899.

44. Ibid., 2.9.1899.
45. *The Irish Builder*, (Dublin)1.1.1901. p. 584; 16.1.1901, p. 603.
46. Ordnance Survey Map. Index to the Curragh. Scale 12.672" = 1 mile. 1902.
47. Records of married quarters, Curragh Camp. 1922. Command Q.M. office, Curragh Camp. Books 1–8.
48. Ordnance Survey Map. Index to the Curragh. 1902.
49. Ibid.
50. In *Leinster Leader*. 17.6.1922.
51. Irish Folklore Collection, County Kildare, Vol.777, p.17. Department of Irish Folklore, U.C.D.

CHAPTER 12

1. In *Kildare Observer*. 14.1.1899.
2. Ibid., 27.5.1899.
3. Ibid., 3.6.1899.
4. Head, Lieut. Col. C.O. *No great shakes*. (London 1943), p. 80–81.
5. In *Leinster Leader*. 15.8.1896, 22.8.1896, 5.9.1896.
6. In the matter of the Defence Act, 1842, the Ordnance Board transfer Act, 1855, and the Acts amending and extending the same; and in the matter of the Glen Imaal Artillery Range. Copy notice to Treat. Chief Secretary's Office, Dublin Castle. 8.3.1899.
7. In *Kildare Observer*. 3.6.1899.
8. Ibid., 3.6.1899.
9. Oral. Mrs. M. Noone (born 1900). 1960.
10. Oral. Nicholas Mac Dermott. 9.12.1993.
11. In *Kildare Observer*. 15.7.1899.
12. Young, M.F. *Letters of a noble woman*. (London 1908), p. 163.
13. In *Kildare Observer* 10.3.1900.
14. Ibid., 3.6.1899.
15. Ibid., 17.6.1899.
16. Ibid., 24.6.1899.
17. Gravestone at Royal Hospital, Kilmainham, Dublin, and unidentified magazine clipping dated 28.6.1899 in scrapbook at Royal Hospital.
18. In *Kildare Observer*, 8.4.1899.
19. Ibid., 1.7.1899.
20. Ibid., 8.7.1899.
21. Ibid., 15.7.1899.
22. Ibid., 29.7.1899.
23. Ibid., 5.8.1899.
24. Ibid., 5.8.1899.
25. Ibid., 12.8.1899.
26. Ibid., 5.8.1899.
27. Ibid., 30.9.1899.
28. Ibid., 30.9.1899, 7.10.1899, 21.10.1899, 28.10.1899.
29. Ibid., 7.10.1899.
30. Ibid., 7.10.1899.
31. Ibid., 14.10.1899.
32. Ibid., 7.10.1899.
33. Ibid., 14.10.1899.
34. Ibid., 21.10.1899.
35. Ibid., 28.10.1899.

36. Ibid., 4.11.1899.
37. Ibid., 21.10.1899.
38. Ibid., 4.11.1899.
39. Ibid., 11.11.1899.
40. Ibid., 25.11.1899.
41. Ibid., 4.11.1899.
42. Ibid., 18.11.1899.
43. Ibid., 4.11.1899.
44. Ibid., 18.11.1899, 16.12.1899, 19.5.1900.
45. Clark, Brian 'The Victoria Cross: a register of awards to Irish-born officers and men', In *Irish Sword* Vol. XVI (1986), p. 194; Wilkins, P.A. *The history of the Victoria Cross.* (London 1904), p. 304; Creagh, General Sir O'Moore, and Humphris, H.M. *The V.C. and D.S.O.* (London 1924), Vol. 1. p. 116.
46. In *Kildare Observer.* 16.12.1899.
47. In *Kildare Observer.* 6.1.1900.
48. Ibid., 20.1.1900, 27.1.1900, 3.2.1900, 17.2.1900, 3.3.1900, 17.3.1900, 14.4.1900, 5.5.1900, 14.7.1900, 27.10.1900, 12.1.1901, 30.3.1901.
49. Ibid., 13.1.1900, 17.3.1900, 24.3.1900.
50. Ibid., 20.1.1900.
51. Ibid., 20.1.1900; *Burke's Irish family records.* (London 1976), p. 342.
52. In *Kildare Observer.* 13.1.1900.
53. In *Leinster Leader.* 13.1.1900.
54. In *Kildare Observer* 27.1.1900.
55. Ibid., 2.2.1901.
56. Ibid., 30.11.1901.
57. Ibid., 10.3.1900.
58. Ibid., 24.3.1900.
59. Ibid., 24.3.1900.
60. Ibid., 14.4.1900.
61. Ibid., 21.4.1900.
62. Ibid., 7.4.1900.
63. Ibid., 14.4.1900.
64. Ibid., 5.5.1900.
65. Ibid., 14.7.1900.
66. Ibid., 9.6.1900.
67. Ibid., 21.9.1901.
68. Ibid., 13.1.1900.
69. Ibid., 28.4.1900.
70. Ibid., 3.8.1901.
71. Ibid., 30.11.1901.
72. Ibid., 8.9.1900, 4.5.1901, 18.5.1901.
73. Ibid., 2.9.1900.
74. Ibid., 17.11.1900.
75. Limb, Sue and Cordingly, Patrick. *Captain Oates, soldier and explorer.* (London 1982), p. 32–3.
76. Ibid., p. 38.
77. In *Kildare Observer.* 23.6.1900.
78. Ibid., 30.6.1900.
79. Ibid., 30.6.1900.
80. Ibid., 7.7.1900.
81. Ibid., 13.10.1900.

82. Ibid., 26.1.1901.
83. Ibid., 23.2.1901.
84. Ibid., 12.1.1901.
85. Ibid., 10.2.1900.
86. Ibid., 2.6.1900.
87. Ibid., 21.7.1900.
88. Ibid., 28.7.1900.
89. Ibid., 24.3.1900.
90. Ibid., 8.9.1900.
91. Ibid., 30.6.1900.
92. Ibid., 6.7.1901.
93. Ibid., 6.1.1900.
94. Ibid., 7.7.1900.
95. Ibid., 4.8.1900.
96. Ibid., 22.12.1900, 2.2.1901.
97. Ibid., 4.5.1901.
98. Ibid., 15.6.1901.
99. Ibid., 26.1.1901, 9.2.1901.
100. Ibid., 10.8.1901.
101. Ibid., 28.12.1901.
102. Ibid., 27.7.1901.
103. Ibid., 7.12.1901.
104. Ibid., 31.8.1901.
105. Ibid., 28.9.1901.
106. Ibid., 13.10.1901.
107. In *Nationalist and Leinster Times.* Centenary Issue. 1883–1983. p. 48.
108. *Census of Ireland 1891*, p. 263; *Census of Ireland 1901*, p. 68, 79.
109. In *Kildare Observer.* 19.10.1901.
110. Ibid., 14.12.1901.
111. In *Leinster Leader.* 29.4.1993.
112. In *Kildare Observer.* 14.12.1901.
113. Ibid., 28.12.1901.
114. Ibid., 14.12.1901.
115. Ibid., 8.9. 1900.
116. Ibid., 10.11.1900.
117. Ibid., 7.7.1900.
118. Ibid., 22.9.1900; Military archives. Plans B36–1152a dated 3.4.1902, 26.6.1902, 9.6.1903.
119. In *Kildare Observer.* 8.9.1900.
120. Ibid., 15.9.1900.
121. Ibid., 10.3.1900.
122. Ibid., 10.3.1900.
123. Ibid., 14.4.1900.
124. Ibid., 15.9.1900.
125. Ibid., 1.12.1900, 15.12.1900.
126. Ibid., 9.3.1901, 16.11.1901.
127. Ibid., 29.9.1900.
128. Ibid., 16.2.1901, 9.3.1901.
129. Ibid., 19.10.1901.
130. Ibid., 14.12.1901.
131. In *The Irish Times.* 19.3.1902.

132. In *Leinster Leader.* 7.6.1902.
133. Limb and Cordingly, Ibid., p. 49.
134. Ibid., p. 59.
135. Ibid., p. 52.
136. Ibid., p. 161.
137. In *Leinster Leader.* 26.7.1902.
138. Ibid., 20 12.1902.
139. Ibid., 10.8.1902.
140. Ibid., 20.12.1902.
141. Ibid., 27.12.1902.
142. Ibid., 20.12.1902.
143. Ibid., 1.1.1910.
144 Ibid., 20.12.1902.
145. Ibid., 27.12.1902.
146 Ibid., 20.12.1902.
147 Ibid., 6.5.1903.
148. Ibid., 28.3.1903.
149. D'Arcy, Fergus *Horses, lords and racing men.* (The Curragh 1991), p. 228, 229.

CHAPTER 13

1. *The Barrack Book, Home Stations.* (H.M.S.O., London 1904), p. 46–7.
2. Hogan, Brig. Gen. P. D. M.S. Tables of Service of British Regiments in Ireland. N.D. Military History Society of Ireland. p. 16, 22, 41A, 59.
3. Census of Ireland. Leinster. 1911, p. 59,66.
4. *Nás na Ríogh from Poorhouse Road to Fairy Flax . . . an illustrated history of Naas.* (Naas 1990), p. 76–7.
5. Kilmainham Papers. Vol. 303. Irish Command Orders. No. 69. 22.3.1907. No. 72. 26.3.1907.
6. Lyttelton, Gen. Sir Neville *Eight years soldiering, politics and games.* (London 1927), p. 286.
7. In *Leinster Leader.* 10.3.1906.
8. Ibid., 3.11.1906
   Kilmainham Papers. Vol. 351. 16.2.1909, 30.3.1909.
9. In *The Irish Times.* 4.3.1911.
10. In *Leinster Leader.* 23.6.1906.
11. Ibid., 5.5.1906.
12. Ibid., 22.12.1906.
13. Ibid., 5.5.1906.
14. Ibid., 4.12. 1909.
15. Ibid., 6.12.1913.
16. Ibid., 15.3.1913, 8.11.1913.
17. Ibid., 9.5.1914.
18. In *The Irish Times.* 15.1.1911.
19. Vanderveen, B.H. (ed). *The Observer's army vehicles directory to 1940.* (London 1940), p. 174.
20. In *Leinster Leader.* 29.9.1906.
21. Ibid., 15.12.1906.
22. Ibid., 26.5.1906.
23. *Porter's Post Office guide and directory for the counties Carlow and Kildare.* (Dublin 1910).
24. Oral. Cummins, Conleth. Newbridge. 1994.
25. Berney, Thomas. Account Book. 1907–1909. p. 63, 148, 162, 188, 348.
26. In *The Irish Times.* 7.6.1911.

27. Newbridge, local history. Social Employment Scheme, Folklore Project. 1988–89. Kildare County Library.
28. Cooke, J. *Bord na Móna, Peat Research Centre.* (Newbridge 1991), p. 52.
29. In *Leinster Leader.* 5.5.1906.
30. *Curragh Camp and district illustrated and described.* Dublin *c.* 1908.
31. In *Leinster Leader.* 13.1.1906.
32. Ibid., 17.3.1906.
33. Ibid., 21.7.1906.
34. Ibid., 1.9.1906.
35. Ibid., 20.3.1909.
36. Ibid., 25.5.1907.
37. Ibid., 22.12.1906, 5.1.1907.
38. Ibid., 30.11.1907.
39. In *Leinster Leader.* 7.1.1911.
40. In *The Irish Times.* 11.1.1911.
41. In *Leinster Leader.* 25.5.1907.
42. Ibid., 18.9.1909.
43. Ibid., 4.9.1909.
44. In *Irish Times.* 8.5.1911.
45. Letter from P. Boland, Newbridge. 17.7.1984.
46. Bell, Pat *Long shies and slow twisters.* (Celbridge 1993), p. 64.
47. Gibson, W. H. *Early Irish golf.* (Naas 1988), p. 80, p. 155, p. 240.
48. In *Leinster Leader.* 28.5.1910.
49. Carson, Patricia 'Sandes Soldiers home', In *Journal of the County Kildare Archaeological Society* Vol. XIV (1970), p. 473.
50. In *Irish Times.* 1.6.1911, 30.10. 1911.
51. In *Kildare Observer.* 22.9.1900.
52. Limb, Sue and Cordingly, Patrick. *Capt. Oates, soldier and explorer.* (London 1982), p. 49.
53. Lyttelton, Gen. Sir Neville. Ibid., p. 286.
54. In *The Irish Times.* 30.3.1911.
55. Ibid., 29.3.1911.
56. Ibid., 18.5.1911, 29.5.1911.
57. Ibid., 4.7.1911.
58. Ibid., 12.7.1911.
59. Ibid., 10.7.1911.
60. Ibid., 5.8.1911.
61. Ibid., 9.6.1911.
62. Ibid., 24.7.1911.
63. Ibid., 24.7.1911.
64. Ibid., 10.8.1911.
65. Ibid., 18.8.1911.
66. Ibid., 28.8.1911.
67. Ibid., 11.9.1911.
68. Ibid., 23.10.1911.
69. Ibid., 20.11.1911, 25.11.1911.
70. In *Kildare Observer.* 29.4.1911.
71. Gough, Sir Hubert *Soldiering on.* (London 1954), p. 98.
72. Smyth, Senator M. Irish Volunteers. 1914–1916. M.S. Military Archives, Dublin. p. 1.
73. In *Leinster Leader.* Centenary Supplement 1980. Section 3 p. 37.
74. In *Leinster Leader.* 15.3.1913.

75. *Burke's Irish Family Records.* (London 1976). p. 342.
76. O'Faoláin, Sean. *Vive Moi!. An autobiography.* (London 1967), p. 71.
77. In *Leinster Leader.* 20.12.1913.
78. Ibid., 19.4.1913.
79. Ibid., 5.7.1913.
80. Ibid., 5.7.1913, 20.12.1913.
81. Ibid., 8.11.1913.
82. O'Faolain, Ibid., p. 71–2.
83. In *Leinster Leader* 31.1.1914.
84. Beckett, I.F.W. *The Army and the Curragh incident 1914.* (London 1986), p. 439–440. Swan, Comdt. D.A. 'The Curragh of Kildare', In *An Cosantoir. The Irish Defence Journal.* (May 1972), p. 57–8.
85. Muenger, E.A. *British Military Dilemma in Ireland. 1886–1914.* (Dublin 1991), p. 177.
86. Fergusson, Sir James *The Curragh incident.* (London 1964), p. 41–2
87. Farrar-Hockley, A. *Goughie.* (London 1975), p. 91.
88. In *The Irish Times.* 3.3.1914; Fergusson, Ibid., p. 69.
89. Daniell, D.S. *4th Hussars. The story of a British cavalry regiment.* (Aldershot 1959), p. 239.
90. Beckett, Ibid., p. 79–80.
91. Fergusson, Ibid., p. 109–111.
92. Beckett, Ibid., p. 189.
93. Fergusson, Ibid., p. 115.
94. In *Illustrated London News.* 28.3.1914. p. 505.
95. Verner, Col. Willoughby *The military life of H.R.H. George, Duke of Cambridge.* (London 1905), Vol. 2. p. 381.
96. Fergusson, Ibid., p. 149–152.
97. Beckett, Ibid., p. 344–5.
98. Muenger, Ibid., p. 203.
99. Ibid., p. 165,202.
100. In *The Irish Times.* 24.3.1914.
101. Muenger, Ibid., p. 199; Johnstone, T. *Orange, green and khaki.* (Dublin 1992), p. 389.
102. Fergusson, Ibid., p. 194.
103. Macready. Rt. Hon. Sir Nevil. *Annals of an active life.* (London 1942), Vol. 1. p. 177.
104. In *The Freeman's Journal.* 2.10.1843.
105. Beckett, Ibid., p. 112.
106. Ibid., p. 155.
107. Daniell, D. S. 4th Hussars. The story of a cavalry regiment. (Aldershot 1959), p. 240. Fergusson, Ibid., p. 116.
108. Beckett, Ibid., p. 269.
109. Fergusson, p. 160.
110. Beckett, Ibid., p. 112.
111. Fergusson, Ibid., p. 160.
112. Beckett, Ibid., p. 156.
113. Ibid., p. 13.
114. Muenger, Ibid., p. 146–7.
115. Hogan, Brig. Gen. P.D. M.S. Tables of service of British regiments in Ireland. N.D. Military History Society of Ireland.

Chapter 14

1. *Buro Staire Mileata: 1913–1921. Chronology Part 1.* p. 69–70.
2. In *Leinster Leader.* 6.6.1914.
3. Smyth, Senator M. M.S. Irish Volunteers 1914–1916. p. 1. Military Archives.
4. In *Leinster Leader* 27.6.1914.
5. Ibid., 11.7.1914.
6. Ibid., 27.6.1914.
7. Ibid., 4.7.1914.
8. Ibid., 1.8.1914.
9. Ibid., 11.7.1914.
10. Ibid., 25.7.1914.
11. *Buro Staire Mileata,* Ibid., p. 89, 98, 102
12. In *Leinster Leader,* 18.7.1914.
13. *Buro Staire Mileata,* Ibid., p. 105, 106.
14. Ibid., p. 95.
15. Jeffrey, M.H. *The trumpet call obey.* (London 1968), p. 91–2.
16. In *Leinster Leader.* 1.8.1914, 8.8.1914.
17. Ibid., 8.8.1914.
18. Ibid., 15.8.1914.
19. Ibid., 8.8.1914.
20. Ibid., 15.8.1914.
21. Jeffrey, Ibid., p. 92.
22. Oral. Tom Garrett. 1992.
23. In *Leinster Leader.* 8.8.1914.
24. Ibid.
25. De Montmorency, Hervey *Sword and stirrup.* (London 1936), p. 242.
26. Ibid., p. 243–4.
27. Ibid., p. 245.
28. Head, Lieut. Col. C.O. *No great shakes:an autobiography.* (London 1943), p. 154.
29. Ibid., p. 158.
30. Ibid., p. 159.
31. Ibid., p. 221–2.
32. In *Leinster Leader.* 15.8.1914.
33. Department of Defence. Property Management Section. Crown Lands. File 1–66: 105–9. *Truth.* London. 12.8.1914. p. 387.
34. D'Arcy, Fergus *Horses, lords and racing men.* (The Curragh 1991), p. 243–4.
35. In *Leinster Leader.* 8.8.1914.
36. Ibid., 15.8.1914.
37. Ibid.
38. Swan, Comdt. D.A. 'The Curragh of Kildare', In *An Cosantoir. The Irish Defence Journal* (May 1972), p. 62.
39. In *Leinster Leader.* 22.8.1914.
40. Ibid., 12.9.1914.
41. Ibid., 19.9.1914.
42. Ibid., 12.9.1914.
43. Ibid., 8.8.1914.
44. Ibid., 12.9.1914.
45. Ibid., 3.10.1914.
46. Ibid., 3.10.1914.
47. Ibid., 19.12.1914.

48. Ibid., 28.11.1914.
49. Ibid., 12.9.1914.
50. Ibid., 3.10.1914.
51. Ibid., 10.10.1914.
52. O' Faoláin, Sean. *Vive Moi! An Autobiography.* (London 1967), p. 72.
53. In *Leinster Leader.* 15.8.1914.
54. Ibid.
55. Ibid.,3.10.1914.
56. Ibid., 17.10.1914.
57. Ibid., 7.11.1914.
58. Ibid., 28.8.1914, 10.10.1914, 28.11.1914, 8.6.1918.
59. Ibid., 7.11.1914, 14.11.1914.
60. Ibid., 7.11.1914.
61. Ibid., 31.10.1914.
62. Ibid., 21.11.1914.
63. Ibid., 20.5.1916.
64. Mosley, Sir Oswald *My life.* (London 1968), p. 45.
65. Ibid., p. 46–7.
66. Ibid., p. 48.
67. Craigie, Eric *An Irish sporting life.* (Dublin 1994), p. 108; Oral. Anne Crookshank. Dublin. 28.7.1994.
68. In *Leinster Leader.* 31.10.1914; Head, Lieut. Col. Ibid., p. 159.
69. Ibid., 17.10.1914.
70. Ibid., 5.12.1914.
71. Ibid., 5.12.1914.
72. Ibid., 15.8.1914.
73. Ibid., 7.11.1914.
74. Ibid., 5.12.1914.
75. Ibid., 28.11.1914.
76. Gibson. W.H. *Early Irish golf.* (Naas 1988), p. 155.
77. In *Leinster Leader.* 21.11.1914.
78. Ibid., 24.10.1914.
79. Laird, Frank M. *Personal experiences of the Great War. (An unfinished manuscript).* (Dublin 1925), p. 7–8, 10.
80. In *Leinster Leader.* 21.1.1922.
81. Oral. Baron de Robeck. 1992.
82. *Everyman's Encyclopaedia.* (London 1961), Vol.7, p. 501.
83. Oral. Mr. T. Garrett. 1992.
84. Callan, Patrick 'Recruiting for the British army in Ireland during the First World War', In *Irish Sword.* Vol. XVII (1990), p. 43–5.
85. Johnstone, Tom *Orange, green and khaki. The story of the Irish regiments in the Great War, 1914–18.* (Dublin 1992), p. 89–90.
86. Cooper, Bryan *The tenth (Irish) division in Gallipoli.* (Dublin 1989), p. 28.
87. Johnstone, Ibid., p. 7–8.
88. Ibid., p. 92.
89. In *Leinster Leader.* 24.4.1915.
90. In *Leinster Leader.* 22.8.1914, 11.12.1915, 11.8.1917.
91. Oral. Casey, P. Western Front Association. 10.1.1994
92. In *Leinster Leader.* 15.1.1916.
93. Ibid., 15.1. 1916, 8.7.1916

94. Ibid., 15.7.1916.
95. Ibid., 22.1.1916.
96. Ibid., 12.2.1916.
97. Beckett, I.F.W. *The Army and the Curragh incident 1914.* (London 1986), p. 425.
98. In *Leinster Leader.* 22.1.1916.
99. Ibid., 5.2.1916.
100. Ibid., 29.7.1916.
101. Ibid., 19.8.1916.
102. Ibid., 30.9.1916.
103. Ibid., 28.10.1916, 16.12.1916.
104. *Walford's the County Families of the UK.* (London 1920), p. 907.
105. Callan, Ibid.
106. In *Leinster Leader.* 12.11.1988.
107. Laird, Frank M. Ibid., p. 65.
108. Hayes, K.E. *A history of the R.A.F. and U.S.N.A.S. in Ireland.* 1913–1923. (Dublin 1988), p. 8–11.
109. Pyne Clarke, Olga. *A horse in my kit bag.* (London 1988), p. 126.
110. Oral. Mr T. Garrett. 1992.
111. Ibid.
112. Smyth, M. M.S. Military Archives. p. 3.
113. Ibid., p. 3–4.
114. Ibid., p. 6–7.
115. MacGiolla Choille, B.M. (editor) *C.S.O. Judicial Division. Intelligence Notes. 1913–16.* (Dublin 1966), p. 86.
116. *Buro Staire Mileata,* Ibid., Part. 1. p. 105.
117. MacGiolla Choille, Ibid., p. 86.
118. Ibid., p. 111.
119. Ibid., p. 110.
120. Ibid., p. 140.
121. Ibid., p. 160.
122. In *Leinster Leader.* 22.1.1916.
123. In *Leinster Leader.* 19.2.1916.
124. Ibid., 11.11.1916.
125. Ibid., 26.7.1916.
126. Ibid., 19.2.1916.
127. In *Kildare Observer.* 11.3.1916.
128. In *Leinster Leader.* 25.3.1916.
129. Ibid., 4.3.1916.
130. Smyth, Senator M. MS. Military Archives. Irish Volunteers 1914–16. p. 8.
131. Oral. Louis Moran, Ballysax. 27.6.1994.
132. MacGiolla Coille, Ibid., p. 204.
133. In *Leinster Leader.* 15.4.1916.
134. MacGiolla Coille, Ibid., p. 232.
135. Ibid., p. 233.
136. Caulfield, Max. *The Easter Rebellion.* (London 1964), p. 188.
137. Smyth, Ibid., p. 8–9.
138. Ibid., p. 11.
139. Smyth, Ibid., p. 12–14.
140. MacGiolla Coille, Ibid., p. 233–4; Hally, Col. P.J. 'The easter 1916 rising in Dublin: the military aspects', In *Irish Sword.* Vol.VII (1966), p. 314.
141. *Sinn Féin Rebellion Handbook.* Dublin 1917. p. 159.

142. Caulfield, Ibid., p. 350.
143. *Burke's Irish Family Records.* (London 1976), p. 1204; Sadleir, T.U. 'Kildare members of parliament 1559–1800', In *Journal of the County Kildare Archaeological Society.* Vol.VII (1914), p. 114.
144. Mac Giolla Coille, Ibid., p. 204.
145. Caulfield, Ibid., p. 110.
146. In *Kildare Observer.* 6.5.1916.
147. In *Leinster Leader.* 28.3.1991.
148. Smyth, Ibid., p. 15–17.
149. In *Leinster Leader* 20.5.1916.
150. Ibid., 28.3.1991.
151. Ibid., Special Supplement 1980. p. 37.
152. In *Leinster Leader.* 8.4.1916, 15.4.1916.
153. Ibid., 17.6.1916.
154. Ibid., 29.4.1916.
155. Ibid., 20.5.1916.
156. Bence-Jones, Mark *Twilight of the ascendancy.* (London 1987), p. 178.
157. Oral. Lieut. Col. C.M.L. Clements. 1992.
158. In *Leinster Leader.* 24.6.1916.
159. Ibid., 8.4.1916.
160. Ibid., 20.5.1916.
161. Ibid., 27.5.1916.
162. Ibid., 1.7.1916.
163. Ibid., 25.11.1916.
164. Ibid., 12.2.1916.
165. Ibid., 1.7.1916.
166. Ibid., 22.7.1916.
167. Ibid., 15.7.1916, 12.8.1916, 19.8.1916.
168. Ibid., 22.7.1916, 14.10.1916.
169. Ibid., 19.8.1916.
170. Ibid., 6.5.1916, 2.12.1916.
171. Ibid., 4.11.1916.
172. Ibid., 11.11.1916.
173. Ibid., 18.11.1916.
174. Ibid., 25.11.1916.
175. Ibid., 11.11.1916.
176. Ibid., 1.1.1916.
177. Ibid., 8.1.1916, 15.1.1916, 26.2.1916, 13.5.1916, 10.6.1916, 17.6.1916, 11.11.1916, 18.11.1916.
178. Ibid., 29.1.1916.
179. Ibid., 8.7.1916.
180. Ibid., 26.8.1916.
181. Ibid., 5.2.1916.
182. Ibid., 24.6.1916.
183. Ibid., 15.7.1916.
184. Ibid., 22.7.1916.
185. Ibid., 22.7.1916.
186. Ibid., 28.10.1916.
187. Ibid., 4.11. 1916.
188. Ibid., 19.8.1916.

189. Ibid., 5.8.1916.
190. Ibid., 16.9.1916.
191. Ibid., 18.11.1916.
192. Ibid., 2.12.1916.
193. Ibid., 2.9.1916.
194. Ibid., 15.7.1916.
195. Ibid., 4.11.1916.
196. Ibid., 9.12.1916.
197. Ibid., 24.6.1916.
198. Ibid., 22.7.1916.
199. Ibid., 9.9.1916.
200. Ibid., 23.12.1916.

CHAPTER 15

1. In *Leinster Leader.* 6.1.1917.
2. War Office 35–182A. Notes etc. (5974-G).
3. Mosley, Oswald *My life.* (London 1968), p. 85.
4 In *Leinster Leader.* 16.2.1918.
5 Ibid., 30.3.1918.
6 Ibid., 4.5.1918, 9.11.1918, 14.12.1918.
7 Ibid., 2.2.1918.
8 Ibid., 4.5.1918.
9 Ibid., 25.5.1918.
10. Ibid., 15.6.1918, 29.6.1918.
11. Hickey, D.J. and Doherty, J.E. *A dictionary of Irish history. 1800–1980.* (Dublin 1980), p. 89.
12. Ibid., p. 90.
13. In *Leinster Leader.* 23.3.1918.
14. Ibid., 8.6.1918.
15. Ibid., 31.8.1918.
16. Ibid., 27.7.1918.
17. Ibid., 20.4.1918.
18. Ibid., 8.6.1918.
19. Ibid., 9.3.1918.
20. Ibid., 3.6.1916.
21. Ibid., 3.10.1914.
22. Ibid., 2.2.1918.
23. Ibid., 17.8.1918.
24. In *The Irish Times.* 7.1.1988.
25. In *Leinster Leader.* 27.7.1918.
26. Ibid., 20.7.1918.
27. Ibid., 1.6.1918, 8.6.1918.
28. Ibid., 22.6.1918.
29. Ibid., 20.4.1918.
30. Ibid., 21.9.1918.
31. Ibid., 4.5.1918.
32. Ibid., 25.5.1918.
33. Ibid., 16.3.1918.
34. *Porter's Post Office guide and directory for the Counties Carlow and Kildare.* (Dublin 1910), p. 83.

35. O'Faoláin, Sean. *Vive Moi! An autobiography.* (London 1968), p. 75–6
36. In *Leinster Leader.* 7.9.1918.
37. Ibid., 26.10.1918, 30.11.1918, 7.12.1918.
38. Ibid., 19.10.1918.
39. Ibid., 21.9.1918.
40. Ibid., 16.11.1918.
41. In *Kildare Observer.* 16.11.1918.
42. Hickey and Doherty, Ibid., p. 90. Revision of the numbers killed continues; Tom Johnstone in *Orange, Green and Khaki* (Dublin 1992, p. 428) quotes casualties as 49,400 from the Irish National Memorial Garden at Islandbridge, Dublin. Pat Casey of The Western Front Association estimates the number of Irishmen who died in the Great War as 35,000.
43. In *Leinster Leader.* 3.6.1916.
44. Ibid., 12.11.1988.
45. Oral. Casey, Pat. 1993.
46. Creagh, General Sir O'Moore and Humphris, H.M. *The V.C. and D.S.O. Vol. 1.* (London 1924), p. 222.
47. *Nás na Ríogh: from Poorhouse Road to Fairy Flax . . . an illustrated history of Naas.* (Naas 1990). p. 97.
48. In *Leinster Leader.* 16.11.1918.
49. Burke, Sir Bernard. *A genealogical and heraldic dictionary of the peerage and baronetage etc.* (London 1953), p. 1260–61.
50. *Burke's Irish family records.* (London 1976), p. 343.
51. Ibid., p. 339.
52. In *Leinster Leader.* 20.8.1988.
53. *Burke's Irish family records.* p. 824.
54. In *Leinster Leader.* 8.7.1916.
55. Ibid., 15.7.1916.
56. Bence-Jones, Mark *Twilight of the ascendancy.* (London 1987), p. 184.
57. Gibson, W.H. *Early Irish golf.* (Naas 1988), p. 155.
58. In *Leinster Leader.* 4.11.1916.
59. Oral. Sheehan, in Newbridge Local History. SES Folklore Project, Kildare County Library. Newbridge. 1988–9. p. 12.
60. Gibson, W.H., Ibid., p. 155.
61. Jeffrey, M.H. *The trumpet call obey.* (London 1968), p. 97–8.
62. In *Kildare Observer.* 18.1.1919.
63. Ibid., 11.1.1919.
64. Ibid., 25.1.1919.
65. Ibid., 18.1.1919.
66. Ibid., 1.2.1919.
67. Ibid., 1.2.1919.
68. In *The Irish Field.* 25.1.1919.
69. In *Kildare Observer.* 8.2.1919.
70. Ibid., 15.2.1919.
71. Ibid., 15.3.1919.
72. Parliamentary debates. 115 H.C. D.E.B. 55. (London 1919).
73. In *Kildare Observer.* 1.3.1919.
74. Ibid., 8.3.1919.
75. Ibid., 22.3.1919.
76. Ibid., 12.4.1919.
77. Ibid., 3.5.1919.

78. *Walford's the County families of the U.K.* (London 1920), p. 1027.
79. In *Kildare Observer*. 13.5.1919.
80. *Buro Staire Mileata 1913–1921*. Part 3, Section I, p. 82.
81. In *Kildare Observer*. 26.7.1919.
82. In *Leinster Leader*. 26.7.1919.
83. In *Kildare Observer*. 12.7.1919.
84. Ibid., 16.8.1919.
85. Ibid., 23.8.1919.
86. Ibid., 13.9.1919.
87. Ibid., 1.10.1919.
88. Ibid., 20.9.1919.
89. War Office 35–182A. Survey of Ordnance Services 1919–1921. p. 2–3.
90. War Office 35–182A. Survey of Ordnance Services 1919–1921. p. 3–5, 9–11.
91. In *Kildare Observer*. 1.10.1919.
92. Ibid., 8.11.1919.
93. Ibid., 6.12.1919.
94. Ibid., 21.6.1919.
95. Ibid., 3.5.1919.
96. Ibid., 22.2.1919.
97. Ibid., 6.12.1919.
98. *Buro Staire Mileata 1913–1921*. Part 3 Section II, p. 102.
99. McCarthy, P. J. 'The R.A.F. and Ireland, 1920–22', In *Irish Sword*. Vol. XVII (1990), p. 179.
100. Ibid., p. 176.
101. Bye-Laws for the Glen Imaal Artillery Range. W.D. 1920.
102. Smyth, Senator M. *The Capuchin Annual*. (Dublin 1970), p. 564.
103. Ibid., p. 570.
104. Ibid., p. 565.
105. Ibid., p. 567.
106. Puirséal, P. *The G.A.A. in its times*. (Dublin 1982), p. 72.
107. Begley, Lieut. Col. M. History of Devoy Barracks. Naas, M.S, N.D.
108. Watson, S.J. *Between the flags*. (Dublin 1969), p. 171.
109. In *Leinster Leader*. 22.10.1920.
110. Crozier, F.P. *Ireland for ever*. (1932), p. 108.
111. Smyth, Senator M. Ibid., p. 567–9.
112. Ibid., p. 565, 570.
113. *Nás na Ríogh: from Poorhouse Road to Fairy Flax . . . an illustrated history of Naas*. (Naas 1990), p. 38.
114. Smyth, Ibid., p. 566,572.
115. In *Kildare Observer*. 22.7.1919.
116. Smyth, Ibid., p. 565,571.
117. Ibid., p. 573.
118. In *Leinster Leader*. 20.6.1920.
119. Ibid., 27.11.1920.
120. Ibid., 20.6.1920
121. Kildare County Council. Minute book Vol. No. 9. November 1920-May 1924.
122. In *Leinster Leader*. 4.9.1920.
123. Hickey, D.J. and Doherty, J.E. *A dictionary of Irish history 1800–1980*. (Dublin 1980), p. 180, 597.
124. In *Leinster Leader*. 18.12.1920.
125. In *Leinster Leader*. 1.1.1921

126. Ibid., 1.1.1921.
127. Ibid., 8.1.1921.
128. Ibid., 15.1.1921.
129. In *Kildare Observer.* 15.1.1921.
130. In *Leinster Leader.* 12.2.1921.
131. In *Kildare Observer.* 19.3.1921.
132. In *Leinster Leader.* 5.2.1921.
133. Ibid., 12.2.1921.
134. Ibid., 12.3.1921.
135. Swan, Comdt. D.A. 'The Curragh of Kildare', In *An Cosantoir. The Irish Defence Journal.* (May 1972), p. 63.
136. Delahunt, Sylvester. Straffan. Autograph book. Rath camp, Curragh, 1921.
137. In *Leinster Leader.* 9.4.1921, 21.5.1921.
138. In *Leinster Leader.* 19.2.1921.
139. Ibid., 2.4.1921.
140. McCarthy, Ibid., p. 181.
141. In *Leinster Leader.* 21.5.1921, 28.5.1921, 23.11.1985.
142. *Buro Staire Mileata 1913–1921.* Chronology. Part 3, Section III: p. 490, 499.
143. In *Leinster Leader.* 25.6.1921.
144. Ibid., 11.6.1921.
145. Ibid., 4.6.1921.
146. Ibid., 2.7.1921.
147. Ibid., 9.7.1921.
148. Ibid., 2.7.1921.
149. Ibid., 12.2.1921.
150. Ibid., 5.3.1921.
151. Watson, Ibid., p. 176.
152. Ibid., p. 176
153. Bence-Jones, Mark. *Twilight of the ascendancy.* (London 1987), p. 232.
154 Kildare County Council Minute Book, No. 9, November 1920-May 1924.

CHAPTER 16

1. Hickey, D.J. and Doherty, J.E. *A dictionary of Irish history 1800–1980.* (Dublin 1980), p. 572.
2. In *Leinster Leader.* 10.12.1921.
3. Ibid., 16.7.1921, 23.7.1921.
4. Ibid., 23.7.1921.
5. Ibid., 21.5.1921.
6. Ibid., 29.10.1921.
7. Ibid., 22.10.1921.
8. Ibid., 3.9.1921, 10.9.1921.
9. Ibid., 27.8.1921.
10. Ibid., 10.9.1921.
11. Ibid., 30.7. 1921.
12. Carroll, Mary. *Genealogy of the Weldons of Kilmorony.* Kildare County Library (1993).
13. In *Leinster Leader.* 8.10.1921.
14. Ibid., 12.11.1921.
15. War Office. 35–180B, 5.10.1921.
16. War Office 35–182 pt.1, 1.4.1922; War Office 35–182A, 'Notes on Guerilla Warfare in Ireland'. In *The Irish Times.* 29–30.12.1993.

17. In *The Irish Times.* 12.11.1921.
18. In *Leinster Leader.* 3.12.1921.
19. Ibid., 17.12.1921.
20. Ibid., 17.12.1921.
21. Ibid., 10.12.1921.
22. Ibid., 17.12.1921.
23. Ibid., 17.12.1921.
24. In *The Irish Times.* 9.12.1921.
25. War Office 35–182. Part 1. 10.12.1921. Ibid., 12.12.1921.
26. D'Arcy, Fergus *Horses, lords and racing men.* (The Curragh 1991), p. 255–6.
27. In *The Irish Times.* 24.12.1921.
28. Robertson, Nora *Crowned Harp.* (Dublin 1960), p. 142.
29. War Office 35–182A. 13.1.1922, 21.1.1922.
30. Brannigan, N.R. Changing the guard: Curragh evacuation 70 years on. In *An Cosantoir. The Irish Defence Journal.* (December 1992), p. 29–30; War Office 35/182A, S.G. 13/23, 13.1.92. General Staff, Dublin; Semple, A.J. The Fenian Infiltration of the British Army in Ireland 1864–7. Unpublished Thesis, T.C.D. (1971).
31. War Office 35–179, 35–182A. 21.12.1921, 24.12.1921.
32. Brannigan, Ibid., p. 29; War Office 35–182A. 21.12.1921, 13.1.1922, 4.4.1922.
33. War Office 35–182. 10.1.1922, 13.1.1922.
34. In *Leinster Leader.* 14.1.1922.
35. Ibid.
36. Ibid.
37. Ibid., 4.2.1922.
38. In *Kildare Observer.* 28.1.1922.
39. In *Freeman's Journal.* 4.2.1922.
40. In *Kildare Observer.* 4.2.1922, 25.2.1922, 4.3.1922, 8.4.1922, 15.4.1922, 22.4.1922, 29.4.1922, 6.5.1922; In *Leinster Leader.* 22.4.1922, 29.4.1922, 17.6.1922, 9.12.1922.
41. In *Leinster Leader.* 21.1.1922.
42. Ibid., 14.1.1922, 4.2.1922.
43. Ibid., 21.1.1922.
44. War Office 35–182. 1.2.1922. Letter from City Hall, Dublin.
45. War Office 35–182 Pt. 1. 3.2.1922. D.Q.M.G.'s Office.
46. In *Leinster Leader.* 11.2.1922.
47. Ibid., 11.2.1922.
48. War Office 93A. Operations War Diary 1922. G.H.Q. Dublin.
49. Costello, Con *Kildare: saints, soldiers and horses.* (Naas 1991), p. 92–4.
50. Begley. Lieut. Col. M. History of Devoy Barracks, Naas. M.S. N.D.; In *Kildare Observer.* 21.1.1922.
51. War Office 35–182 Pt.1, 20.2.1922.
52. War Office 35–182 Pt. 1. 1.4.1922.
53. Military Archives: Collins Papers. A-0768 XVI. 24.2.1922.
54. *Burke's Irish Family Records.* (London 1976), p. 1222.
55. In *Leinster Leader.* 4.2.1922.
56. Ibid., 25.2.1922.
57. Ibid., 25.2.1922.
58. Brannigan, Ibid., p. 29.
59. War Office 35/182: Part 1. 5th Division.
60. War Office 35/199: 14620Q. 21.3.1922, A.A. and Q.M.Q., 5th Division, Curragh; War Office 35/182: O/4060: 22.3.1922. Ordnance Officer. 5th Division, Curragh.

61. Military Archives. Collins Papers. A-0768 XVI. 26.3.1922.
62. War Office 35/199: 14620/Q, Part II. 21.3.1922. HQ 5 Div.; War Office 35/182A 14620/Q, Part I, 4.4.1922. Schedules A-D. H.Q. 5 Div.; War Office 35/189. 13.3.1922. H.Q. 5 Div.
63. In *Leinster Leader.* 15.4.1922.
64. Military Archives. Collins Papers. A-0768 XVI. 9.5.1922.
65. War Office 35–182 Pt.1. 3.5.1922. H.Q. 6 Div.; War Office 35–182A. 18.4.1922. H.Q. 14 Inf. Bde.
66. In *Leinster Leader.* 29.4.1922.
67. Swan, Comdt. D.A. 'The Curragh of Kildare', In *An Cosantoir. The Irish Defence Journal.* (May 1972), p. 66.
68. War Office 35–182A p.5, 4.4.1922. H.Q. 5 Div.
69. War Office 35–182A, p. 2. 4.4.1922. H.Q. 5 Div.
70. War Office 35–182A. p. 6–10. 4–4–1922. H.Q. 5 Div.
71. In *Leinster Leader.* 15.4.1922.
72. Ibid., 15.4.1922.
73. Ibid., 1.4.1922.
74. Ibid., 1.4.1922.
75. Ibid., 15.4.1922.
76. Ibid., 29.4.1922.
77. National Archives: O.P.W. file. Department of Justice H 99–10. Unsigned and undated memo.
78. Oral. Mr P. Harkins, Newbridge. 7.7.1994.
79. National Archives: O.P.W. file. Department of Justice H 99–10. Unsigned and undated memo.
80. In *Leinster Leader.* 6.5.1922.
81. Ibid., 13.5.1922.
82. Ibid., 6.5.1922.
83. War Office 35–182. 10.5.1922. p. 1–2. Evacuation Orders 5 Div.
84. In *Leinster Leader.* 20.5.1922.
85. In *Leinster Leader.* 13.5.1922.
86. Brannigan, Ibid., p. 30; War Office. 35/182. Dublin District. S.G. 915–1. 3.2.1922.
87. War Office 35–182. 10.5.1922. p. 1–3. H.Q. 5 Div.
88. In *Leinster Leader.* 20.5.1922; Swan, Ibid., p. 66.
89. In *Freeman's Journal.* 17.5.1922.
90. In *Leinster Leader.* 20.5.1922.
91. Oral. Lieut. Liam Collins Supply and Transport Corps, who was present in the camp. on the day of the take-over related his memories later to Lieut. Col. Sean Barrett. Recorded from him in 1992.
92. O'Farrell, Padraic. *The Blacksmith of Ballinalee.* (Mullingar 1993), p. 78.
93. Oral. Sr Veronica [Brigid] Tracy. Holy Family Community, Curragh Camp. 1994.
94. Brannigan, Ibid., p. 29–30.
    War Office 35–182. Pt. 1.9.4.1922. General Staff Dublin District.
95. In *Freeman's Journal.* 17.5.1922.
96. In *The Irish Times.* 17.5.1922.
97. In *Kildare Observer.* 21.1.1922, 20.5.1922.
98. In *Leinster Leader.* 20.5.1922.
99. Ibid., 20.5.1922.
100. Oral. Mr Charles Choiseul, Two Mile House, Naas. 1.3.1990.
101. In *Leinster Leader.* 27.5.1922.
102. Ibid., 27.5.1922.
103. Ibid., 3.6.1922, 12.8.1922, 26.8.1922.
104. In *Nationalist and Leinster Times.* Centenary Issue. (1883–1983), p. 48.

105. In *Leinster Leader.* 22.7.1922.
106. Ibid., 19.8.1922.
107. Military Archives: A-6492. 9.8.1922, 28.10.1922, 2.11.1922, 29.11.1922, 18.1.1923, 20.1.1923, 30.1.1923, 14.6.1923.
108. Military Archives. A-6969. 16.9.1922.
109. Military Archives. Ibid., 10.8.1922.
110. Military Archives. A-7689. 16.11.1922.
111. In *Leinster Leader.* 26.8.1922.
112. Ibid., 17.6.1922.
113. Jeffrey, M.H. *The trumpet call obey.* (London 1968), p. 98.
114. In *News of the World.* 18.3.1923.
115. In *Nationalist and Leinster Times.* Centenary Issue 1883–1983. p. 48.
116. Begley, Lieut. Col. M. History of Devoy Barracks, Naas. M.S., N.D.
117. *Defence Forces Handbook.* (Dublin 1982). p. 7.
 In *Nationalist and Leinster Times.* Ibid., p. 48.
118. Military Archives. A-13709. 2.1.1925. p. 2; Cooke, John. *Bord na Móna, Peat Research Centre.* (Newbridge 1991), p. 57.
119. Military Archives. A-13709. 17.4.1923.
120. Ibid., 17.4.1923, p. 5.
121. In *Freeman's Journal.* 5.5.1922.
122. Military Archives. A-13709. 18.12.1924.
123. Military Archives. Ibid., 16.12.1924, 18.12.1924.
124. Military Archives. Ibid., 14.1.1925.
125. Cooke, Ibid., p. 58.
126. Military Archives. A-13709. 2.3.1927, 3.3.1927.
127. The Price family in Newbridge 1895–1989. p. 3. In, Folklore Project 1988–89. SES. County Library, Newbridge.
128. James Dillon, Ballyshannon. Folklore Tapes 1914–1922. Kildare County Library, Newbridge.
129. Oral. Ann Dempsey, Newbridge. May 1992.
130. Tom Corcoran, Newbridge. Folklore Project. County Library, Newbridge.
131. Larragy, Angela *The Leader Advertiser.* 13.9.1986.
132. Oral, Dempsey, Ann. 1992.
133. In *Leinster Leader.* Punchestown Supplement. (April 1960), p. 16.
134. Fleming, Lionel. *Head or Heart.* (London 1965), p. 64.
135. Berman, David and Jill 'Journal of an officer stationed in Naas and Baltinglass, 1832–3', In *Journal of the County Kildare Archaeological Society.* Vol. XV (1976), p. 269.
136. Robertson, Nora *Crowned Harp.* p. 153.
137. Duggan, John P. *A History of the Irish army.* (Dublin 1991), p. 116, 119.

EPILOGUE

1. *Saorstat Eireann. Census of Population 1926. Vol. X. General Report.* (Dublin 1926). p 17.
2. Begley, Lieut. Col. M. Unpublished M.S. History of Devoy barracks, Naas. (No Date).
3. *Kildare Observer.* 19.5.1923.
4. Begley, Ibid.
5. *Kildare Observer.* 2.6.1923.
6. Cooke, John. *Bord na Móna: Peat Research Centre.* (Newbridge 1991). p 58–61.
7. *Defence Forces Handbook.* (Dublin 1982) p 17.
8. Army Census 12/13 November 1922: Curragh to Crosshaven. *Defence Forces Handbook.* (Dublin 1982) p.7.

9. *Handbook of the Curragh Camp.* (Curragh 1930). p 21.
10. *Kildare Observer* 13.1.1923
11. *Kildare Observer* 28.4.1923.
12. *Leinster Leader* 17.3.1923.
13. *Kildare Observer* 24.3.1923.
14. *Kildare Observer* 15.9.1923
15 Beckett, I.F.W. *The Army and the Curragh Incident 1914.* (London 1986). p 439–440
16 *The Irish Times,* 21.3.1996.
17. Valiulis, M.G. *Almost a Rebellion: The Irish Army Mutiny of 1924.* (Cork) 1985 p 44, 51.
18. Bolton, A.D. (ed). *The Irish Reports.* (Dublin 1926). p 266.
19. *The Last Post. National Graves Association.* (Dublin 1976). p. 22, 110.
20 Leinster Leader, 22.10.1921.
21. *Handbook of the Curragh Camp* (Curragh 1930) p. 23.
22 *Oglaigh na hEireann. Irish Defence Forces Handbook* 1968 (Dublin 1968) p. 14.
23 Share, Bernard. *The Emergency. Neutral Ireland 1939–45.* (Dublin 1978) p. 137–138.
24. *Defence Forces Handbook* (Dublin 1982) p 23.
25. Curragh Command Press Officer, 13.3.1996.
26. *Leinster Leader* 21.4.1923.

BIBLIOGRAPHY

PRIMARY SOURCES

ILLUSTRATIONS

Album of photographs put together by Capt. E.D. Fenton 86th Regiment for the Prince of Wales. 'Souvenirs of Soldiering at the Camp Curragh, 1861'. Royal Archives, Windsor Castle.

An Cosantoir. May 1972. Reproductions of illustrations from Lugard; R.E. plans for the rebuilding of the camp, and photographs from Lawrence and of the Rath Camp in 1921 and of the hand-over of the camp in 1922. View of the Foot Camp, Curragh of Kildare. Engraving of *c.* 1800.

Collection of 16 photographs taken in the Curragh during the visit of the Prince of Wales in 1861. Royal Archives. Windsor Castle.

Curragh Camp and District. Eason and Son, (Dublin, *c.* 1908).

Curragh photographs. Irish Architectural Archive, Dublin.

Drawing of Shoemaker's shop, Highland Regt., Curragh Camp 1888. Collection Mr H. Hamilton.

Drawings of buildings proposed for Curragh Camp, Lugard, (Dublin, 1858). National Library.

Drawings, Lady Fremantle. *c.* 1866. Collection Knight of Glin.

Eason photographic collection: Athy, Curragh, Newbridge barracks. National Library.

Lawrence Collection; photographs of the Curragh camp and of Naas and Newbridge barracks. National Library.

The Curragh Camp. Headquarters. *c.* 1857. engraving. Wisenheart and Son, 42 Lr. Sackville St., Dublin.

Collection Mr D. Gillman.

The Curragh, Co. Kildare. Dr J. Lamprey. Oil on canvas 24 x 61. 1861. Gorry Gallery, Dublin.

The Royal Quarters and the Grand Review at the Curragh. *Illustrated London News.* July 13th 1861.

The Curragh—The Camp from Gibbets Rath. Walter Frederick Osbourne, R.H.A. Watercolour 6" x 9.5" N.D. James Adam Salesrooms, Dublin.

The Queen's Birthday at the Curragh. Photographs of the Review. *The Navy and Army Illustrated.* (London, 1897).

Bibliography

MANUSCRIPTS AND PRIVATE PAPERS

Church Registers.

Ballysax/Ballyshannon Church of Ireland 1855–1859.
Curragh Garrison West Church. Church of Ireland. 1856–1922. National Archives. M.S. 5050–55.
Carnalway, Church of Ireland. 1858–1889.
Curragh Garrison East Church. Roman Catholic 1855–1918.
Kilcullen, Church of Ireland. 1860–1883.
Kilcullen. Roman Catholic Church. 1856–1899.
Kildare Cathedral. Church of Ireland. 1802–1897.
Kildare, Roman Catholic Church. 1844–1899.
Naas, Church of Ireland. 1679–1830.
Newbridge, Church of Ireland. 1845–1899.
Newbridge, Garrison Church. Church of Ireland. 1867–1922. R.C.B. Library, Dublin.
Newbridge, Roman Catholic Church. 1842–1899.

Vestry Book.

St David's Church, Naas. Minutes of Select Vestry. 1805–1884.

Curragh Camp.

Command Q.M.'s office: Records of Married Quarters. 1922.
Sandes Soldiers' home, File Q.M. CH/10287. 1949–1984.

Department of Defence, Property Management Section, Dublin.

British Army Terriers giving return of all lands and tenements (including clearance rights), purchased or leased for more than 21 years in the U.K. and Channel Islands. Ireland 1901, and addenda to same.
Copy of Lease, Commissioners of Woods and Forests to Kildare Hunting Club 28.1.1866. Box 289, No. 28.
Correspondence re Curragh Stand House and Rathbride hare park, 1856, 1861, 1866. File No. 43, parts II and V.
Crown Lands: File 1406. 1866–1910.
Curragh of Kildare. Transcript notes of Commission of Inquiry, held at Newbridge. 1866. Parts I and II.
Indenture; R.A. Meekings and Sec. War Dept., 3.7.1860. Bundle 7, Deed 351.
Letter from Curragh Ranger re booth rents, 11.5.1868, and receipts for same.
Letter from Curragh Ranger re trespass money, 22.9.1868.
Letter re provision of house for Wesleyan chaplain, 11.9.1898.
Letter from G.O.C. Curragh to D.A.G. Dublin, re clearing furze from Curragh plain, 31.7.1899.
Leases of War Department Property.
Memorandum of Agreement for forming an Encampment on the Curragh of Kildare, 13.1.1859. (Document, Curragh 97).
Parchment; Act of Consecration of burial place for military encampment, 14.10.1869. (Box 293. No. 26).
Return of properties hired by the War Department for periods of 3 years, and of rents payable on War. Department Freehold Properties, 1919. Irish Command, revised 31.8.1920.
Register of Lettings War Department Property 1878–1921.
War Office list of lands and buildings in the occupation of the War Department Irish Command, 30.11.1917.

Department of Irish Folklore, University College Dublin.
Irish Folklore collection, county Kildare, vol. 777.

Diocesan Archives, Kildare and Leighlin, Carlow.
Haly papers. 1855.

Dublin Corporation Archives.
Mansion House Fund: Local committee no. 565. Kildare, county Kildare.

Kildare County Library, Newbridge.
Board of Guardians. Naas Union. Minutes of Meetings. 1865–1881.
Carroll, Mary. Genealogy of the Weldon's of Kilomorony. 1993.
Hughes, Henry. The History of Newbridge, County Kildare. 1995.
Newbridge, local history. Social Employment Scheme, Folklore Project. 1988/89.

Land Registrar of titles, county Kildare.
Draft award of arbitrator in the matter of bye-laws regulating the Curragh range ground and the closing of the Donnelly's Hollow road, 3.7.1896.

Land Valuation Office.
Valuation lists, county Kildare.

Masonic Grand lodge, Dublin.
Registers Vol. 2 (1906–1914), Vol. 3 (1878–1921), Vol. 4 (1906–1908). 4th Series.
Masonic Lodge No. 397. Curragh, Newbridge. Minute Book 1917–1920.

Military Archives, Dublin.
All Ireland Army Rifle Meeting. 1895. Poster/Programme.
Buro Staire Mileata. 1913–1921. Chronology. Parts I, II and III. *Roinn Cosanta.* (Dublin N.D.).
Collins papers:
Army census 12/13 November 1922, Curragh to Crosshaven.
Evacuation of British Forces 1922. A/0768. group IX.
Droichead Nua Barracks. 11.1.1922. A/0768 II.
Letter from Curragh Superintendent to Minister of Defence. 24.2.1922. A/0768 XVI.
Letter from Curragh Superintendent to Richard Mulcahy. 9.5.1922. A/0768 XVI.
Letter and instructions from M.O.D. to Evacuation Officer. 26.3.1922. A/0768 XVI.
Letter Evacuation Officer to M.O.D. 9.5.1922. Evacuation of Naas barracks A/0768 XV.
Letters re occupation of quarters in Curragh Camp, January 1922 to 14 June 1923. Files A/ 6492, A/ 6969.
Proposal that Curragh camp should be made a military area. 16.11.1922. File A/7689.
Proposal that Protestant church be used by Roman Catholics. 6.11.1923. File A/14870.
Correspondence re number of civilians employed at Curragh camp pre evacuation, and re unemployment in Newbridge, and of possibility of quartering troops in Newbridge barracks. 1923–1927. A/13709
Tables of Service of British Regiments in Ireland, by, Brig. Gen. P.D. Hogan, M.S. compiled for the Military History Society of Ireland. N.D.
Sandes soldiers home. File C42. Letter 25.2.1955.
Smyth, Michael. M.S. Irish Volunteers, 1914–1916. County Kildare. *c.* 1960.

Bibliography

National Archives.

Census Schedules, household returns, county Kildare, 1901, 1911.

Chief Secretary's Office: registered papers, 3/157/1. 8.5.1865.

Dept. of Justice : file H 99/10.

Office of Public Works: files 7/8/13, 7/8/27.

National Library, Dublin.

Kilmainham Papers. M.S. 1266–8, 1855–1862; MS 1269–71, 1862–1879; M.S. 1110, 1854–62; M.S. 1351, 1909.

Lugard, Lieut. Col. H.W. *Narrative of operations in the arrangement and formation of a camp for 10,000 infantry on the Curragh of Kildare.* (Dublin 1858).

Mayo Papers. M.S. 11,222, 1866.

O'Flanagan, Reverend Michael. *O.S. Letters, 1837.* pt II. (Bray, 1930).

Private Papers.

Bardan's Hotel, Kilcullen, Visitor's Book 1897–1902.

Berney, Thomas, Saddler Kilcullen, account books.

Begley, Lieut. Col. M. Unpublished M.S., History of Devoy barracks, Naas. No date.

Boland, P. Letter. 17.7.1984.

Brophy family, Herbertstown, government contractors. nineteenth century accounts.

Curragh camp; Dance Card 1859. (Private Collection).

Delahunt, Sylvester. Tuckmill, Straffan. Autograph book, Rath camp. (Curragh 1921).

Gorry Gallery, Dublin. Exhibition catalogue eighteenth, nineteenth and twentieth-century Irish painters. 28.4.1989.

McGlynn family, Brownstown, coal merchants, nineteenth century accounts.

Morrin, Noel. Letter to C. Costello, re post-mark collection. 1992.

Royal Irish Military Tournament, Ballsbridge, Dublin. Programme.1895. (Private Collection).

Public Record Office, London.

War Office Records.

Correspondence re building of the Curragh Camp 1855. WO/44/611: Courtney and Stephens. 12.3.1855; Arthur and George Holme. 14.4.1855 and 8.6.1855.

Notice to Builders. Command Royal Engineers's office, Dublin. 16.3.1855.

Notes on guerilla warfare in Ireland, ND, :5774/G. WO/35/182 pt 1.

Distribution of Cavalry Regiments and Battalions in Ireland. 1921. Cab.24/126. WO/A/0672, pts I&II.

Fifth Division Location Return: w/e 3 Sept. 1921. 7/9/1921. WO 35/179.

Summary of Ordnance Services in Dublin District and 5th Divisional Areas from the end of 1919 until the end of 1921. WO/ 35/ 182A.

Withdrawal proposals. Secret. Col. J. Brind, G.H.Q. , Parkgate, Dublin. 10.12.1921. WO/35/182.

General Headquarters: Movement Plans, 21.12.1921, 13. 1. 1922. WO 35/182A.

Withdrawal: handwritten note, re codeword 'Finis': Duty Officer, Dublin District. 21.1.1922. WO/35/182A.

Withdrawal policy. Secret. Col. W.J. Maxwell Scott, 5th Division. 24.12.1921. WO/35/182.

Disposal Board. Local Sales. Secret. Lt. Col. F. Dalrymple, A.Q.M.G. 5th Division 10.1.1922. WO/35/182.

Disposal Board. Local Sales, instructions for Lt. Col. F. Dalrymple, A.Q.M.G. 5th Division. 13.1.1922. WO/35/182.

Withdrawal and Reduction of Guards: Secret Message. 13.1.1922. WO/35/182A.

Letter from Michael Collins to commander in chief, Parkgate, Dublin. 1.2.1922. WO/35/182.

Letter from Gen. F.N. Macready to Provisional Government, 2.2.1922. WO/35/182.
Order ref. British and Free State Troops meeting when on the march. 3.2.1922. WO/35/182.
Withdrawal, security precautions. 20.2.1922. WO/35/182 pt I.
Units in 5th Division Area 4.3.1922. WO/35/182.
Evacuation of the Curragh, Position of civilian clerks. 13.3.1922. WO/35/189.
Evacuation instructions. Curragh area. 21.3.1922. Lt. Col. F. Dalrymple A.Q.M.G. WO/35/182.
List of foremen in Curragh to represent Provisional Government. Ordnance Officer H.Q. 5th Division. 22.3.1922. WO/35/182.
Statement showing Personnel required for care of R.A.S.C. Buildings etc. on handing over to the Irish Provisional Government. 22.3.1922. WO/35/182A.
Secret and Urgent Messages from Col. W.J. Maxwell Scott, 5 Div. 1.4.1922, 12.4.1922, and 15.4.1922. WO 35/182.
Evacuation of the Curragh. Order from Maj. Gen. Sir Hugh S. Jeudwine. 4.4.1922. WO/35/182A.
Evacuation of Kildare. Order from Lt. Col. F. Dalrymple. 10.4.1922. WO/35/182. pt 1.
Provisional Orders for the Hasty Evacuation of the Curragh. Brigade Major 14th Inf. Bde. 18.4.1922 WO/35/182A. pt I.
Evacuation, troop withdrawal and accommodation. 19.4.1922. WO/35/182 ptI.
Evacuation of the Curragh: Secret Message from Lt. Col. Sir Francis Elphinstone Dalrymple. 24.4.1922. WO/35/182 pt I.
Evacuation of the Curragh. Order from Col. E. Evans. 3.5.1922. WO/35/182 pt I.
Order from Lt. Col. F. Dalrymple, secret and urgent. 10.5.1922. WO/35/182.
Operations war diary, 1922. G.H.Q. Dublin. WO/93A.

Royal Hospital, Kilmainham, Dublin.
Lord Roberts, press cuttings relative to.

MAPS

Larkin, William. *Map of County Kildare.* 1811. National Library road maps. Vol. 15 A3–16. Map No. 50.
Maps in the Department of Defence, Dublin. Property Management Section:
Crown Lands. 14.7.1855. File 1406.
Curragh. Scale 6":1 mile. Trace from O.S. Map of the County Kildare. No date, but prior to 1855.
Curragh Sheet, made from O.S. Sheets 22, 23, 27, 28. Scale 6":1 mile. 1855. Edition 1863. Certified copy of map lodged with the Clerk of the Peace for the Co. Kildare on 22.12.1867 and a duplicate lodged in the Record and Writ Office, Dublin. 9.10.1868. Amended to show proposed new road from Ballysax to Athgarvan.
Curragh Sheet. 1860. Scale 6":1 mile. Map referred to in the Commissioners Award, dated 30.6.1869. Another copy of the same map showing the existing public roads on the Curragh.
Curragh Sheet. Scale 6":1 mile. 1868 edition of 1855 map. Amended to show proposed branch of G.S. and W.R. to Curragh encampment.
Curragh of Kildare. Scale 6":1 mile. Map deposited for the purposes of the Act of the Oireachtas 1961.
Map: War Department, Index to the Curragh. Scale 12.672": 1 mile. OS 1902.
Maps in the office of the Land Registrar of Titles county Kildare.
O.S. Curragh sheet, 6":1 mile 1867. Manuscript insertion of proposed road to west of old road from Ballysax to Athgarvan. [The road was never constructed].
O.S. Curragh Sheet 6":1 mile, revised 1875. Inscribed: In the matter of bye-laws made by her majesty's principle secretary of state for the War Department for the regulation of the Curragh rifle range ground and the closing of the Donnelly's Hollow road in county Kildare.

Noble and Keenan. *Map of County Kildare*, 1752.
Petty, Sir William. *Map of Co. Kildare* 1685. (Newcastle-upon-Tyne 1968).
Saoirse. A map of the Anglo-Irish War. 1919–1921. Dublin.1986.
Townland Sheets county Kildare. Scale 6":1 statute mile. Sheets 17, 18, 22, 23, 28, O.S. Dublin 1909.
Taylor, Lieut. Alexander. *A Map of the County Kildare.* 1783.
Walker, Henry. *Map of Curragh of Kildare.* 1807.

MONUMENTS

Memorial on Curragh Plain to Capt. Collins. 1860.
Memorial to *Vonolel*, Lord Roberts's charger. Royal Hospital Kilmainham. 1899.
Royal Crest and date on Curragh Water Tower. 1900.

NEWSPAPERS.

Curragh News.
Dublin Chronicle.
Faulkner's Dublin Journal.
Finn's Leinster Journal.
Freeman's Journal.
Irish Field.
Irish Independent.
The Irish Times.
Kildare Observer.
Leinster Express.
Leinster Leader.
Nation.
Nationalist and *Leinster Times.*
News of the World.
The Times, (London).
United Ireland.

OFFICIAL PUBLICATIONS.

Award of Commissioners. Curragh of Kildare Act , 1868. (Dublin 1869).
Barrack Book, the Home Stations. H.M.S.O. (London 1904).
Barrack Book, the Irish Command. 1921. H.M.S.O. (London 1921).
British Labour Statistics Historian Abstracts 1886–1968. H.M.S.O. (London 1971).
Bye-laws for the Regulation of the Curragh. Curragh of Kildare Act 1868. (Dublin December 1868).
Bye-laws for the Regulation of the Curragh 1868; amendments to, 1873. (Dublin 1873).
Bye-laws, Glen Imaal Artillery Range. 1920. (Wicklow 1920).
Bye-laws, Curragh Rifle Ranges, bye-laws as to Donnelly's Hollow Road. (Curragh Camp 1921).
Bye-laws, Curragh 1964. (Dublin 1964).
Census of Ireland for the year 1861: part V, General Report. (Dublin 1864).
Census of Ireland for the year 1871: part I, area, houses and population, also the ages, civil condition, occupations, birthplaces, religion, and education of the people. Vol. 1. Province of Leinster, No. 3 County of Kildare. (Dublin 1872).

Census of Ireland for the year 1881: part I. area, houses, and population, also the ages, civil or conjugal condition, occupations, birthplaces, religion, and education of the people. Vol. I. Province of Leinster. No. 3. County of Kildare. (Dublin 1881).

Census of Ireland for the year 1891: part 1, area, houses, and population, also the ages, civil or conjugal condition, occupations, birthplaces, religion and education of the people. Vol. I. Province of Leinster. No. 3 County of Kildare. (Dublin 1891).

Census of Ireland for the year 1901: part 1, area, etc. Vol. I. Province of Leinster. No. 3. County of Kildare. (Dublin 1901).

Census of Ireland for the year 1911: part 1, area, etc. Vol. I. Province of Leinster. No. 3. County of Kildare. (Dublin 1912).

Extracts of any correspondence that has taken place with reference to the granting of a Lease of the Curragh of Kildare by the Crown to the War Department. 1865. (301) XXXII.339.

Griffith, Richard. *General Valuation of Rateable Property in Ireland. County of Kildare.* (Dublin 1853).

Kildare County Development Plan. Third Revision. Part I. Kildare County Council. (Kildare 1985).

Notice to Treat, in the Matter of the Defence Act, 1842, the Ordnance Board Transfer Act, 1855, and the Acts Amending and Extending the same; and in the matter of the Glen of Imaal artillery range. C.S.O., Dublin Castle. (Dublin 8.3.1899).

Sites and Monuments Record. County Kildare. Archaeological Survey of Ireland. Office of Public Works. (Dublin 1988).

Saorstát Eireann Census of Population 1925. Vol X General Report (Dublin 1926)

Warrant for the Regulation of the Barracks in Great Britain, Ireland and the Colonies. H.M.S.O. (London 1824).

PARLIAMENTARY PAPERS

Acts

An act against swine feeding in the Curragh of Kildare, (27 Ed. 1, 1299).

An act to prohibit horse races in the neighbourhood of the city of Dublin, (31 Geo. 3, *c.* 38, 1791).

An act for the better prevention of contagious diseases. 1866. (29 and 30 Vict. *c.* 35, 1866).

An act to make better provision for the management and use of the Curragh of Kildare, (31 and 32 Vict., *c.* 60, 1868).

An Act to confirm the award under the Curragh of Kildare Act 1868, and for other purposes relating thereto. (33 and 34. Victoria. *c.* 74.1870).

Bills

A Bill for the better prevention of contagious diseases at certain naval and military stations. (1866 (78) II.219).

A Bill for the better prevention of contagious diseases at certain naval and military stations (as Amended by a Select Committee) (1866.(116) II.219).

A Bill to make better provision for the management and use of the Curragh of Kildare. (1866 (136) II.383).

A Bill to confirm the award under the Curragh of Kildare Act 1868 and for other purposes relating thereto. (1870 (175) I.369).

A Bill to provide for the detention in certain hospitals of persons affected with certain diseases, and to repeal the contagious diseases Acts. 1866–1869. (30 and 31 Victoria *c.* 106 (22)1883).

A Bill to repeal the contagious diseases Acts 1866–1869. (1873 (29) I.227).

A Bill for the repeal of the contagious diseases Act 1878. (1878 (59) I.313).

Reports

'Army Medical Department Report for the year 1860'. XXXIII.I.1862.

'Army Medical Department Report for the year 1869'. XXXVIII.379.1871.

'Army Medical Department Report for the year 1878'. XLIV.331.1878–79.

'Army Medical Department Report for the year 1882'. XLIX.299.1884.

'Army Medical Department Report for the year 1887'. XLIX.169. 1889.

'Army Medical Department Report for the year 1896'. LIV.557. 1897.

'Army Medical Department Report for the year 1911'. LII.255. 1912–1913.

'Report from an official committee on barrack accommodation for the army'. War Department 17.7.1855.(405) XXXII. 37–272.

'Report of committee on preparation of army medical statistics for the year 186'. 1862. (3051) XXXIV.I.

'Report of Treasury Commissioners' local inquiry with view to legislation on the subject of the Curragh of Kildare. 1867–1868'. (329) LV. 743.

'Report from the Select Committee on the Curragh of Kildare Bill, together with the proceedings of the committee, minutes of evidence and appendix'. 25.6. 1868. (404) X.1.

'Report from the Select Committee on the Curragh of Kildare Bill, together with the proceedings of the Committee, minutes of moral affects of the C.D.A.' 1873. (209) XL.433.

'Report to War Office by the Inspector of Certified Hospitals on the moral effects of contagious diseases Act 1873'. (209) XL.433.

Returns

'Contract prices of bread (per 4lb loaf) supplied to the troops in each county of Ireland in each half year, in each of the years (ended 30 November) 1871, 1972 and 1873'. House of Commons. LXXX.334.

'Contract prices for fresh meat (per lb) supplied to the troops in each county of Ireland, in each half year, in each of the year (ending on 30 November) 1871, 1872, 1873'. LXXX. 334/5.

'Return of State of Curragh 1833'. Bills and Acts. 1833.(81) XXXV.505.

'Return of Divine Service 1856'. 1858. XXXVII. 916.

'Return showing the gross sum expended in the erection of the camp at the Curragh; number of troops it is capable of accommodating, number quartered there on 1st June 1858, and rate per cent per annum of the mortality among the troops, and the number who died in each year 1857–58'. (496) XXXVII. 235.

'Return of number of persons flogged in army of G.B. and Ireland. 1856'. (17) XXXVII. 307.1857–58.

'Return of expenses connected with the erection of barracks, huts, stabling, etc., including the cost of draining, paving and supply of water on the Curragh of Kildare. 1861'. (270) XXXVI. 309.

'Return relating to Convictions and Arrests for Trespass on Curragh of Kildare, within the last two years 1860 and 1861'. 1862 (141) XXXII.299.

'Return of number of convictions and arrests for trespass on Curragh of Kildare, before Camp Court and Petty Sessions at Newbridge or Kildare 1860 and 1861'. Parliamentary paper, House of Commons. 1862.(141) XXXII.299.

'Return allowance to officiating clergymen to troops. 1865'. XXXII.

PLANS

Curragh camp, Squares C and G. Principal detached blocks of buildings, and detail plans. 1855. Lugard (Dublin 1858).
Naas military barracks. R.E. Chatham. 1886. Officers' Mess, Devoy barracks, Naas.
Plans in Command Engineer's Office, Curragh Camp:
Curragh Camp. 'C' Square. Plans: 9/A3, 16/14,42/15,34/15 No. 2,73/21,33/15,54/18.
Curragh Camp. Plans of Gough and Keane barracks: Plans 15/21,5/23,2/3.
Standard Plan No. 23. Serjeants' Mess One Battalion Infantry, to accommodate 58 members, and Serjeants' Mess for One Brigade Division of R.F.A. to accommodate 28 members. Director of Fortifications and Works. 1905.
Plans in Military Archives, Dublin:
Athy barracks, Perambulation Plan. No date.
Curragh camp. Plan of prison. 1882. Plan 24/24 No. 3.
Curragh camp. Plan of Garrison Bakery. 1883. Plan 8/28 No. 1.
Curragh camp. Cavalry barracks. 1893. Plans: 164/13 No.2, 125/13, 18/13. 109/13.
Curragh camp. Plans of hospital. 12/6, 3/17, 17/17, 3/6.
Kildare barracks, Record of Reappropriation of Lock Hospital. 1906.
Kildare barracks. Site Plan. R.E. 1902. Mil.
Kildare Lock Hospital, plan of well and tank tower. 1869.
Lock Hospital, site and plan. 1867.
Lock Hospital, County Kildare. 1882.
Newbridge barracks,county Kildare. R.E. Chatham. 1863.
Record Site Plan, Kildare Lock Hospital. 1876.

WORKS OF REFERENCE

Baily's hunting directory. (London 1896–)
Boylan, Henry. *A dictionary of Irish biography.* (Dublin 1978).
Burke, Sir Bernard, *A genealogical* and *heraldic dictionary of the peerage* and *baronetage etc.* (London 1887).
Burke's Irish family records. (London 1976).
Burke's peerage. (London 1953).
Dictionary of National biography. (London 1917).
Dod, Charles R. *The peerage, baronetage* and *knightage of Great Britain* and *Ireland.* (London 1855, 1880).
Everyman's encyclopaedia. (London 1961).
Hickey, D.J. and Doherty, J.E. *A dictionary of Irish history 1800–1980.* (Dublin 1980).
Lewis, Samuel. *A topographical dictionary of Ireland. Part II.* (London 1837).
Loeber, R.A *Biographical dictionary of architects in Ireland. 1600–1720.* (London 1981).
Porter's Post Office guide and *directory for the counties Carlow* and *Kildare.* (Dublin 1910).
Thom's Irish almanac and *official directory.* (Dublin 1846).
Vanderveen, Bart. H. ed. *The observers army vehicles directory to 1940.* (London 1974).
Walford's the county families of the UK. (London 1920).

SECONDARY SOURCES

THESES

Haire, D.N. British Army in Ireland 1868–1890. (M. Litt., T.C.D. 1973).
Semple, A.J. The Fenian infiltration of the British Army in Ireland, 1864–7. (M. Litt., T.C.D. 1971).

VERBAL REPORTS

Bradley, Garda C., Artillery Place, Newbridge.
Brophy, Mr D., Baltracey, Naas.
Casey, Mr P., Western Front Association, Dublin.
Choiseul, Mr C., Two-Mile-House, Naas.
Clements, Lt. Col. C.M.L., Carnalway, Kilcullen.
Crookshank, Miss Anne, Dublin.
Cummins, Mr C., Main St., Newbridge.
Dempsey, Mrs Ann, Ballysax, Curragh
Dunne, Mr K., Kildare.
Garrett, Mr T., Maddenstown, Curragh.
Garrett, Mrs T., Maddenstown, Curragh.
Gillman, Mr D., Bray.
Harkins, Mr P., Newbridge.
Molloy, Mr J., Newtown, Suncroft.
Moran, Mr L., Ballysax.
Noone, Mrs Margaret, Dunlavin.
Orford, Mr B., Suncroft, Curragh.
Rafferty, Mrs Margaret, Brownstown, Curragh.
Robeck, Baron de, Gowran Grange, Naas.
Tracy, Sister Veronica, Curragh camp.

LATER PUBLICATIONS

Books

Andrews, J.H. *Irish historic towns atlas. No.1. Kildare.* R.I.A. (Dublin. 1986).
Annals of the Four Masters. Part I. (Dublin. 1990).
Annals of the Sisters of Mercy. (New York. 1881).
Annual register (London. 1861).
Ascoli, David. A *companion to the British army 1660–1983.* (London. 1983).
Atkinson, A. *The christian tourist.* (Dublin. 1815).
Ballymore Eustace Development Group. *History of Ballymore Eustace.* (Ballymore Eustace. 1992)
Barton, Derick. *Memories of ninety years.* (Dublin. 1989).
Beckett, I.F.W. *The army* and *the Curragh incident 1914.* (London. 1986).
Beckett, I.F.W and Simpson, K. *A nation in arms. A social study of the British army in the First World War.* (London. 1990).
Bell, Pat. *Long shies and slow twisters.* (Celbridge. 1993).
Bence-Jones, Mark. *Twilight of the ascendancy.* (London. 1987).
Berry, Henry F. *Statutes and Ordinances, and Acts of the Parliament of Ireland. King John to Henry V.* (Dublin. 1907).

Bodkin, Thomas. *My Uncle Frank.* (London. 1947).
Bolton, A.D. (ed) *The Irish Reports.* (Dublin 1926).
Boulger, D.C. *Battle of the Boyne.* (London. 1911).
Boylan, Lena. *Castletown and its owners.* (Leixlip. 1968).
Bredin, Brig. Gen. A.E.C. *A history of the Irish soldier.* (Belfast. 1987).
Brewer, J.N. *The beauties of Ireland.* (London. 1825).
Burke, Helen. *The people* and *the Poor Law in 19th century Ireland.* (Littlehampton. 1987).
Carr, J.C. (ed) *A Century of Medical Radiation in Ireland. An Anthology.* (Dublin 1995).
Caulfield, Max. *The Easter Rebellion.* (London. 1964).
Comerford, R.V. *The Fenians in context. Irish politics and society 1848–82.* (Dublin. 1985).
Condren, Mary. *The serpent and the goddess.* (San Francisco. 1989).
Cooke, John. *Bord na Mona: Peat Research Centre.* (Newbridge. 1991).
Cooper, Bryan. *The tenth (Irish) division in Gallipoli.* (Dublin. 1993).
Costello, Con. *Guide to Kildare and West Wicklow.* (Naas. 1991).
Costello, Con. *Kildare: saints, soldiers and horses.* (Naas. 1991).
Coyne, W.P. *Ireland, industrial and agricultural.* (Dublin. 1901).
Creagh, Gen. Sir O'Moore and Humphris, H.M. *The VC and DSO. Vol. 1.* (London. 1925).
Crooks, Maj. J.J. *History of the Royal Irish Regiment of Artillery.* (Dublin. 1914).
Crozier, F.P. *Ireland for ever.* (Bath. 1932).
Cunliffe, M.F. *The Royal Irish Fusiliers 1793–1950.* (London. 1952).
Curragh Camp and District. Illustrated and Described. (Dublin. *c.* 1908).
Daniell, D.S. *4th Hussars. The story of a Cavalry regiment.* (Aldershot. 1959).
D'Arcy, F.A. *Horses, lords and racing men.* (The Curragh. 1991).
Delany, Ruth. *The Grand Canal of Ireland.* (Newton Abbot. 1972).
De Montmorency, Harvey. *Sword and Stirrup.* (London. 1936).
Devoy, John. *Recollections of an Irish Rebel.* (New York. 1929).
Duggan, John P. *A history of the Irish Army.* (Dublin. 1991).
Dunlop, Robert. *Plantation of renown.* (Kildare. 1970).
Farrar-Hockley, Antony. *Goughie.* (London. 1975).
Fergusson, Sir James. *The Curragh Incident.* (London. 1964).
Fitzpatrick, W.J. *Times and correspondence of Rt. Rev. Dr Doyle. Vol. II.* (Dublin. 1861).
Fleming. Lionel. *Head or harp.* (Dublin. 1965).
Fothergill, G.A. *The National Stud: A Gift to the State.* (Edinburgh. 1916).
Gibson, W.H. *Early Irish golf.* (Naas. 1988).
Gilbert, Sir John. *History of the Viceroys of Ireland.* (Dublin. 1865).
Gilbert, Sir John. *History of the Irish confederation and war in Ireland. Vol. II.* (Dublin. 1882–85).
Godley, Gen. Sir Alexander. *Life of an Irish soldier.* (London. 1939).
Gough, Gen. Sir Hubert. *Soldiering on.* (London. 1954).
Gould, R. W. *Locations of British cavalry, infantry and machine gun units 1914–1924.*
(South Woodford. 1977).
Grant, Elizabeth. *The highland lady in Ireland.* (Edinburgh. 1991).
Hall, Mr and Mrs S.C. *Ireland, its Scenery, Character, etc. Part II.* (London. 1837).
Haldane, General Sir Aylmer. *A Soldier's Saga.* (Edinburgh & London 1948).
Hamilton, I.B.M. *The happy warrior.* (London. 1966).
Handbook of the Curragh Camp. (Curragh. 1930).
Harris, Henry. *The Royal Irish Fusiliers.* (London. 1972).
Hart, Col. H.G. *The Annual Army List, etc.* (London. 1872–)
Hayes, K.E. *A history of the RAF and the United States Naval Air Service in Ireland. 1913–1923.* (Dublin. 1988).
Hayes-McCoy, G. A. *Irish battles.* (Dublin. 1969).

Head, Lt. Col. C.O. *No great shakes.* (London. 1943).
Hynes, Captain Rory. *History of Clarke Barracks 1855–1930.* (Curragh Camp. 1994). (Dublin 1940).
Jackson, Maj. E.S. *The Inniskilling Dragoons.* (London. 1909).
James, E.A. *British regiments 1914–1918.* (London. 1978).
James, N.D.G. *Plain soldiering. A history of the armed forces on Salisbury Plain.* (Salisbury. 1987).
Jeffrey, M.H. *The trumpet call obey.* (London. 1968).
Johnstone, Tom. *Orange, green* and *khaki. The story of the Irish regiments in the Great War, 1914–18.* (Dublin. 1992).
Joyce, P.W. *Irish names of places.* (Dublin. 1895).
Kiely, K., Newman, M., and Ruddy, J. *Tracing your ancestors in Co. Kildare.* (Newbridge. 1992).
Kildare County infirmary. (Carlow. N.D.)
Lacy, Thomas. *Sights and scenes in our fatherland.* (London. 1863).
Laird, Frank M. *Personal experiences of the Great War.* (Dublin. 1925).
Lawless, Emily. *With Essex in Ireland.* (London. 1902).
Levenson, Samuel. *Maud Gonne.* (London. 1976).
Limb, Sue and Cordingly Patrick. *Captain Oates, soldier and explorer* (London. 1982).
Lock hospitals. (London. 1882).
Luddy, M. and Murphy, C. (Ed) *Women surviving.* (Dublin. 1990).
Lynch, P. and Vaizey, J. *Guinness's brewery in the Irish economy 1759–1876.* (Cambridge. 1960).
Lyons, F.S.L. and Hawkins R. (editors). *Ireland under the Union, varieties of tension. Essays in honour of T.W. Moody.* (Oxford. 1980).
Lyons, M. R. *Naas Lawn Tennis Club, a short history. 1881–1992.* (Naas. 1992).
Lyttelton, Gen. Sir Neville, *Eight years soldiering, politics* and *games.* (London. 1927).
MacBride, Maud Gonne. *A servant of the Queen.* (London. 1938).
McCance, Capt. S. *History of the Royal Munster Fusiliers.* (Aldershot. 1927).
McEvoy, John., *Carlow College 1793–1993* (Carlow. 1993).
Mac Giolla Coille, Brendain. (Editor). *Intelligence Notes 1913–1916, preserved in the S.P.O.* (Dublin. 1966).
Mac Lysaght, Edward. *Irish life in the 17th Century.* (Cork. 1950).
Macready, Rt. Hon. Sir Nevil. *Annals of an active life.* (London. 1942).
McAnally, Sir Henry. *The Irish Militia 1793–1816.* (Dublin. 1949).
Mayo, Earl of and Boulton, W.B. *A history of the Kildare hunt.* (London. 1913).
Mercer, Maj. J.D. *Record of the North Cork regiment of militia. 1793–1880.* (Dublin. 1886).
Mitchell, Frank. *The Irish landscape.* (Dublin. 1976).
Moody, T.W. and Martin F.X. (editors). *The course of Irish History.* (Cork. 1987).
Moral Utility of a Lock hospital. (Dublin. 1872).
Mosley, Sir Oswald. *My life.* (London. 1968).
Muenger, E.A. *British military dilemma in Ireland. 1886–1914.* (Dublin. 1991).
Murray , K.A and McNeill, D.B. *The Great Southern* and *Western Railway.* (Dublin. 1976).
Myler, Patrick. *Regency Rogue.* (Dublin. 1976).
Naas Local History Group. *Nás na Riogh.* (Naas. 1990).
Nelson, Thomas. *The Land War in County Kildare. Maynooth Historical Series No. 3, 5–31.* (Maynooth.1985).
Norton, James E. *History of the South County Dublin harriers etc.* (Dublin. 1991).
O'Connor Morris, M. *Hibernia Venatica.* (London. 1878).
O'Connor, Tommy. (editor). *Ardfert. A hurling history.* (Ardfert. 1988).
O'Farrell, Padraic. *The blacksmith of Ballinalee.* (Mullingar. 1993).
O'Faoláin, Sean. *Vive Moi! An autobiography.* (London. 1967).
O Muineog, Micháel. *Kilcock GAA history.* (Clane. 1987).
Historical Manuscripts Commission Publications. *Ormonde manuscripts.* (London. 1920).

O'Toole, Jimmy. *Carlow gentry.* (Carlow. 1993).
Paterson, John. *Meath and Kildare. An historical guide.* (Kingscourt, Co. Cavan. 1981).
Pelly, Patricia and Tod, Andrew. (Editors) *The highland lady in Ireland.* (Edinburgh. 1991).
Pochin Mould, D.D.C. *Saint Brigid.* (Dublin. 1964).
Puirseal, Padraig. *The G.A.A. in its time.* (Dublin. 1982).
Purcell, Mary. *Matt Talbot.* (Dublin. 1954).
Pyne Clarke, Olga. *A horse in my kit bag.* (London. 1988).
Rawson, T.J. *Statistical survey of Co. Kildare.* (Dublin. 1807).
Richards, Frank. *Old Soldier's Never Die.* (London. 1964).
Richards, Frank. *Old-Soldier Sahib.* (London. 1965).
Robertson, Nora. *Crowned harp.* (Dublin. 1960).
Robins, Joseph. *The lost children.* (Dublin. 1980).
Repington, Lt. Col. Charles A. Court. *Vestigia.* (London. 1919).
Ryan, A.P. *Mutiny at the Curragh.* (London. 1956).
St Aubyn, Giles. *Edward VII: Prince and King.* (London. 1979).
Scott Daniel, David. *4th Hussars. The story of a British cavalry regiment.* (Aldershot. 1959).
Share, Bernard *The Emergency: Neutral Ireland 1935–45.* (Dublin 1978).
Songhurst, W.J. (editor). *Ars quatuor coronatorum.* (Lodge No. 2976. London). (Margate. 1925).
Spiers, E.M. *The late Victorian army.1868–1902.* (Manchester. 1992).
Sutcliffe, Sheila. *Martello towers.* (Newton Abbot. 1972).
The Last Post. National Graves Association. (Dublin 1976).
Tulloch, Maj. Gen. A.B. *Recollections of forty years' service.* (London. 1903).
Turner, Maj. Gen. Sir. A.E. *Sixty years of a soldier's life.* (London. 1912).
Valiulis, M. A. Almost a Rebellion: The Irish Army Mutiny of 1924. (Cork 1924).
Verner, Col. Willoughby. *The military life of H.R.H. George, Duke of Cambridge.* (London. 1905).
Watson, S.J. *Between the flags.* (Dublin. 1969).
Welcome, John. *The sporting empress.* (London. 1975).
Welcome, John. *Irish horse racing.* (Dublin. 1982).
Wilkins, P.A. *The history of the Victoria Cross.* (London. 1904).
Willes, Lord Chief Baron Edward. *Letters 1757–1762.* (Aberystwyth. 1990).
Wren of the Curragh. (London. 1867).
Wyndham, Horace. *The Queen's service.* (London. 1899).
Young, Arthur. *A tour in Ireland.* (Cambridge. 1925).
Young, M.F. *Letters of a noble woman.* (London. 1908).

ARTICLES

Abbreviations used:
An Cosantoir: the Irish Defence Journal.
I.H.S.: Irish Historical Studies.
I.N.S.: Irish Naturalists' Journal.
Irish Sword: The Journal of the Military History Society of Ireland.
J.K.A.S.: Journal of the county Kildare Archaeological Society.
J.R.S.A.I.: Journal of the Royal Society of Antiquaries of Ireland.
J.S.A.H.R.: Journal of the Society for Army Historical Research.
P.R.I.A.: Proceedings of the Royal Irish Academy.

Berman, David and Jill, 'Journal of an officer stationed at Naas and Baltinglass 1832–3', in *J.K.A.S*, vol. XV (1976).

Blanco, R.L., 'The attempted control of V.D. in the army in mid-Victorian England', in the *J.S.A.H.R.*, vol. 45, no. 184 (1967).
Boylan, Lena, 'Celbridge in Vanessa's time', in *J.K.A.S.*, vol. XVI (1986).
Brannigan, N.R.,'Changing the guard: Curragh evacuation 70 years on', in *An Cosantoir* , (December. 1992).
Bryan, Col. Dan, 'The Curragh Training Camp', in *The Irish Sword*, vol. I (1953).
Byran, Col. Dan, 'Ballyshannon fort county Kildare 1642–1650', in *The Irish Sword* vol. IV (1960).
Callan, Patrick, 'Recruiting for the British army in Ireland during the First World War', in *The Irish Sword*, vol. XVII (1990).
Carson, Patricia, 'Sandes Soldiers' Home', in *J.K.A.S.*, vol.XIV (1970).
Clark, B.D.H.,'The Victoria Cross, a register of wards to Irish-born officers and men', in *The Irish Sword*, vol. XVI (1986).
Connoly, Sean and Picard, J.M., 'Cogitosus: Life of St. Brigit', in *J.R.S.A.I.*, vol. 117 (1987).
Connolly, Sean, 'Vita Prima Sanctae Brigitae', in *J.R.S.A.I.* ,vol. 119 (1989).
D'Arcy, Fergus, 'The ranger of the Curragh of Kildare', in *J.R.S.A.I.*, vol. 122 (1992).
De Courcy-Wheeler, Capt.H.E., The tower on the hill of Allen, in *J.K.A.S.*, vol VII (1914).
D'hOir, Siobhain, 'Guns in medieval and tudor Ireland', in *The Irish Sword*, vol. XV. (1983).
Fehan, John and McHugh, Roland, 'The Curragh of Kildare as a hygrocybe grassland', in *I.N.J.*, vol. 24, no.1 (1992).
Ferguson, Kenneth, 'The development of a standing army in Ireland', in *The Irish Sword*, vol. XV (1983).
Fitzgerald, Lord Walter, 'The rangers of the Curragh of Kildare', in *J.R.S.A.I.*, vol. VII (1897).
Fitzgerald, Lord Walter, 'The Curragh; its history and traditions', in *J.K.A.S.*, vol. III (1902).
Fox, Capt. S.E., 'First use of gun powder in Irish history', in *The Irish Sword*, vol. XI (1974).
Gibson, W. H., 'The north and south moats, Naas', in *J.K.A.S.* vol. XVII (1991).
Haire, D.N., 'In aid of the civil power 1868–90', in *Ireland under the Union, essays in honour of T.W. Moody*, (ed) F.S.L. Lyons and R.A.J. Hawkins (Oxford, 1980).
Hally, Col. P.J.'The Easter Rising 1916 in Dublin: the military aspects', in *The Irish Sword*, vol. VII (1966).
Hawkins, Richard, 'An army on police work 1881–2' in *The Irish Sword*, vol. XI (1974).
Hennessey, W.M., 'The Curragh of Kildare', in *P.R.I.A.*, vol. IX (1867).
Heffron, Col. Michael, 'Collins barracks, Dublin', in *An Cosantoir* (May 1968).
Johnston, Major S.H.F., 'The Irish Establishment', in *The Irish Sword*, vol. I (1953).
Kerrigan, Paul, 'The defences of Ireland', in *An Cosantoir*, (August 1974).
Kerrigan, P. M., 'A military map of Ireland in the late 1790s', in *The Irish Sword*, vol. XII (1976).
Kerrigan, P. M., 'Seventeenth-century fortifications, forts and garrisons in Ireland', in *The Irish Sword*, vol. XIV (1981).
Kerrigan, P. M., 'A return of barracks in Ireland 1811', in *The Irish Sword*, vol. XV (1983).
Loeber, Rolf, 'Biographical dictionary of engineers in Ireland 1600–1730', in *The Irish Sword*, vol. XIII (1979).
Luddy, Maria, 'Prostitution and rescue work in nineteenth-century Ireland', in *Women Surviving: Studies in Irish Women's History in the 19th and 20th centuries*, (Dublin, 1990).
Luddy, Maria, 'Women and the Contagious Diseases Acts 1864–1886', in *History Ireland*, (Spring 1993).
McCarthy, P.J. 'The R.A.F. and Ireland 1920–2', in *The Irish Sword*, vol. XVII (1990).
MacCauley , J.A., 'The Dublin Fusiliers', in *The Irish Sword*, vol VI (1964).
Mooney, Fr Canice. O.F.M., 'Quarters of the army in Ireland 1769', in *The Irish Sword*, vol.II (1956).
Moore, Capt. A.M., quoted in proceedings, in *J.R.S.A.I.*, vol. V (1859).
Ni Chinneide, Sheila, 'An 18th century French traveller in Kildare', in *J.K.A.S.*, vol. XV (1976).

O'Grady, J.S., 'History and antiquities of the Hill of Allen', in *J.K.A.S.* vol. IV (1905).
O Riordáin, S.P., 'Excavations of some earthworks on the Curragh, county Kildare', in *P.R.I.A.* vol. LIII, sec. C., no. 2 (1950).
Otway-Ruthven, Jocelyn, 'The medieval county of Kildare', in *I.H.S.*, vol. XI, no. 43 (1959).
Perira, H.P.E., 'Barracks in Ireland, 1729, 1769', in *The Irish Sword*, vol. I (1953).
Sadleir, T.U., 'Kildare members of parliament 1559–1800', in *J.K.A.S.*, vol. VII (1914).
Semple, A.J., 'The Fenian infiltration of the British army', in *J.S.A.H.R.*, vol. 52, no. 211 (1974).
Simms, J.G., 'Collins barracks, Dublin', in *The Irish Sword*, vol. III (1958).
Smyth, Senator Michael, 'Kildare battalions, 1920', in *The Capuchin Annual*, (1970).
Swan, Comdt. D.A.., 'The Curragh of Kildare', in *An Cosantoir* , (May, 1972).
Tully, Col. F.J., 'letter to the editor', in *An Cosantoir*, (September, 1965).
Wailes, Bernard, 'Excavations at Dún Ailinne', in *J.K.A.S.*, vol. XV (1962).
Wall, Thomas, 'Two Fenians in Kildare', in *J.K.A.S.*, vol. XIV (1970).
Wolfe, George, 'The Wolfe family of county Kildare', in *J.K.A.S.*, vol. III (1902).
Wyse Jackson, R., 'Queen Anne`s Irish army establishment in 1704', in *The Irish Sword*, vol. I (1953).

PERIODICALS.

All the Year Round, (London. 1864, 1867).
An Baile Seo Ghainne. (Kilcullen. 1977).
Defence Forces Handbook. (Dublin. 1982).
Dublin Gazette. (January 10th 1873)
Golfers' Companion. (Dublin. 1983).
Green Howard's Gazette. (London, July 1899).
Handbook of the Curragh Command. (Dublin. 1984).
Illustrated London News. (London. 1861, 1914).
Irish Builder. (Dublin. 1901).
Irish Life. (Dublin. 1912).
Irish Military Guide. (Dublin. 1891–1919).
Monthly Official Directory of the Curragh Camp and Newbridge. (Dublin. December 1887, February 1891).
Navy and Army Illustrated. (London. 1897).
Oglaigh na hEireann: Irish Defence Forces Handbook 1968. (Dublin. 1968).
Soldier. (London. August, 1954).
The Regiment. An Illustrated Military Journal for Everybody. (London. 25.5.1896).

INDEX

Numbers in bold refer to illustrations

Index

C

WATER

PLAN SHEWING THE MANNER IN

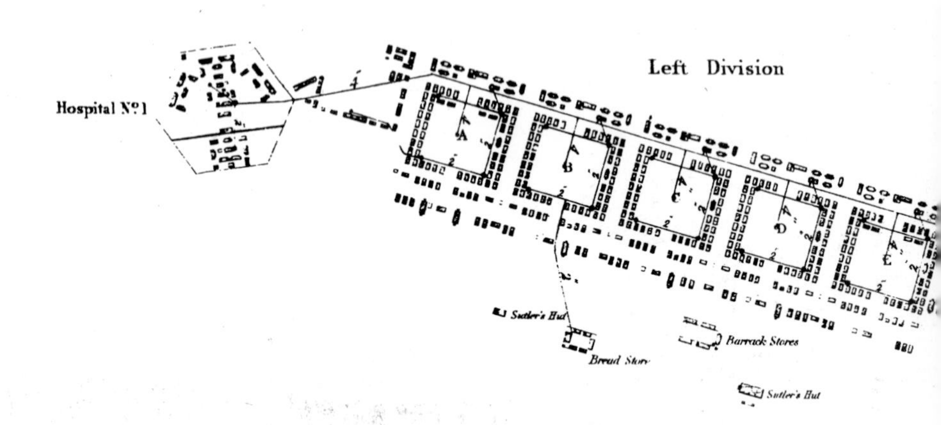

Map of Curragh Camp layout in 1858: (from Lieut. Col. H.W. Lugard's *Narrative of operations in the arrangement and formation of a Camp for 10,000 Infantry on the Curragh of Kildare. 1858.)*

Note _ The sizes of the Cast I
Ascending pipes from the
Clock Tower, -6 inches in
Pipes from the valve box
Hare Park, &c. and to the
diameter.
Distribution pipes round
buildings 2 and 3 inches

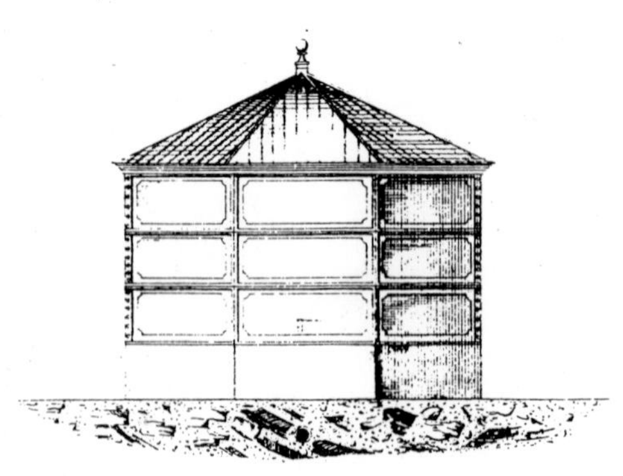

IRON TANK FOR 10,000 GALLONS